Springer Water

The book series Springer Water comprises a broad portfolio of multi- and interdisciplinary scientific books, aiming at researchers, students, and everyone interested in water-related science. The series includes peer-reviewed monographs, edited volumes, textbooks, and conference proceedings. Its volumes combine all kinds of water-related research areas, such as: the movement, distribution and quality of freshwater; water resources; the quality and pollution of water and its influence on health; the water industry including drinking water, wastewater, and desalination services and technologies; water history; as well as water management and the governmental, political, developmental, and ethical aspects of water.

Suhaib A. Bandh · Basharat Mushtaq

Wastewater Treatment Technology

Technical Progressions and Advances

Springer

Suhaib A. Bandh
Sri Pratap College
Cluster University
Srinagar, Jammu and Kashmir, India

Basharat Mushtaq
Jaknap Projects
Environmental Engineering
and Green Solutions
New Delhi, Delhi, India

ISSN 2364-6934 ISSN 2364-8198 (electronic)
Springer Water
ISBN 978-3-031-86683-8 ISBN 978-3-031-86684-5 (eBook)
https://doi.org/10.1007/978-3-031-86684-5

This Springer imprint is published by the registered company Springer Nature Switzerland AG
The registered company address is: Gewerbestrasse 11, 6330 Cham, Switzerland

Introduction

Water is an indispensable resource that sustains life and drives global economic and social development. However, with rapid urbanization, industrialization, and population growth, the quality and availability of this critical resource faces unprecedented challenges. The increasing prevalence of water pollution, scarcity, and mismanagement has underscored the urgent need for innovative strategies and technologies to preserve and purify water. Wastewater treatment emerges as a cornerstone in addressing these challenges, providing not only a means to mitigate environmental contamination but also opportunities for resource recovery and sustainable water management.

The book, *Wastewater Treatment Technology*, is designed to serve as a comprehensive resource for researchers, engineers, policymakers, and students engaged in environmental science and engineering. It delves into the multifaceted aspects of water and wastewater treatment, bridging fundamental concepts with advanced technologies and practical applications. The book begins by elucidating the key parameters and principles governing water quality, underscoring the complex interplay between natural processes and anthropogenic influences, and shedding light on how land use, pollution, and climate change shape water systems. By integrating cutting-edge analytical techniques and data-driven approaches, it emphasizes the critical importance of monitoring and managing water quality to ensure ecosystem and human health.

Transitioning to the methodologies employed to safeguard water resources, the book offers a thorough examination of both traditional and innovative treatment processes, ranging from coagulation and filtration to advanced oxidation and membrane technologies. By highlighting the evolution and application of these techniques, it addresses the growing demand for resource-efficient, sustainable, and cost-effective solutions. The focus then shifts to one of the most pressing global concerns—providing safe and potable drinking water. The intricacies of treatment processes aimed at removing contaminants, pathogens, and impurities are explored with an emphasis on regulatory compliance and public health, ensuring the delivery of high-quality drinking water from source to tap.

Attention is then directed to the critical domain of wastewater management, comprehensively exploring the sequential stages of wastewater treatment, from preliminary and primary treatments to advanced biological and chemical methods. By integrating case studies and real-world examples, the text highlights the transformative potential of innovative technologies in mitigating pollution and promoting environmental sustainability. Additionally, the book addresses the often-overlooked yet essential aspect of wastewater treatment—the management of sludge. Various treatment methods and disposal strategies are examined, emphasizing the dual objectives of minimizing environmental impacts and harnessing the resource potential of sludge byproducts. From anaerobic digestion to sustainable reuse options, it underscores the significance of responsible sludge management in a circular economy framework.

Wastewater treatment technologies are not only technical solutions but also instruments of social, economic, and environmental transformation. Effective treatment and management of wastewater can lead to a reduction in public health risks, the restoration of polluted water bodies, and the mitigation of greenhouse gas emissions. Furthermore, the recovery of energy, nutrients, and clean water from wastewater exemplifies the potential for turning waste streams into valuable resources, fostering a circular economy. The integration of emerging technologies such as artificial intelligence, machine learning, and the Internet of Things (IoT) in treatment systems holds the promise of revolutionizing the monitoring, control, and optimization of processes, making them more efficient and adaptive to dynamic challenges.

The overarching goal of this book is to foster a deeper understanding of wastewater treatment technologies and their pivotal role in addressing contemporary water challenges. By bridging theory with practical insights, it equips readers with the knowledge and tools needed to design, implement, and optimize treatment processes. Furthermore, it emphasizes the need for interdisciplinary collaboration and innovation to achieve water security and sustainability in an era marked by increasing environmental complexity. Through the exploration of policy frameworks, economic considerations, and technological advancements, this book provides a holistic perspective on how wastewater treatment can contribute to resilient water management systems.

As the global community strives to meet the United Nations Sustainable Development Goals (SDGs), particularly SDG 6 (Clean Water and Sanitation), this book aims to contribute to the collective effort by offering a valuable repository of knowledge. It invites readers to explore the dynamic field of wastewater treatment, inspiring them to pioneer solutions that safeguard this vital resource for generations to come.

Contents

Chapter 1
Concepts of Water Quality

1.1 Introduction

Water quality encompasses a spectrum of characteristics that define its suitability for various purposes, ranging from drinking to recreational activities like swimming. It serves as a gauge of its condition concerning the requirements of both living organisms and human endeavors. This multifaceted concept is defined by an array of physical, chemical, biological, and microbiological properties, ensuring that it meets the necessary standards for diverse applications across sectors such as domestic, agricultural, and industrial.

Indicators of water quality include parameters like dissolved oxygen, bacterial concentrations, salinity, and turbidity. Moreover, microscopic algae and contaminants such as pesticides, herbicides, and heavy metals are also assessed to ensure the integrity of water quality (Tyagi et al., 2013). Through vigilant monitoring and proactive maintenance, the safety and suitability of water for its intended uses can be assured.

It is critical to recognize that while various users may have distinct demands for water quality, these needs may not always align harmoniously. The activities of one user can impinge upon those of another, either by requiring water quality outside the range acceptable to the other user or by diminishing quality during its use (Meybeck et al., 1992).

Natural ecosystems possess intrinsic value and serve as sensitive indicators of changes or deteriorations in overall water quality. They offer invaluable insights that complement conventional physical, chemical, and biological data (Roy et al. 2011). Various natural factors, including geology, topography, meteorology, hydrology, and biology, exert profound influences on water quality, shaping its characteristics both on the surface and underground. Human interventions, such as dam construction, wetland drainage, and flow diversion, can significantly alter water quality dynamics.

S. A. Bandh and B. Mushtaq, *Wastewater Treatment Technology*, Springer Water,
https://doi.org/10.1007/978-3-031-86684-5_1

Additionally, pollution from domestic, industrial, and urban sources, as well as agricultural chemicals, can degrade water quality, particularly in times of low water availability (World Health Organization, 2004).

Even when water is abundant, its unsuitable quality can curtail its utility. Natural ecosystems are finely attuned to water quality, and disruptions to this balance can have far-reaching consequences. Human activities are primary contributors to a range of water quality issues, capable of disrupting entire aquatic ecosystems and limiting water use. Single influences can spawn multiple problems, and conversely, a single problem can stem from multiple contributing factors.

1.2 Water Quality of Different Sources

The issue of water quality is multifaceted, demanding a nuanced understanding of the composition and status of water bodies. To comprehensively evaluate water quality across different sources, it is imperative to scrutinize its chemical composition, biological constituents, particulate matter, and physical attributes through a structured three-step approach:

1. Conduct chemical analyses of water, particulate matter, and aquatic organisms.
2. Perform biological tests such as toxicity assessments and measurements of enzyme activities.
3. Record descriptions of aquatic organisms, encompassing their distribution, density, biomass, physiology, and diversity.

Furthermore, physical measurements including water temperature, pH levels, conductivity, light penetration, particle size of suspended and deposited materials, dimensions of the water body, flow velocity, and hydrological balance should be recorded.

According to GESAMP (1988), pollution of aquatic ecosystems stems from human activities introducing substances or energy directly or indirectly through diverse practices. This can lead to adverse effects such as risks to human health, depletion of living resources, disruption of aquatic activities such as fishing and degradation of water quality for agricultural, industrial and other purposes.

The significance of water quality hinges on its intended usage. Water designated for drinking must be devoid of any chemicals or microorganisms that could jeopardize human health. It is vital to recognize that pollution of aquatic ecosystems has widespread repercussions, impacting not only the environment but also economies and human well-being. Hence, proactive measures are essential to prevent pollution and safeguard our aquatic ecosystems. This can be achieved through adopting sustainable practices and implementing policies that prioritize the conservation of our natural resources. By doing so, we can ensure the continued safety and health of our water for future generations.

1.2.1 *Water Quality of Surface Waters*

Surface water, encompassing both saltwater in oceans and freshwater in rivers, streams, and lakes, is a pivotal element of Earth's hydrological cycle (Winter and Kickuth 1989). This cycle involves the continuous movement of water to and from the Earth's surface, driven by processes such as precipitation, runoff, evaporation, and seepage. While some bodies of surface water persist year-round, others are seasonal, fluctuating in response to changing environmental conditions. The seepage of water into the ground gives rise to groundwater, an integral component of the hydrological cycle.

Surface water exists in various forms, including perennial bodies that endure year-round, ephemeral watercourses that flow seasonally, and man-made reservoirs created by structures like dams and constructed wetlands. Understanding the dynamics of surface water is paramount for effective water resource management and environmental conservation.

By understanding the movement and behavior of surface water, we can enhance our capacity to anticipate and mitigate water-related disasters such as floods and droughts. Additionally, man-made surface water structures serve as vital resources for irrigation, drinking water supply, and other human needs, contributing significantly to societal well-being.

The significance of surface water extends beyond practical utility—it is an indispensable component of the Earth's water cycle, sustaining ecosystems and supporting biodiversity. Through the study of different surface water types and their behaviors, we can refine our management strategies and ensure the preservation of water resources for future generations. Thus, a comprehensive understanding of surface water dynamics is essential for safeguarding the water security and ecological integrity of our planet.

1.2.2 *Water Quality of Lakes and Reservoirs*

Stratification, a key process influencing water quality in deep, still bodies like lakes and reservoirs, occurs when the water separates into distinct layers with differing densities, with one layer overlaying the other. While temperature disparities primarily drive this phenomenon by causing density variations, differences in solute concentration can also induce stratification.

The stratified layers in lakes or reservoirs are subject to distinct influences, impacting water quality. The surface layer, receiving more sunlight, contrasts with the lower layer, physically isolated from the atmosphere, which serves as a source of gases like oxygen. Moreover, the lower layer may come into contact with decomposing sediments, exerting an oxygen demand and resulting in significantly lower oxygen concentrations compared to the upper layer. Anoxic conditions in bottom sediments can lead to the diffusion of various compounds, including ammonia, nitrate, phosphate, sulfide, silicate, iron, and manganese compounds, into the lower water layer.

Understanding the ramifications of stratification on water quality is paramount for the management and preservation of lakes and reservoirs. Monitoring and mitigating stratification ensure the continued health and sustainability of these water bodies, critical for generations to come. Lake water serves as a vital resource for various life forms and ecosystems, providing numerous benefits in both natural and anthropogenic settings. Lakes are pivotal in maintaining water quality and quantity for diverse human sectors, such as drinking water supply and irrigation.

Monitoring water quality in lakes and reservoirs is essential for maintaining safe water for various activities, including drinking, bathing, fishing, agriculture, and aquaculture. Long-term trends and short-term changes serve as indicators of environmental health and alterations in the water catchment area. Thus, proactive monitoring and management of water quality are essential for safeguarding these invaluable resources and ensuring their continued availability and health for present and future generations (Edokpayi et al., 2018; Gasim, 2015).

To facilitate this monitoring endeavor, Lake Water Products offer a semi-continuous observation record for a substantial number of medium- and large-sized lakes, as identified by the Global Lakes and Wetlands Database (GLWD) or other specific environmental monitoring initiatives. This record encompasses three key water quality parameters, in addition to separately provided data on lake surface water temperature.

1. **Turbidity**: This parameter measures the clarity of a lake, indicating its ability to allow sunlight to penetrate deeper parts of the water column. Turbidity levels may vary seasonally due to factors such as river discharge and the proliferation of phytoplankton, including algae and cyanobacteria.
2. **Trophic State Index (TSI)**: This index serves as a vital indicator of a lake's productivity, indirectly reflecting its eutrophication status over extended time scales. TSI is based on the abundance of phytoplankton present in the water body, providing insights into nutrient availability and ecological balance.
3. **Lake Surface Reflectance**: For scientific users interested in algorithm development, lake surface reflectance data offer valuable information regarding the apparent color of the water body. These reflectance bands can be utilized to generate true-color images by combining visual wavebands, aiding in the interpretation and visualization of water quality dynamics.

These parameters collectively contribute to a comprehensive understanding of lake water quality, enabling informed decision-making and proactive management strategies. By leveraging these monitoring tools, stakeholders can effectively safeguard the health and sustainability of lakes, ensuring their continued vitality for present and future generations.

The unique character of each lake ecosystem stems from a multitude of fundamental factors, including climate, atmospheric inputs, geological composition, physiography, and land use practices. A comprehensive understanding of these factors is imperative for effectively protecting and preserving the distinctiveness of each lake ecosystem.

- **Climate**: Climate exerts a profound influence on the hydrological and chemical attributes of a lake. Factors such as temperature, wind patterns, precipitation, and solar radiation shape the composition of the biological community within the lake. Precipitation is particularly crucial as it governs runoff and the transport of nutrients and sediments into the lake.
- **Atmospheric Inputs**: Atmospheric inputs, including acid rain and airborne particles, serve as significant sources of contaminants for lakes. Additionally, indirect atmospheric inputs through watershed runoff contribute to the accumulation of pollutants in lake ecosystems.
- **Geological Composition**: The soil type in the watershed dictates the potential for runoff and erosion, while the geological characteristics of underground rock formations influence the quantity and quality of groundwater inflows and outflows.
- **Physiography**: Physiographic features such as the area, surface topography, presence of upstream lakes and wetlands, altitude, and land slope of the lake's watershed impact surface-water runoff and the nature of chemicals and sediments entering the lake. Human land use practices can substantially alter these factors, modifying runoff patterns and nutrient/sediment inputs.
- **Land Use**: The type, location, extent, and history of land cover and land use, including agricultural, rural, and urban development, significantly affect surface-water and groundwater inflows and outflows. Land use practices influence the quantity and types of sediment, nutrients, and chemicals transported into the lake from the surrounding watershed.
- **Morphometry**: The morphometry of a lake, comprising its size, shape, and depth, plays a pivotal role in determining lake currents and mixing patterns. It also influences thermal and chemical stratification characteristics, highlighting the importance of understanding Lake Morphometry for ecosystem study and informed management decisions.

By comprehensively assessing and understanding these basic factors, stakeholders can develop effective management strategies tailored to the unique characteristics of each lake ecosystem. This holistic approach is essential for conserving and enhancing the health and resilience of lake ecosystems, ensuring their continued ecological integrity and societal benefits.

1.2.3 Water Quality of Rivers

Rivers, the natural watercourses flowing into oceans, lakes, or other bodies of water, typically receive contributions from converging tributaries. These vital waterways serve as conduits for draining water from upland areas, with their dynamic currents facilitating the rapid dilution and decomposition of pollutants, a process less efficient in standing water bodies. Despite their significance, many rivers and streams worldwide suffer from substantial pollution.

Multiple interconnected factors influence the natural water quality of rivers and reservoirs. As water traverses the Earth's surface, it interacts with minerals in soil and rocks, dissolving a diverse array of substances and never remaining entirely pure. Natural water invariably contains soluble inorganic and organic compounds, along with substantial amounts of suspended minerals. The composition and quantity of impurities in natural water vary by location and time of year, influenced by factors such as geology, climate, topography, biological processes, and land use practices. These impurities dictate the characteristics of water bodies, with dissolved oxygen emerging as a primary indicator of water quality.

Dissolved oxygen is indispensable for the survival of aquatic organisms and influences numerous other water quality indicators, including biochemical and aesthetic aspects such as odor, clarity, and taste. Hence, monitoring and preserving the quality of rivers and streams are imperative to safeguard the health and balance of our planet's ecosystems. By ensuring adequate oxygen levels and mitigating pollutant inputs, we can uphold the vitality and resilience of these crucial watercourses, benefiting both aquatic life and human communities dependent on them.

1.2.4 Water Quality of Reservoirs

Reservoirs, whether naturally occurring or human-made, serve specific functions by providing a dependable and regulated water source. These bodies of water fulfill diverse purposes, including municipal water supply, industrial usage, power generation, agricultural irrigation, river regulation, flood control, fisheries, recreational activities, navigation, and waste disposal in certain cases. Reservoirs are strategically located in areas where water availability is limited or excessive, or where controlled water management is essential for agricultural or industrial needs.

In regions facing water scarcity, reservoirs are crucial for conserving available water resources, ensuring their availability during periods of high demand for irrigation or drinking purposes. They serve as indispensable assets in meeting the escalating demand for water resources, particularly in areas lacking natural water sources. Reservoirs provide a stable and modifiable water supply for various applications, ranging from essential drinking water provision to industrial processes.

Furthermore, reservoirs contribute to environmental management by regulating river flow and mitigating flood risks, thereby enhancing water resource sustainability. Additionally, they offer recreational opportunities such as fishing, boating, and other leisure activities, augmenting their societal value.

As global population growth and water resource demands intensify, the significance of reservoirs in sustaining water availability and managing water resources will only amplify. By fulfilling critical functions in water supply, regulation, and recreation, reservoirs play an indispensable role in ensuring the prudent and sustainable utilization of water resources in an increasingly water-stressed world.

1.2.5 Key Factor Influencing the Surface Water Quality

The quality of water in lakes, rivers, and reservoirs undergoes seasonal and geographic variations, irrespective of pollution. It is crucial to recognize that there isn't a singular measure defining good water quality; suitability for one purpose, such as drinking, may not align with guidelines for irrigation, highlighting the multifaceted nature of water quality assessment (Meybeck & Helmer, 1996; WHO, 2004).

Water quality guidelines furnish vital scientific insights regarding parameters and toxicological thresholds relevant to safeguarding specific water uses. Factors profoundly influencing surface water quality encompass sedimentation, runoff, erosion, dissolved oxygen, pH, temperature, organic materials, pesticides, and various contaminants like oils, greases, detergents, and litter.

Monitoring and managing these factors are imperative to uphold the safety and health of water resources. By doing so, we can preserve the environment and foster sustainable development for future generations. Moreover, substances present in the atmosphere significantly impact rainfall quality. While natural gases like carbon dioxide, oxygen, and nitrogen are dissolved in rain, pollutants such as dust, volcanic gases, sulfur dioxide, and toxic chemicals contaminate rainfall as it descends (Roy & Majumder, 2017).

Runoff, traversing soil and rocks, dissolves and transports substances, affecting surface water quality for varying durations based on constituent concentrations. For instance, runoff over limestone-rich soil carries elevated calcium carbonate levels, altering water chemistry. Conversely, flow over metal-rich rocks dissolves metals, further modifying water properties.

Industrial farming, mining, and forestry significantly impact river, lake, and groundwater quality by elevating nutrient, pesticide, and sediment concentrations. Industrial activities intensify metal and toxic chemical levels, elevate sediment content, raise temperatures, and deplete dissolved oxygen, adversely impacting aquatic ecosystems and rendering water unsuitable for various uses.

Urban runoff exacerbates water quality issues by transporting debris and pollutants from streets to receiving water bodies, augmenting nutrient, sediment, animal waste, petroleum products, and road salt concentrations. Addressing the adverse effects of industrial and urban activities on aquatic ecosystems is imperative to ensure water suitability for established purposes (Edokpayi et al., 2017; Roy & Majumder, 2017). Thus, concerted efforts are necessary to mitigate anthropogenic impacts on water quality and safeguard aquatic ecosystems for present and future generations.

1.2.6 Water Quality of Underground Sources

Groundwater, residing beneath the Earth's surface within soil, sand, and rock crevices, constitutes a crucial resource stored and slowly moving through geological formations known as aquifers (Mohamed et al., 2018). These aquifers function as natural filters, gradually purifying water as it percolates through them. However, various factors, including contamination by substances like petroleum, hydrocarbons, pesticides, minerals, and pathogenic microorganisms, can compromise the groundwater quality (Lee, 2004).

The quality of groundwater is intricately influenced by several factors, including the composition of recharge water, interactions between water and soil/rocks in the saturated zone, residence time, and the reactions occurring within the aquifer. Consequently, significant variations in groundwater quality may arise, particularly in regions featuring diverse rock compositions and solubilities (Chilton, 1996; Gary, 2005; Verma et al., 2020).

Principal processes shaping water quality within aquifers encompass physical, geochemical, and biochemical mechanisms:

- Physical processes such as dispersion/dilution, filtration, and gas movement aid in impurity removal from water.
- Geochemical processes, including complexation, acid–base reactions, oxidation–reduction, precipitation-solution, and adsorption–desorption, play pivotal roles in water purification.
- Biochemical processes, such as microbial respiration and decay, cell synthesis, and other biological activities, contribute to contaminant breakdown and removal from water.

These intricate processes collectively govern groundwater quality, highlighting the dynamic interplay between natural phenomena and human-induced influences. Understanding and managing these processes are imperative for safeguarding groundwater resources and ensuring their sustained availability for various societal needs.

1.2.7 Contamination of Ground Water

Groundwater, being shielded beneath the ground, is generally less prone to contamination compared to surface water, as contaminants must permeate through soil and rock layers, which often act as natural filters (Satapathy & Syed, 2015). However, the risk of contamination persists, especially if there are fractures or fissures in the overlying soil and rocks, facilitating the flow of toxic compounds (Mohamed et al., 2018). Human activities exert a significant influence on groundwater quality, given that most groundwater originates from rainwater infiltration from the surface. Contaminants can infiltrate groundwater directly from sources such as abandoned wells, mines, quarries, and buried sewerage pipes, bypassing natural decontamination processes.

Monitoring and safeguarding groundwater quality are imperative to ensure safe drinking water and maintain ecosystem health. Understanding the origins and pathways of contamination enables proactive measures to prevent and mitigate its impacts. Improperly constructed wells pose a common risk for groundwater contamination. Cracked or incorrectly constructed wells allow surface contaminants to seep into the groundwater (Chandra et al., 2015). This risk intensifies if wells are situated near potential contamination sources such as septic tanks, livestock areas, agricultural fields, manure piles, or petroleum storage tanks. Hence, meticulous well construction and regular monitoring are essential for preserving groundwater quality and safeguarding public health and environmental integrity.

1.2.8 Natural Contamination of Groundwater Quality

Groundwater, a vital resource traversing through aquifers at a slow pace, exhibits extensive interaction with surrounding rocks or sediments, unlike surface water. Typically, geological materials constituting aquifers are relatively inert or composed of minerals that dissolve slowly into groundwater. Nevertheless, over time, groundwater gradually accumulates dissolved material as it maintains contact with the aquifer. Unfortunately, certain regions harbor rocks or sediments containing minerals that may contaminate water with elements rendering it unsuitable for human consumption or agricultural use (WHO, 2004). Contaminants such as copper, arsenic, mercury, fluorine, sodium, and boron pose significant threats.

Contamination can manifest in two ways:

1. Elevated levels of the contaminating element within the aquifer material.
2. Normal aquifer material encountering specific water or aquifer attributes facilitating contaminant accumulation to significant levels.

Regardless of the cause, natural groundwater contamination poses a serious concern warranting attention. Ensuring water supply safety necessitates regular groundwater monitoring and testing. This proactive approach enables the identification of potential contaminants, facilitating the implementation of suitable measures to prevent their escalation to harmful levels. Effective filtration systems and treatment methods play crucial roles in removing contaminants from water, safeguarding public health and environmental integrity.

1.2.9 Anthropogenic Contamination of Groundwater Quality

The vulnerability of aquifers to pollution hinges on various factors, including the depth to the water table, the permeability of surface-to-aquifer materials, aquifer permeability, surface slope, and precipitation volume (Verma et al., 2020). Confined aquifers generally exhibit lower vulnerability compared to unconfined ones, with deeper aquifers being less susceptible than shallow ones. Steeper surface slopes typically lead to surface water runoff rather than infiltration, potentially reducing contamination risks. Furthermore, areas with lower precipitation tend to have lower contamination risks compared to regions with heavy rainfall.

Anthropogenic sources pose significant threats to groundwater quality, including agricultural chemicals and animal waste, industrial discharges, landfill leachate, mining and quarry activities, fuel storage tank leaks (especially at gas stations), septic system effluents, road runoff, and chemical spills during transportation.

Protecting groundwater from contamination necessitates implementing appropriate measures. This includes adopting best practices in agriculture and industrial operations, ensuring proper disposal of hazardous materials, and maintaining fuel storage tanks to prevent leaks. By undertaking these actions, we can safeguard groundwater as a safe and dependable resource for present and future generations.

1.2.10 Water Quality of a Pond

The quality of water in a pond is subject to the influence of various natural and anthropogenic factors. Natural factors, such as the origin of the pond water and the geological composition of the surrounding watershed, play a role in determining water quality characteristics. While these factors are often beyond human control, they typically do not pose significant issues. However, more severe water quality challenges often stem from human activities within or near the pond (Swistock and Clark, 2015). Effective management practices and early detection of problems through regular testing are essential to mitigate the impacts of these activities.

When evaluating water quality concerns, it is crucial to consider the primary uses of the pond. Different parameters should be tested based on the intended use of the water body (McComas, 2003). For example, a pond intended for animal drinking water requires different testing parameters compared to a pond primarily used for recreational fishing. By comprehending the interplay of natural and human factors affecting water quality in ponds, appropriate management approaches can be implemented to preserve a balanced ecosystem. Routine testing and monitoring serve as vital tools for identifying and addressing potential issues before they escalate into significant problems.

1.2.11 Key Parameters Influencing Pond Water Quality

The quality of pond water is influenced by various factors, each of which plays a crucial role in determining its overall health and suitability for different purposes:

1. **Temperature**: Pond water temperature fluctuates seasonally, with surface water being more influenced by air temperature than deeper water. These fluctuations impact the metabolic rates of aquatic organisms and the distribution of dissolved gases, which in turn affect the overall pond ecology.
2. **Dissolved Oxygen**: Adequate levels of dissolved oxygen are vital for the survival of aquatic life. Biological decay of organic matter can deplete oxygen levels, particularly in warmer months or when ponds are treated with herbicides, leading to oxygen-deficient conditions detrimental to aquatic organisms.
3. **Coliform Bacteria**: Coliform bacteria serve as indicators of water quality, with elevated levels often indicating contamination from sources like animal waste or septic systems. Monitoring coliform levels helps assess the potential health risks associated with pond water use.
4. **Nitrogen and Phosphorous**: Excessive nitrogen and phosphorus inputs, typically from agricultural runoff, septic systems, or fertilizers, can lead to eutrophication and algal blooms in ponds. Elevated nutrient levels can disrupt aquatic ecosystems and degrade water quality.
5. **Light Availability**: Availability of light energy influences pond productivity and photosynthesis, with shallow ponds supporting abundant aquatic plant growth due to light penetration. However, high turbidity can limit light penetration, affecting plant growth and leading to organic matter decay and gas production.
6. **Water Color**: The color of pond water can provide insights into its quality and use. Clear water is desirable for fish culture, while specific colors may indicate the presence of phytoplankton, zooplankton, sand particles, or organic matter. Unfavorable colors, such as black or red, may signify poor water quality conditions unsuitable for aquatic life or certain uses.

By understanding and monitoring these factors, pond managers can implement appropriate management strategies to maintain water quality, support aquatic ecosystems, and ensure the suitability of pond water for various purposes. Regular testing and proactive measures are essential for sustaining healthy pond environments.

Prevention of Pond Water Quality

Proactive management techniques are essential for preventing water quality problems in ponds and maintaining a healthy aquatic environment. The effective strategies include:

1. **Regular Water Testing**: Periodic testing of pond water helps assess bacterial levels and identify any hidden issues affecting water quality. Monitoring parameters such as pH, dissolved oxygen, and nutrient levels can guide corrective actions.
2. **Fish Stocking**: Selecting fish species adapted to the natural temperature regime of the pond helps maintain ecological balance and control excessive growth of aquatic plants and algae, which can degrade water quality.
3. **Pollution Prevention**: Limiting polluting activities near the pond and managing land use practices to minimize runoff can prevent contamination from entering

the water. This includes proper waste management, avoiding the use of harmful chemicals, and controlling erosion.

4. **Herbicide Management**: When using aquatic herbicides, it is crucial to carefully follow label directions and avoid over-treatment. Treating only a portion of the pond at a time allows for gradual control of vegetation while minimizing impacts on water quality and aquatic life.
5. **Vegetated Buffer Zones**: Vegetation acts as a natural barrier, trapping contaminants and promoting infiltration. Maintaining vegetated buffer strips around the pond perimeter, with wider buffers on steeper slopes, helps filter runoff, reduce sedimentation, and prevent pollutants from entering the water.
6. **Surface Water Management**: Using ditches, grading techniques, and proper drainage infrastructure can redirect polluted surface water away from the ponds, preventing contamination and preserving water quality.

By implementing these management techniques, pond managers can effectively safeguard water quality, support diverse aquatic ecosystems, and ensure the long-term health and sustainability of the ponds.

1.3 Water Quality Requirements for Different Users

In order for water bodies to function properly and meet the needs of their users, they must adhere to specific quality standards. For instance, drinking water must meet stringent purity criteria. As demands for water availability and supply continue to rise, managing water quality has become as critical as managing its quantity. Water serves various purposes, each with its own quality requirements. Thus, it is essential to identify the intended uses of water in a particular body to establish appropriate quality standards. In India, the Central Pollution Control Board (CPCB) has introduced the concept of "designated best use," where the highest quality requirement among the various uses determines the standard. The CPCB has classified water uses into five designated best uses (Table 1.1), aiding water quality planners and managers in setting quality targets and designing suitable restoration programs. By understanding the specific needs of each user, efficient and effective utilization of water resources can be ensured.

Water, a vital resource, serves diverse purposes such as drinking, irrigation, and industrial processes. However, the safety and availability of water for each use hinges on water quality standards. These standards are established through rigorous scientific examinations to safeguard against health hazards and usage restrictions (Drexler et al., 2014). The process of setting water quality standards is complex and varies based on factors like nature, society, culture, and economic conditions of each region. Once established, it is imperative to enforce these standards effectively, necessitating continuous monitoring to ensure compliance with regulations.

Table 1.1 Five designated best uses of water as per CPCB

Designated best use	Class	Criteria
Drinking water source without conventional treatment but after disinfection	A	1. Total coliforms organism MPN/100 mL shall be 50 or less 2. pH between 6.5 and 8.5 3. Dissolved oxygen 6 mg/L or more 4. Biochemical oxygen demand 5 days 20 °C, 2 mg/L or less
Outdoor bathing (organised)	B	1. Total coliforms organism MPN/100 mL shall be 500 or less 2. pH between 6.5 and 8.5 3. Dissolved oxygen 5 mg/L or more 4. Biochemical oxygen demand 5 days 20 °C, 3 mg/L or less
Drinking water source after conventional treatment and disinfection	C	1. Total coliforms organism MPN/100 mL shall be 5000 or less 2. pH between 6 and 9 3. Dissolved oxygen 4 mg/L or more 4. Biochemical oxygen demand 5 days 20 °C, 3 mg/L or less
Propagation of wild life and fisheries	D	1. pH between 6.5 and 8.5 2. Dissolved oxygen 4 mg/L or more 3. Free ammonia (as N) 4. Biochemical oxygen demand 5 days 20 °C, 2 mg/L or less
Irrigation, industrial cooling, controlled waste disposal	E	1. pH between 6.0 and 8.5 2. Electrical conductivity at 25 °C micro mhos/cm, maximum 2250 3. Sodium absorption ratio max. 26 4. Boron max. 2 mg/L
	Below-E	Not meeting any of the A, B, C, D and E criteria

Source CPCB

Monitoring plays a crucial role in identifying hazards within specified concentration limits. If hazards are detected, revisions may be required, including potential amendments to regulations (Tyagi et al., 2013). Regulatory standards should mandate monitoring to verify compliance. Conversely, guidelines serve as references for achieving desired water quality goals or standards. It's essential to note that guidelines lack legal binding and carry less weight than regulatory standards.

1.4 Use of Water for Different Requirements

Water is a versatile resource utilized for various purposes, each with specific quality requirements. The main categories of water use include:

1. **Domestic Water Requirement**: Used for household purposes like drinking, cooking, cleaning, and bathing.
2. **Industrial Water Requirement**: Utilized in manufacturing, cooling, and cleaning processes in industries.
3. **Irrigation Requirements**: Essential for agricultural activities to ensure crop growth.
4. **Animal Water Supply**: Required for animals for drinking and other purposes.
5. **Preservation and Breeding of Aquatic Species**: Vital for the preservation and breeding of fish and other marine life.
6. **Recreation and Leisure**: Lakes, rivers, and oceans offer recreational activities like swimming, boating, and fishing.
7. **Generation of Electricity**: Used in hydroelectric power plants for electricity generation.
8. **Navigation**: Rivers and canals serve as waterways for transportation of goods and people.
9. **Landscape Harmony**: Enhances landscapes through fountains, waterfalls, and other water features.
10. **Dilution and Transport of Wastes**: Utilized to dilute and transport waste materials such as sewage and industrial effluents.

Water requirements for domestic and industrial use often necessitate prior treatment to meet high-quality standards (Drexler et al., 2014; Sharma et al., 2019). Domestic water requirements are particularly stringent, requiring compliance with various quality criteria. Conversely, low-demand water uses like waste dilution and transportation have less specific quality criteria. However, when multiple uses are assigned to a water body, it must meet diverse quality criteria (Marcos et al. 2017; Van Wyk & Scarpa, 1999). For example, reservoirs may serve purposes such as water supply, electricity generation, recreation, and irrigation (Table 1.2).

Water is an indispensable resource involved in various internal water cycles, as well as the Earth's hydrological cycle, leading to modifications in its characteristics (Van Wyk & Scarpa, 1999) (Fig. 2).

During these cycles, water quality evolves at each stage of its journey. Water and wastewater follow several routes, including raw water, treated water, raw wastewater, treated wastewater, stormwater, and receiving bodies. Initially abstracted from water bodies like rivers, lakes, or groundwater, raw water possesses a specific quality. Through treatment processes to meet domestic or industrial water supply needs, raw water transforms into treated water. Following usage, this treated water becomes raw wastewater, undergoing further quality transformations and becoming liquid waste.

These stages demonstrate the dynamic nature of water quality as it progresses through different stages of use and treatment, emphasizing the importance of managing and maintaining water quality throughout its journey.

Treatment of wastewater is crucial for removing contaminants before discharge into receiving bodies. This process significantly impacts the quality of the liquid (Yotova et al., 2019). Similarly, stormwater, which carries pollutants, is collected

Table 1.2 Association between water use and quality requirements

General requirement	Specific use	Required quality
Domestic requirement	Drinking, bathing, cleaning, washing etc.	• Free from chemical substances (harmful to health) • Free from organisms (harmful to health) • Low aggressiveness and hardness • Aesthetically pleasant (low turbidity, colour, taste and odour and absence of microorganism)
Industrial requirement	Water incorporated into the manufacture of various products (e.g. medicines)	• Free from chemical substances (harmful to health) • Free from organisms (harmful to health) • Aesthetically pleasant (low turbidity, colour, taste and odour and absence of microorganism)
	Water that enters into contact with the products	• Variable with the product
	Water that does not enter into contact with the product (e.g. refrigeration units, boilers)	• Low hardness • Low aggressiveness
Irrigation requirement	Horticulture products ingested raw or with skin	• Free from chemical substances (harmful to health) • Free from organisms (harmful to health) • Non-excessive salinity
	Other plantations	• Free from chemical substances (harmful to the soil and plantation) • Free from organisms (harmful to animal health)
Animal water supply	Drinking, bathing	• Free from chemical substances (harmful to animals health) • Free from organisms (harmful to animals health)

(continued)

Table 1.2 (continued)

General requirement	Specific use	Required quality
Preservation and breeding of aquatic species	Aquaculture and animal breeding	• Variable with the environmental requirements of the aquatic species to be preserved • Free from chemical substances (harmful to animals, workers and consumer health) • Free from organisms (harmful to animals, workers and consumer health) • Availability of nutrients
	Vegetable growing	• Free from chemical substances (toxic to vegetables and consumers) • Availability of nutrients
Recreation and leisure	Primary contact (direct contact with the liquid medium e.g. swimming, water skiing, surfing)	• Free from chemical substances (harmful to health) • Free from organisms (harmful to health) • Low level of suspended solids and oils and greases
	Secondary contact (without direct contact with the liquid medium; e.g. leisure navigation, fishing, contemplative viewing)	• Pleasant appearance • Low turbidity • No odour
Generation of electricity	Hydroelectric power plants	• Low aggressiveness
	Nuclear or thermoelectric power plants (e.g. cooling towers)	• Low hardness
Navigation	–	• Low pressure of course material that could be dangerous to vessels
Dilution and transport of wastes	–	• No required quality

and treated before discharge into receiving bodies, contributing to changes in water quality (EPA, 2017).

Upon reaching the receiving body, a combination of dilution and self-purification mechanisms further modifies water quality (Yotova et al., 2019). Understanding these routes is essential for effective management and maintenance of water resources. Through the implementation of robust treatment processes, we can ensure the preservation of water as a valuable resource for future generations.

1.4.1 Domestic Water Requirements

Domestic water is vital for various household activities, including drinking, food preparation, and hygiene. Access to clean water is essential for sustaining life and preventing the spread of diseases, particularly among vulnerable populations such as children under 5 years old and adults over 50 years old (Prüss-Ustün et al., 2019). Domestic water requirements encompass both in-house needs and outdoor uses. In-house requirements entail activities like drinking, cooking, sanitation, cleaning, and laundry, while outdoor needs include watering gardens, sprinkling lawns, and washing cars.

The average domestic water consumption in Indian cities, as per the National Building Code, is estimated to be 135 L per capita per day (Table 1.3) (Fan et al., 2014). However, actual water requirements can vary based on household size, location, and living conditions (Howard et al., 2020). Developed countries tend to have higher per capita water usage compared to less developed ones, and urban areas generally exhibit higher water demands than rural areas. Per capita water demand can range from as low as 75 L per day to as high as 500–600 L per day (Gleick & Iwra, 1996).

The factors influencing per capita water requirements are indeed multifaceted and can vary significantly based on various determinants. Some key factors that impact per capita water requirements are:

1. **Climatic Conditions**: Arid regions typically require more water due to increased needs for bathing, cleaning, and gardening, especially in hot and dry climates where there is a higher demand for drinking water.
2. **Socioeconomic Status and Habits**: Affluent households often have more water-intensive lifestyles, such as using bathing tubs or engaging in water-consuming recreational activities, leading to higher water consumption.
3. **Urbanization and City Size**: Larger cities generally have higher per capita water demands due to greater population density and associated infrastructure needs, such as firefighting and wastewater disposal systems. Even smaller industrial towns may have significant water demands due to indirect usage.

Table 1.3 Average water use per person per day in urban areas

Requirement of water	Consumption in litres per capita per day
Drinking	5
Cooking	5
Bathing	50
Toilet flushing	30
Washing of clothes	20
Washing of utensils	15
Washing/cleaning of houses	10

Source Manual on water supply and treatment 2019

4. **Water Tax Policies**: The method of water tax collection, such as through water meters, can influence consumption behavior. While water meters enable controlled consumption and accurate billing, their installation and maintenance can incur additional costs and technical challenges.
5. **Policy Implementation**: Municipalities or water supply agencies may assess water taxes based on factors like property size, quality, and basic consumption principles to ensure fair and efficient tax collection.
6. **Efficiency Measures**: Streamlining water tax collection processes and ensuring accurate assessments can help optimize revenue collection without unduly burdening consumers or causing inconvenience.

By considering these factors and implementing appropriate policies and measures, municipalities can better manage water resources and ensure equitable access to clean water while promoting sustainable usage practices.

1.4.2 Institutional Water Requirements

Assessing the water requirements of various institutions (Table 1.4) in a town or city is essential for effective water supply system planning. Some key institutions and their average additional per capita water demand, along with an approximate estimate of their water requirements is as follows:

1. **Hospitals**: Hospitals require a significant amount of water for various purposes such as patient care, sanitation, cleaning, and medical equipment sterilization. On average, the additional per capita demand for hospitals can range from 25 to 60 L per day.
2. **Schools**: Schools need water for drinking fountains, restrooms, cleaning, and other educational activities. The additional per capita demand for schools typically ranges from 25 to 40 L per day.
3. **Restaurants**: Restaurants require water for food preparation, cleaning, dishwashing, and restroom facilities. The additional per capita demand for restaurants can vary widely based on their size and capacity, ranging from 30 to 50 L per day.
4. **Hotels**: Hotels have diverse water needs, including guest rooms, dining facilities, laundry services, swimming pools, and landscaping. The additional per capita demand for hotels depends on their size and amenities, ranging from 40 to 60 L per day.
5. **Railway Stations and Bus Terminals**: Public transportation hubs require water for restroom facilities, cleaning, and drinking fountains. The additional per capita demand for railway stations and bus terminals can range from 20 to 40 L per day.
6. **Government Offices**: Government offices need water for administrative purposes, restroom facilities, cleaning, and landscaping. The additional per capita demand for government offices varies based on the size and functions of the office, typically ranging from 20 to 30 L per day.

Table 1.4 Recommended water requirement for various institutions

S. No.	Institutions	Liters per head per day
1.	Hospital (including laundry)	
(a) No. of beds exceeding 100	450 (per bed)	
(b) No. of beds not exceeding 100	340 (per bed)	
2.	Hotels	180 (per bed)
3.	Hostels	135
4.	Nurses homes and medical quarters	135
5.	Boarding schools/colleges	135
6.	Restaurants	70 (per seat)
7.	Air ports and sea ports	10
8.	Junction stations and intermediate stations where mail or express stoppage (both railways and bus stations) is provided	70
9.	Terminal station	45
10.	Intermediate stations (excluding mail and express stops)	45 (could be reduced to 25 where bathing facilities are not provided)
11.	Day schools/colleges	45
12.	Offices	45
13.	Factories	45 (could be reduced to 25 where bathing facilities are not provided)
14.	Cinema, concert halls and theatres	15

Source Manual on water supply and treatment 2019

Proper assessment of the water requirements for these institutions is crucial for designing an efficient and sustainable water supply scheme for urban areas. By considering these factors, planners can ensure that adequate water resources are allocated to meet the needs of both residential and institutional users.

1.4.3 Industrial Water Requirements

Understanding the water requirements of industries that includes the underlying factors is crucial for effective planning and management of industrial water supply schemes.

1. **Type and Size of Industry**: The water requirements of an industry depend on its type and size. Heavy industries such as steel manufacturing or chemical processing typically require large quantities of water for cooling, processing, and other industrial operations. On the other hand, light industries such as textiles or food processing may have lower water demands.
2. **Number of Industries**: The cumulative water demand of industries in a particular area depends on the number and type of industries present. Industrial clusters or zones may require a dedicated water supply scheme to meet the collective water needs of all industries.
3. **Specific Industry Requirements**: Certain industries may have unique water requirements based on their processes and operations. For example, industries that involve water-intensive processes such as paper manufacturing or semiconductor fabrication may require specialized water treatment facilities to meet quality standards and regulatory requirements.
4. **Housing and Amenities**: In some cases, large industries may require additional water supply for employee housing, recreational facilities, and other amenities associated with the industry.
5. **Sustainable Water Use Practices**: Promoting sustainable water use practices is essential for managing industrial water supply schemes effectively. This may include implementing water conservation measures, recycling and reuse of wastewater, and adopting technologies for efficient water use.

By considering these factors and incorporating them into the planning and management of industrial water supply schemes, authorities can ensure that the water needs of industries are met efficiently while also promoting sustainable water management practices.

Table 1.5 represents the water usage and wastewater generation of different industries in India. It is important to consider these factors while planning and managing industrial water supply schemes to ensure that the needs of all industries are met while also promoting sustainable water use practices.

1.4.4 Water Requirements for Public Use

This pertains to the necessary water requirements for:

- Traffic terminals
- Fire services
- Public Gardens and more.

Prioritizing water allocation to public areas ensures that essential services can continue uninterrupted, even during times of water scarcity or emergencies. By investing in water infrastructure and management strategies that support these facilities, communities can enhance resilience and improve overall public health and well-being.

Table 1.5 Water use by different industries in India (Aggarwal & Kumar, 2011; Goldar, 2003)

Industrial sector	Annual consumption (million m^3)	Annual wastewater discharge (million m^3)	Proportion of total water consumed (percent)
Thermal power plant	35,157.4	27,000.9	87.87
Engineering	2019.9	1551.3	5.05
Pulp and paper	905.8	695.7	2.26
Textiles	829.8	637.3	2.07
Steel	516.6	396.8	1.29
Sugar	194.9	149.7	0.49
Fertilizers	73.5	56.4	0.18
Others	314.2	241.3	0.78
Total	40,012.0	30,729.2	100.0

1.4.5 Water Requirements for Traffic Terminals and Stations

Water requirements for traffic terminals and stations, including railway stations, bus stations, harbors, and airports, are essential for ensuring the comfort and safety of passengers. These requirements encompass various facilities within these terminals, such as waiting rooms, waiting halls, and retiring rooms. The National Building Code of 2016 provides guidelines for determining the water supply requirements for traffic terminals and stations (Table 1.6).

To estimate the water requirements of these facilities, it is essential to consider factors such as the average number of passengers handled by the station daily, as well as the personnel and vendors who utilize the facilities. This information is crucial for effective planning and allocation of resources to meet the water needs of these terminals and stations.

Table 1.6 Guidelines for determining the water supply requirements for traffic terminals and stations

Nature of station/terminal	With bathing facilities	Without bathing facilities
Litres/capita/day		
Intermediate stations (excluding mail and express stops)	45	23
Junction stations and intermediate stations where mail or express stoppage is provided	70	45
Terminal stations	45	45
International and domestic airports	70	70

Source National Building Code 2016, BIS

1.4.6 Requirement for Fire Fighting

As population density increases and hazardous activities like fire risks become more prevalent. Industries, particularly those handling explosive materials, require special firefighting arrangements, while towns and cities must ensure additional water supplies for firefighting in emergencies (Hall, 2011). Municipalities and corporations install fire hydrants, and high-rise buildings in densely populated areas are mandated to have firefighting provisions.

Fire hydrants are strategically positioned approximately 150 m apart and linked to water mains. When a fire occurs, firefighting pumps are activated to dispense water at high pressure. A minimum water pressure of 1–5 kg/cm^2 is required for 4–5 h of continuous firefighting (Nema, 2012). According to regulations, one-third of the firefighting water requirement should come from storage reservoirs, with the remainder allocated to static tanks at strategic points. These tanks can be filled from the nearest water sources, such as streams, ponds, or water tankers.

For cities with populations exceeding 50,000 the required water volume in m^3 is calculated using Eq. (1.1) and implementing these measures is crucial for safeguarding public safety and minimizing the impact of fires.

$$Q = 100\sqrt{P} \quad (1.1)$$

where,

Q is water required in m^3 and
P is population in thousands.

Rate of fire demands is calculated by Kuichling's formula which gives amount of water required in L/min

$$Q = 3182\sqrt{P}$$

where

Q is water required in L/min and
P is population in thousands.

1.4.7 Agricultural Requirement

In India, public water supplies are typically not used for agricultural purposes. However, in some farmhouses, water may be utilized for kitchen gardens or minor agricultural activities. In contrast, in foreign countries, private water resources are often utilized extensively for farming, especially for crop irrigation. Public supply is commonly used for dairies, cattle troughs, farmhouses, horticulture, and greenhouse cultivation (Nema, 2012) (Table 1.7).

Table 1.7 Agricultural usage and estimated consumptions

Usage	Estimated consumption
Intensive dairy farming	80 L/day per hectare of grazing fully utilized or 1350 L/day per cow
Average agricultural demand for mixed farming	57 L/day per hectare of farm land
Glass houses	12,400 L/day per hectare in winter and three times of this value in summer

It is important to recognize that while public water supplies may not be commonly used for agriculture in India, they still play a crucial role in other aspects of farming. Additionally, the utilization of private water resources for farming in foreign countries underscores the significance of water management and conservation efforts on a global scale. By responsibly and sustainably utilizing various water sources, we can ensure the ongoing success and development of agriculture, both domestically and internationally.

1.4.8 General Usage of Water

In certain countries, the distinction between water usage for public consumption and that for trade and industry is not always clear. Public use encompasses water for various purposes such as fountains, parks, governmental buildings, and their surrounding areas, often provided free of charge. The amount of water utilized for these purposes can vary significantly, ranging from zero to approximately 100 L per person per day, depending on factors such as the city's classification and climate.

On the other hand, the demand for water for firefighting, routine fire hydrant testing, temporary building supplies, and sewer flushing is typically minimal and does not necessitate separate estimation when expressed as an average daily per capita figure throughout the year.

Accurately measuring water usage is crucial for effective management and conservation of this vital resource. Therefore, countries must implement systems to differentiate between public, trade, and industrial water usage to ensure proper allocation and management. This differentiation aids in identifying areas where conservation efforts can be focused and ensures that water resources are utilized efficiently and sustainably.

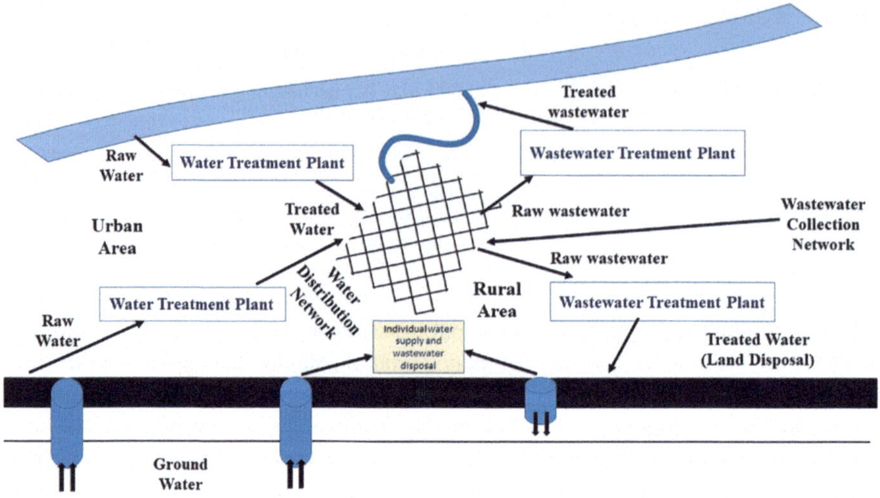

Fig. 1.1 Routes of water use and disposal

1.5 Per Capita Water Demand

Per capita water demand refers to the average daily requirement of water for one person, encompassing various purposes such as household needs, institutional uses, public uses like street cleaning, sewer flushing, as well as industrial, commercial, and firefighting needs. It is calculated as the average daily per capita demand, reflecting the amount of water an individual requires on a daily basis for different activities. This calculation is essential for effective water management and conservation efforts. By understanding the average daily water needs of individuals, we can optimize the allocation of water resources to meet the population's requirements while ensuring sustainability and conservation (Fig. 1.1).

1.6 Water Demand and Its Types

The global water crisis has become a significant concern, with many countries facing severe water shortages. Urban water authorities and planners are grappling with the challenge of meeting the increasing demand for water while striving to establish sustainable urban water systems (Ramulongo et al., 2017). Water demand refers to the quantity of water needed by users to fulfill various purposes such as drinking, washing, cooking, and irrigation.

Water usage has been steadily rising globally, increasing by approximately 1% annually since the 1980s. This upward trend is attributed to factors such as population growth, socio-economic development, and changing consumption patterns. Experts predict that this trend will persist until 2050, with a projected increase of 20–30% above current water usage levels, primarily driven by rising demand in the industrial and domestic sectors (WWAP (UNESCO World Water Assessment Programme), 2019).

Alarmingly, over 2 billion people live in countries experiencing high water stress, and approximately 4 billion people face severe water scarcity during at least one month of the year (Hoo, 2019; Min et al., 2021; Wang et al., 2018). These stress levels are expected to exacerbate as water demand grows and the impacts of climate change intensify. As the global population continues to expand, the demand for water increases daily. The quantity of water required for specific purposes, such as millimeters per crop or liters per person per day, serves as a metric for this demand (Arbués et al., 2020). Water demand can be categorized as either consumptive (for households, industries, and agriculture) or non-consumptive (for habitat preservation, fisheries navigation, and salinity control at river mouths). Additionally, a distinction can be made between in-stream and off-stream demand (Min et al., 2021).

Various factors influence water demand, including continuous population growth, changes in lifestyle, demographic structure, and the potential effects of climate change. At the global level, economic development, industrialization, agricultural practices, urbanization, energy production, and water management policies are among the significant factors shaping water demand dynamics. Addressing these factors is crucial for effectively managing water resources and mitigating the impacts of the global water crisis.

1.7 Types of Water Demand

The demand for water varies significantly across different sectors, locations, and time periods, presenting challenges in accurately assessing the exact quantity of water required by the public. Factors such as climate, population density, and economic activity can all influence water consumption patterns, leading to notable variability in demand across various areas. To tackle this challenge, experts often rely on thumb rules and empirical formulas to estimate water demand, which can yield reasonably accurate results. However, it is important to acknowledge that these estimates may not always be precise and may require adjustments based on local conditions and other factors.

In general, a city may face several types of water demand, including residential, commercial, industrial, and agricultural, each with unique water requirements and necessitating different strategies for effective water resource management. Understanding the complex and dynamic nature of water demand is crucial for developing sustainable water management practices that can meet the needs of both current and future generations. By collaborating to address this critical issue, we can ensure that our communities have access to clean and reliable water resources essential for their well-being.

To accurately determine the total water demand for a specific area of the community, it is essential to consider all individual demands and make appropriate provisions based on the unique needs of the population for whom the water supply system is being designed. This entails conducting a thorough analysis of various factors such as population size, water usage patterns, and environmental conditions. By adopting a comprehensive approach to water supply planning, we can ensure that the community's needs are met sustainably and efficiently. The following are some important types of water demands:

1. Domestic water demand
2. Industrial and commercial water demand
3. Demand for public uses
4. Agricultural water demand
5. Fire demand.

1.7.1 Domestic Water Demand

Water is a vital resource supplied to residential households for both interior and exterior use, encompassing activities such as drinking, cooking, bathing, washing, lawn sprinkling, gardening, and sanitation (Wang et al., 2018). The amount of water required varies depending on various factors such as living conditions, household size, income levels, cultural practices, climate variations, and the efficiency of water fittings installed (Mohandas, 2013). Efficient and effective supply of water to meet domestic demand necessitates careful planning and management of water resources to ensure adequacy and sustainability.

In urban areas, the average daily water consumption per person is estimated to be 135 L under normal conditions (U.S. Geological Survey 2015). However, in developed countries with advanced lifestyles, the demand for water in the domestic sector is notably higher. Globally, household water usage constitutes approximately 8% of total water usage, with a basic household water requirement estimated to be around 50 L per person per day, excluding water usage for gardens.

Ensuring the quality of drinking water is paramount, as it must meet stringent standards to prevent any risk of immediate or long-term harm to human health (Safe Drinking Water Foundation, 2015). Various factors influence domestic water demand, including household income, water prices, family size, age, composition, and weather variables such as temperature and precipitation (Arbués et al. 2003).

According to estimates, different demographic groups have varying water consumption requirements. For instance, the US Geological Survey suggests that a female should use 273.07 L per day, a male 164 L per day, a child 161.21 L, and a retired person 401.38 L per day (Mohandas, 2013). However, consumption patterns can vary significantly across regions and income levels. In India, households with high incomes may consume 250–600 L per person per day, while low-income households may use only 40 L per person per day. Discrepancies in access to water resources are evident globally, with many people in Africa receiving as little as 20 L of water per day, insufficient to meet basic hygiene and cooking needs (Min et al., 2021).

Efforts to address these disparities and ensure equitable access to clean water are crucial for promoting public health and well-being worldwide.

1.7.2 Industrial and Commercial Water Demand

The provision of water to various establishments such as offices, factories, hospitals, hostels, and industries is a crucial aspect of urban planning. The amount of water required varies depending on the nature of the city and the number and types of industries and commercial establishments present.

Water serves as a versatile resource in industrial processes, including food and drink production, washing, concrete batch mixing, and various others (Gleick, 2003). Industrial water use encompasses a wide array of activities, such as fabrication processing, washing and cooling, mining, hydropower generation, and thermal power generation. Different industries may require different types of water sources, ranging from portable water to recycled water or treated effluent from sewage treatment plants, for their manufacturing processes.

Despite the potential benefits of recycled water, its current utilization remains relatively low, constituting only a minor component of total industrial water use (Gleick, 2003). The demand for industrial water varies significantly from city to city, highlighting the importance of matching water supply with specific industrial purposes. Additionally, considerations such as system reliability and the disposal of cooling tower blowdown must be taken into account to ensure efficient and sustainable water management practices in industrial settings.

The industrial sector's need for water can be categorized into four distinct groups, each with its own characteristics and implications for water management:

1. **Cooling Water Demand**: This category involves the extraction of water from rivers or estuaries for cooling purposes, with minimal loss upon return. While some commercial and office buildings may use water-cooled air conditioning systems, this demand is typically not met by the public supply.
2. **Major Industrial Demand**: Industries with consumption exceeding 1000 m^3 per day fall into this category. Examples include papermaking, chemical manufacturing, iron and steel production, and oil refining.
3. **Large Industrial Demand**: Factories falling within this category utilize between 100 and 500 m^3 of water per day. Activities such as food processing, vegetable washing, drink bottling, and chemical production are common in this group. Typically, these demands are met by the public water supply.
4. **Medium to Small Industrial Demand**: This group comprises factories and small manufacturers with water usage below 50 m^3 per day. The majority of these businesses source their water from the public supply.

It is important to recognize that the industrial sector's demand for water can have significant environmental and public supply implications. Therefore, it is crucial for industries to adopt sustainable water management practices to ensure the long-term availability of this vital resource.

1.7.3 Water Demand for Public Use

Public water demand encompasses the total volume of water withdrawn and delivered by both public and private water suppliers to fulfill the needs of various users. These users include domestic, commercial, and industrial entities, each with distinct requirements for water (Hoo, 2019; Min et al., 2021; Nagarajan, 2005). A public water supply system is characterized by serving at least 25 people or having a minimum of 15 service connections. Examples of such systems include those serving cities, towns, military bases, apartment complexes, and large mobile home parks. Additionally, water supplied to public places such as parks, golf courses, swimming pools, municipal buildings, firefighting, and public works activities like street cleaning and sewer flushing are considered part of the public water supply system (Nagarajan, 2005). Public water demand also encompasses the water requirements of various public entities and activities. In summary, the human demand for public water encompasses a broad spectrum of activities and purposes, highlighting the importance of ensuring that public water suppliers can adequately meet users' needs while upholding the safety and quality of the water supply (Fig. 1.2).

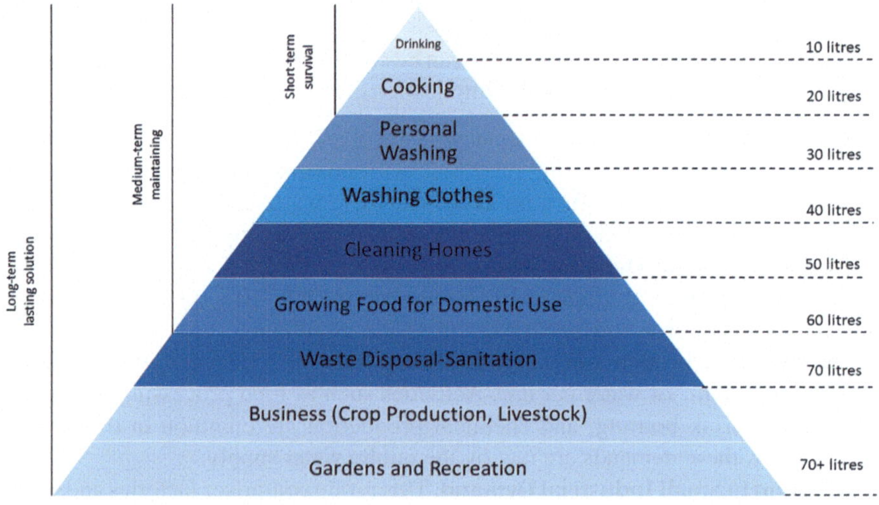

Fig. 1.2 Human requirement of water demand

1.7.4 Fire Demand

In urban areas, the occurrence of a fire outbreak necessitates a reliable and sufficient water supply for firefighting purposes. However, surface water sources like ponds or reservoirs may not always be readily available in busy areas such as towns, or cities (Hoo, 2019; Metcalf & Eddy, 2007; Min et al., 2021).

To address this challenge, it is essential to implement proactive measures, including the storage of water in underground reservoirs strategically located throughout the area. Additionally, the installation of fire hydrants at regular intervals along main pipelines, typically spaced approximately 100–150 m apart, ensures that firefighting teams have immediate access to water when needed (Metcalf & Eddy, 2007).

These measures are critical not only for safeguarding public safety but also for protecting property and infrastructure from the devastating impacts of fires. By investing in these preventative measures, communities can enhance their preparedness and response capabilities, reducing the severity of fire-related incidents and minimizing potential damage.

1.8 Water Quality Standards

Water quality standards play a crucial role in ensuring the safety and sustainability of our water resources for various purposes, including drinking, recreation, and ecological health. These standards are established based on scientific evidences of the potential adverse effects of water contaminants on human health and the environment (WHO, 2017).

Regulatory bodies at the state, territorial, tribal, or federal levels are responsible for setting and enforcing water quality standards to protect human health and aquatic ecosystems (USEPA, 2017). These standards define the desired condition of a water body and outline the necessary measures to achieve or maintain that condition.

The establishment of water quality standards is essential for preserving the integrity of water bodies and ensuring their suitability for different uses, such as swimming, fishing, and drinking water supply (Soltani et al., 2021). By adhering to these standards, authorities can effectively manage water resources and mitigate potential risks to public health and the environment.

Water quality standards serve as a fundamental framework for managing and protecting our water resources. By maintaining high-quality standards, we can safeguard the health of ecosystems and communities that depend on clean and safe water for their well-being and prosperity.

Water quality standards consist of four essential elements crucial for maintaining the health and safety of water bodies:

1. **Designated Uses**: These specify the intended purposes of the water body, such as drinking water supply, aquatic habitat, or recreational activities. Identifying designated uses is essential for establishing appropriate water quality criteria.
2. **Water Quality Criteria**: These are guidelines that limit the concentrations of various pollutants in the water to protect the designated uses. Criteria can be numeric, specifying maximum allowable concentrations of contaminants, or narrative, describing desired water quality conditions.
3. **Antidegradation Policy**: This policy aims to prevent the degradation of water quality in high-quality waters and protect existing uses. It ensures that water quality does not decline over time and that designated uses are maintained without compromise.
4. **General Policies**: These address implementation issues such as low flows, variances, and mixing zones. They provide guidance on how to manage and maintain water quality standards effectively.

States, territories, and authorized tribes may include additional components in their water quality standards as needed. Adhering to these standards is crucial for ensuring that water bodies remain safe and healthy for all designated uses, both now and in the future, in a sustainable manner.

When developing water quality standards, states consider various uses of water and may designate specific use classifications based on the intended purposes. These classifications encompass a range of activities and functions, including.

1.8.1 Public Water Supplies

Public water supplies are crucial sources of water used for drinking and food processing, among other purposes. These supplies may include surface water sources such as rivers, lakes, and reservoirs, as well as groundwater sources such as wells and aquifers. Before being distributed to the public, water from these sources often undergoes treatment processes to remove contaminants and ensure its safety for consumption.

Treatment processes may include filtration, disinfection (such as chlorination), sedimentation, and pH adjustment, among others. These processes are designed to remove or reduce various contaminants, including bacteria, viruses, parasites, chemical pollutants, and other impurities that may be present in the water.

Ensuring the safety and quality of public water supplies is paramount for protecting public health and preventing waterborne diseases. Regulatory agencies, such as the Environmental Protection Agency (EPA) in the United States, set standards and regulations to govern the quality of public water supplies and require regular monitoring and testing to ensure compliance with these standards.

By maintaining high standards for water treatment and distribution, authorities can provide communities with access to clean, safe drinking water that is essential for sustaining life and promoting public health.

1.8.2 Protection and Propagation of Fish Shellfish and Wildlife

The protection and propagation of fish, shellfish, and aquatic flora is an important aspect of water quality standards. This category encompasses activities such as raising cold water fish, warm water fish, shellfish, and oyster propagation, as well as the protection of aquatic flora and wildlife.

In this context, protection refers to measures taken to safeguard the habitats and ecosystems that support fish, shellfish, and aquatic plants. This includes preserving water quality, maintaining adequate flow levels, and minimizing pollution and habitat destruction.

Propagation involves activities aimed at promoting the growth and reproduction of fish, shellfish, and aquatic flora. This may include stocking water bodies with fish species, creating and maintaining spawning grounds, and implementing measures to enhance water quality and habitat suitability for aquatic life.

Wildlife protection within this category extends to waterfowl, shorebirds, and other water-oriented wildlife species that depend on healthy aquatic ecosystems for food, shelter, and breeding. Measures to protect these species may include habitat conservation, pollution control, and the establishment of wildlife reserves and protected areas.

Overall, the protection and propagation of fish, shellfish, and aquatic flora, as well as the conservation of associated wildlife, are critical components of water quality standards aimed at maintaining healthy aquatic ecosystems and promoting biodiversity.

1.8.3 Recreation

Recreational uses of water are an important consideration in water quality standards, encompassing a variety of activities that people engage in for leisure and enjoyment. These recreational uses can be categorized into primary and secondary activities, each involving different levels of immersion in water.

(a) **Primary Recreation**: This category includes activities where individuals are fully immersed in water. Examples include swimming, waterskiing, skin-diving, and surfing. These activities typically require direct contact with water and are often associated with swimming areas, beaches, and water parks.

(b) **Secondary Recreation**: Secondary recreational activities are those where immersion in water is less likely or not required. Examples include boating, wading, rowing, and fishing. While participants may come into contact with water, it is not the primary focus of the activity. Secondary recreation activities are often enjoyed in lakes, rivers, and other water bodies suitable for boating and fishing.

These categories can be further subdivided into various subcategories based on specific recreational pursuits, such as fishing, sailing, power boating, and rafting. Each subcategory may have its own set of water quality considerations and standards to ensure the safety and enjoyment of participants.

Additionally, it's important to recognize and protect recreational uses that do not involve direct contact with water. Activities such as hiking, camping, and bird watching contribute to the overall recreational value of water bodies and can enhance public enjoyment of natural areas. States are encouraged to include these non-contact recreational uses in their water quality standards to promote the conservation and enjoyment of natural resources (USEPA, 2017).

1.8.4 Agriculture and Industry

The classification of water for agricultural and industrial use plays a vital role in assessing its suitability for various purposes, including irrigation, livestock consumption, range grazing, and industrial processes. These classifications help ensure that water quality meets the specific needs of agriculture and industry while protecting crops, livestock, and industrial equipment from potential damage.

(a) **Agricultural Use Classification**: Water classified for agricultural use is evaluated based on its suitability for irrigation, livestock consumption, and range grazing. This classification considers factors such as nutrient content, salinity, pH levels, and the presence of contaminants that could affect soil fertility and crop health. Water quality standards for agricultural use aim to safeguard crops and livestock from damage caused by irrigation and other exposures. These standards help farmers make informed decisions about water management practices to optimize agricultural productivity while minimizing environmental impacts.

(b) **Industrial Use Classification**: Water designated for industrial use is primarily used for cooling and various industrial processes. This classification ensures that water quality meets the specific requirements of industrial applications, including cooling systems and manufacturing processes. Industrial water must be free from contaminants and impurities that could damage industrial equipment or interfere with production processes. Water quality standards for industrial use focus on protecting equipment from corrosion, scaling, and fouling, thereby ensuring efficient operation and prolonging equipment lifespan.

While specific criteria may not have been developed exclusively for agricultural and industrial water uses, existing water quality standards for human health and aquatic life are typically stringent enough to protect these uses. Regulatory agencies may adapt and apply these standards to agricultural and industrial contexts, taking into account the unique needs and characteristics of each sector. By ensuring that water quality meets the requirements of agricultural and industrial users, these classifications contribute to sustainable water management practices and support the continued growth and development of agriculture and industry.

1.8.5 Other Uses

States have the authority to adopt additional designated uses for water bodies beyond those specified in federal regulations, such as coral reef preservation, marinas, groundwater recharge, aquifer protection, and hydroelectric power generation (Labaster & Lloyd, 2013). These additional designated uses may require the establishment of specific water quality criteria to protect them effectively. For instance, in India, the Central Pollution Control Board (CPCB) has introduced the concept of "designated best use," which identifies the highest-quality use of water for a particular body of water among several potential uses (Soltani et al., 2021). This classification aids water quality managers and planners in setting water quality targets and developing appropriate restoration programs for various water bodies.

Water quality criteria are fundamental components of state water quality standards, defined as the elements expressed as constituent concentrations, levels, or narrative statements (USEPA, 2017). They represent the conditions necessary to maintain the physical, chemical, and biological integrity of water bodies and protect their designated uses. If water quality criteria are exceeded, there may be risks to human health or adverse effects on ecological integrity.

Water quality criteria can take two forms: numerical and narrative. Numerical criteria specify the maximum allowable pollutant levels in a water body, while narrative criteria describe the desired conditions for a water body to be free from specific contaminants or adverse conditions (Soltani et al., 2021). These criteria are essential tools for establishing water quality standards that protect public health and welfare, enhance water quality, and fulfill the objectives of the Clean Water Act. By adhering to water quality criteria, regulatory agencies can ensure the integrity of water bodies and promote the sustainable management of water resources.

The narrative criterion for water quality serves as a crucial set of guidelines to ensure the safety and purity of our water supply. These guidelines encompass various aspects to safeguard against potential contaminants and maintain the integrity of water bodies. Some examples of the narrative criteria include:

- Ensuring that water is free from toxic, radioactive, conventional, non-conventional, deleterious, or other polluting substances that could hinder the fulfillment of specified designated uses.
- Preventing the presence of substances in water that are attributable to wastewater discharges or other pollutant sources, which could lead to issues such as objectionable deposit settling, debris floating, or the formation of nuisances like scum, oil, or other matter.
- Avoiding the occurrence of undesirable characteristics in water, such as objectionable color, odor, taste, or turbidity, which could affect its suitability for consumption or use.
- Ensuring that water does not contain substances that may cause injury, toxicity, or adverse physiological responses in humans, animals, or plants.
- Preventing the proliferation of undesirable or nuisance aquatic life in water bodies.

These guidelines are established to maintain the quality of our water supply and protect both human health and the environment. Adherence to these narrative criteria is essential to ensure that our water remains safe for consumption and use, promoting the well-being of individuals and ecosystems alike.

1.8.6 Antidegradation Requirements

One of the fundamental principles of the Clean Water Act is to preserve the physical, chemical, and biological integrity of water bodies. This is achieved through the implementation of antidegradation requirements, which serve as a framework for maintaining and safeguarding the quality of water that has already been attained. The key components of antidegradation requirements include designated uses and water quality criteria, which outline the objectives for preserving the existing uses of different water bodies, preventing deterioration and loss of aesthetic value, safeguarding against threats to indigenous aquatic life, and conserving national water resources.

It is imperative for each nation to formulate, adopt, and uphold national and state-specific antidegradation policies related to water quality standards. These policies must be incorporated into water quality management practices to ensure effective implementation (Labaster & Lloyd, 2013; Soltani et al., 2021). If antidegradation requirements are not explicitly addressed in the standard regulations of a state or nation, they must be specifically referenced in the water quality standards to establish a clear functional relationship between the policy and the standards. State and central water regulatory authorities should implement more stringent and mandatory antidegradation statements to establish water quality standards. The implementation procedures for antidegradation must delineate, on a case-by-case basis, whether and to what extent the water quality may be permitted to decline (Tables 1.8 and 1.9).

1.9 Water Quality Monitoring

Freshwater is a vital resource essential for agriculture, industry, and human survival, playing a critical role in achieving sustainable development (Chan & Han, 2018). It sustains all life forms, including humans, animals, and plants (Demetillo et al., 2019). However, water quality can either support life and good health or pose risks of disease and illness. Unfortunately, rapid environmental degradation driven by population growth, development, and climate change has heightened the urgency for researchers to address the negative impact on water sources (Bakker, 2012; Demetillo et al., 2019). Water pollution in different water bodies like lakes, streams, rivers, and oceans—has led to a global demand for advanced environmental monitoring systems, particularly in water quality monitoring (Demetillo et al., 2019).

Table 1.8 International comparison of drinking water quality standards and guidelines for chemicals (maximum allowable concentration)

Parameters	Unit	Country standards					Health effects
		WHO	E.U.	USEPA	CANADA	NESREA	
Chemical parameters							
Conductivity	μS/cm	1000				1000	None
Total dissolved solids (TDS)	mg/L	500	500	500	500	500	Based on taste; TDS > 500 mg/L results in excessive scaling in water pipes, water heaters, boilers and appliances
Total suspended solids (TSS)	mg/L	–	–	–	–	–	
Total solids	mg/L	500	500	500	500	500	
Total alkalinity	mg/L	100					
Total hardness	mg/L	100			150		
Calcium hardness	mg/L	75					None
Magnesium hardness	mg/L	30					
Aluminium	mg/L	0.2	0.2	0.2	0.2	0.2	Potential neuro-degenerative disorders
Arsenic	mg/L	0.01	0.01	0.01	0.01	0.01	Cancer
Barium	mg/L	0.7	–	2	1		Hypertension
Bromate	mg/L	0.01	0.01	0.01	0.01		Renal cell tumors (classified as probable carcinogen)

(continued)

Table 1.8 (continued)

Parameters	Unit	Country standards					Health effects
		WHO	E.U.	USEPA	CANADA	NESREA	
Cadmium	mg/L	0.003	0.005	0.005	0.005	0.003	Toxic to kidney
Chloride	mg/L	100	250	250	250	250	None
Copper	mg/L	2	2	1	1	1	Gastro-intestinal disorder
Chromium	mg/L	0.05	0.05	0.1	0.05		Cancer
Cyanide	mg/L	0.07	0.05	0.2	0.2	0.07	Very toxic to the thyroid and the nervous system
2,4-dichlorophenol	mg/L	–	0.0001	–	0.9		Possibly carcinogenic
Fluoride	mg/L	1.5	1.5	4	1.5	1.5	Fluorosis, skeletal tissue (bones and teeth) morbidity
Iron	mg/L	0.3	0.2	0.3	0.3	0.3	None
Lead	mg/L	0.01	0.01	0.015	0.01	0.01	Cancer, interference with vitamin D metabolism, affects mental development in infants, toxic to the central and peripheral nervous systems
Manganese	mg/L	0.05	0.05	0.05	0.05	0.05	Neurological disorder
Mercury	mg/L	0.001	0.001	0.002	0.001		Affects kidney and central nervous system

(continued)

Table 1.8 (continued)

Parameters	Unit	Country standards					Health effects
		WHO	E.U.	USEPA	CANADA	NESREA	
Metolachlor	mg/L	0.01	0.001	–	0.05		
Nitrate	mg/L	11	11	10	10	50	Cyanosis, and asphyxia (blue-baby syndrome) in infants under 3 months
Potassium	mg/L	10	10	10	10	10	People with kidney infections or other conditions, such as heart disease, coronary artery disease, hypertension etc., and those who are taking medications that interfere with normal body potassium handling should avoid the consumption of water treated by water softeners using potassium chloride
Sulphate	mg/L	200	250	250	500	100	500 mg/L has potential to cause physiological effects

(continued)

Table 1.8 (continued)

Parameters	Unit	Country standards					Health effects
		WHO	E.U.	USEPA	CANADA	NESREA	
Sodium	mg/L	200	200	200	200	200	Sodium-based water softener is used, a separate unsoftened supply for cooking and drinking purposes is recommended
2,4,6-trichlorophenol	mg/L	0.2	0.0001	–	0.005		Cancer
Trihalomethanes	mg/L	–	0.1	0.08	0.1		Cancer
Physical parameters							
Color	TCU	3	20	15	≤ 15	15	None
Odour	Threshold odor no	Unobjectinable	≤ 1 @ 25 °C	Unobjectionable	Inoffensive	Unobjectionable	None
Temperature	Degree Celsius				≤ 15	Ambient	None
Turbidity	NTU	≤ 5	4	0.5–1.0	0.1–1.0	≤ 5	None
Taste	Dilution number	Unobjectinable	≤ 1	Unobjectinable	Inoffensive	Unobjectinable	None
pH		6.5–8.5	6.5–8.5	6.5–8.5	6.5–8.5	6.5–8.5	None

(continued)

Table 1.8 (continued)

Parameters	Unit	Country standards					Health effects
		WHO	E.U.	USEPA	CANADA	NESREA	
Biological parameters							
E. coli	cfu/mL	0	0	0	0	0	Urinary tract bacteremia, meningitis, diarrhea (one of the main cause of infections, morbidity and mortality among children), acute renal failure and haemolytic anaemia
Faecal streptococcus	cfu/mL	0	0	0	0	0	Indication of recent faecal contamination
Presumptive	cfu/mL	0	0	0	0	0	
Aerobic mesophillic	cfu/mL	100	100	100	100	100	
Fungi/yeast	cfu/mL	0	0	0	0	0	
Total coliform count	cfu/mL					10	Indication of faecal contamination
Clostridium perfringes spore	cfu/mL	0	0	0	0	0	Index of intermittent faecal contamination

Note All standards and guidelines in mg/L
A dash (–) indicates that no standard or guideline has been established for a given parameter
Source Guidelines for Canadian Drinking Water Quality (2006)
US EPA National Primary Drinking Water Standards
E.U. Council Directive on the quality of water intended for human consumption
Australian Drinking Water Guidelines, 2004
World Health Organization Drinking Water Quality Guidelines (2004)

Table 1.9 Comparison of selected parameters listed by different agencies and countries

Parameter	Symbol	WHO	European Union	United States	China	Canada	India (BIS)
1,2-dichloroethane		–	3.0 µg/L	5 µg/L	–	–	
Acrylamide		–	0.10 µg/L	TT[b]	–	–	
Aluminium	Al		0.2 mg/L			No limit listed	0.03 mg/L
Antimony	Sb	ns	5.0 µg/L	6.0 µg/L	–	6.00 µg/L	
Arsenic	As	10 µg/L	10 µg/L	10 µg/L	50 µg/L	10.0 µg/L	0.05 mg/L
Barium	Ba	700 µg/L	ns	2 mg/L	–	1.00 mg/L	
Benzene		10 µg/L	1.0 µg/L	5 µg/L	–	–	
Benzo(a)pyrene		–	0.010 µg/L	0.2 µg/L	0.0028 µg/L	–	
Beryllium	Be					–	
Boron	B	2.4 mg/L	1.0 mg/L	–	–	5.00 mg/L	1.0 mg/L
Bromate		–	10 µg/L	10 µg/L	–	–	
Cadmium	Cd	3 µg/L	5 µg/L	5 µg/L	5 µg/L	5.00 µg/L	0.01 mg/L
Calcium	Ca					200 mg/L	75 mg/L
Chromium	Cr	50 µg/L	50 µg/L	0.1 mg/L	50 µg/L (Cr6)	0.050 mg/L	0.05 mg/L
Cobalt	Co					–	
Copper	Cu	–	2.0 mg/L	1.3 mg/L[a]	1 mg/L	1.00 mg/L	0.05 mg/L
Cyanide		–	50 µg/L	0.2 mg/L	50 µg/L	–	0.05 mg/L
Epichlorohydrin		–	0.10 µg/L	TT[b]	–	–	
Fluoride		1.5 mg/L	1.5 mg/L	4 mg/L	1 mg/L	–	1.0 mg/L
Gold	Au					No limit listed	
Hardness	$CaCO_3$					0–75 mg/L = soft	300 mg/L
Iron	Fe		0.2 mg/L			0.300 mg/L	0.3 mg/L
Lanthanum	La					No limit listed	
Lead	Pb	–	10 µg/L	15 µg/L[a]	10 µg/L	10.0 µg/L	0.05 mg/L

(continued)

Table 1.9 (continued)

Parameter	Symbol	WHO	European Union	United States	China	Canada	India (BIS)
Magnesium	Mg					50.0 mg/L	30 mg/L
Manganese	Mn		0.05 mg/L			0.050 mg/L	0.1 mg/L
Mercury	Hg	6 μg/L	1 μg/L	2 μg/L	0.05 μg/L	1.00 μg/L	0.001 mg/L
Molybdenum	Mo					No limit listed	
Nickel	Ni	–	20 μg/L	–	–	No limit listed	
Nitrate		50 mg/L	50 mg/L	10 mg/L (as N)	10 mg/L (as N)	–	45 mg/L
Nitrite		3 mg/L	0.50 mg/L	1 mg/L (as N)	–	–	
Pesticides (total)		–	0.50 μg/L	–	–	–	Absent
Pesticides (individual)		–	0.10 μg/L	–	–	–	
pH						6.5–8.5	6.5–8.5
Phosphorus	P					No limit listed	
Polycyclic aromatic hydrocarbons		–	0.10 μg/L	–	–	–	
Potassium	K					No limit listed	
Scandium	Sc					No limit listed	
Selenium	Se	40 μg/L	10 μg/L	50 μg/L	10 μg/L	10.0 μg/L	0.01 mg/L
Silicon	Si					No limit listed	
Silver	Ag					0.050 mg/L	
Sodium	Na					200 mg/L	
Strontium	Sr					No limit listed	
Tetrachloroethene and trichloroethene		40 μg/L	10 μg/L	–	–	–	
Tin	Sn					No limit listed	
Titanium	Ti					No limit listed	
Tungsten	W					No limit listed	
Uranium	U					0.10 mg/L	

(continued)

Table 1.9 (continued)

Parameter	Symbol	WHO	European Union	United States	China	Canada	India (BIS)
Vanadium	V					No limit listed	
Zinc	Zn					5.00 mg/L	5.0 mg/L
Vinyl chloride			0.50 μg/L				
Chlorides			250 mg/L				250 mg/L
Electrical conductivity			2500 μS/cm at 20 °C				
Total dissolved solids							500 mg/L
Sulphate							200 mg/L

Note – indicates that no standard has been identified by editors of this article
ns indicates that no standard exists
μg/L = micrograms per litre, or 0.001 ppm; mg/L = 1 ppm, or 1000 μg/L
[a] Means action level; not a concentration standard. A public water system exceeding the action level must implement "treatment techniques" which are enforceable procedures
[b] TT (treatment technique). The public water system must certify that the combination of dose and monomer level does not exceed: acrylamide = 0.05% dosed at 1 mg/L (or equivalent); epichlorohydrin = 0.01% dosed at 20 mg/L (or equivalent)
Source

- Guidelines for Drinking-water Quality, Fourth Edition; World Health Organization; 2011
- US EPA, OP (2013-02-22). "Summary of the Safe Drinking Water Act" (US EPA)
- Task Force of Canadian Council of Resource and Envir. Ministers Guidelines for Canadian Drinking Water Quality, 1996. Amend. Health Canada (2006)
- "Indian Standard Drinking Water Specification". Central Ground Water Board
- Australian Drinking Water Guidelines (2011) (http://www.nhmrc.gov.au/guidelines/publications/eh52). "Council Directive 98/83/EC of 3 November 1998 on the quality of water intended for human consumption, ANNEX I: PARAMETERS AND PARAMETRIC VALUES, PART B: Chemical parameters" EUR-Lex
- "Primacy Enforcement Responsibility for Public Water Systems". Drinking Water Requirements for States and Public Water Systems. Washington, D.C.: United States Environmental Protection Agency (EPA)

The wasteful use of freshwater and water pollution threaten the developmental projects, necessitating water treatment to produce safe drinking water. Increased chemical use in various manufacturing processes, such as fertilizers, farming, construction, and industrial production, as well as direct discharge of polluted water into nearby water bodies by industries, has significantly contributed to the global decline in water quality (Cloete et al., 2016). Discharge of toxic chemicals, over-pumping of aquifers, long-range atmospheric transport of pollutants, and contamination promoting algal growth (potentially leading to eutrophication) are major causes of water quality degradation today. It is imperative to take action to protect freshwater resources and ensure their sustainability for future generations. Monitoring water quality is crucial for environmental protection, involving detection and control of contaminant substances harmful to aquatic resources. The Water Act of 1974 established pollution control boards responsible for enforcing provisions including maintaining and restoring the "wholesomeness of our water resources." To achieve this, a system of water use characterization was developed to determine the degree of wholesomeness to be maintained or restored (Moparthi et al., 2018).

Water quality monitoring is integral to the rational development and management of water resources. Over time, there has been ongoing evolution in methods for gathering information on changes in water quality. Monitoring entails obtaining information on water systems, including chemistry, biology, statistics, and economics (Pasika & Gandla, 2020; Sugapriyan et al., 2018). It spans various types of water uses and functions and the nature of water resources, whether surface water, groundwater, or seawater.

Monitoring water quality in the twenty-first century presents challenges due to numerous chemicals used daily that can enter water bodies. It involves sampling and analyzing water components and conditions, including dissolved oxygen, bacteria, nutrients, and pollutants like pesticides, metals, and oils (Central Pollution Control Board (CPCB), 2008). Water quality monitoring requires a comprehensive, multidisciplinary approach to ensure the health and safety of aquatic resources. Factors such as pH and temperature significantly influence water quality; temperature can affect dissolved oxygen levels, while pH can impact the toxicity of ammonia (Bartram & Ballance, 1996). Monitoring, defined by the International Organization for Standardization (ISO), involves programmed processes of sampling, measurement, and recording or signaling various water characteristics to assess conformity to specified objectives.

Monitoring activities include long-term, short-term, and continuous programs standardized for measuring and observing the status and trends of aquatic environment. Surveys are finite duration, intensive programs measuring and observing aquatic environment quality for specific purposes, while surveillance is continuous, specific measurement, and observation for water quality management and operational activities. Implementing these monitoring activities ensures water quality maintenance and prompt addressing of potential issues.

1.9.1 Elements of Water Quality Monitoring

Water quality monitoring is vital for safeguarding the safety and health of our water resources. However, before initiating any water sampling and analysis, it is crucial to clearly define the required information and what is already available. This helps establish the main objective of the monitoring program and address any existing gaps. To achieve this, preparing a "water quality monitoring program document" or "study plan" is useful, outlining the objectives and potential limitations of the program.

The study plan should commence with a clearly defined objective for the monitoring program and a comprehensive description of the study area. This involves identifying the geographical boundaries of the area, existing and planned water uses for various purposes, and current and anticipated point and non-point sources of contamination (Pasika & Gandla, 2020). By doing so, the study plan can effectively cater to the specific needs of the study area. The key elements of a study plan include a clear statement of aims and objectives, description of the concerned study area, characterization of the sampling sites, identification of the water quality variables to be measured, proposed frequency and timing of sampling, estimation of the required resources for implementing the design, and a plan for quality control and assurance. Adhering to these guidelines enables the design and implementation of water quality monitoring programs effectively, ensuring the safety and health of our water resources (Fig. 1.3).

1.9.2 Purpose of Monitoring

Water quality monitoring has traditionally served to ensure water suitability for its intended purposes. Over time, it has evolved to include impact monitoring, which examines trends in aquatic environment quality and assesses how human activities and waste treatment operations affect it. Moreover, monitoring now extends to estimating nutrient or pollutant fluxes discharged into lakes, oceans and other available ecosystems. Background quality monitoring is also common to establish a basis for comparing impact monitoring results.

Prior to implementing a monitoring and assessment program, preliminary surveys are crucial for determining the program's focus. Water quality monitoring and assessment rely on fundamental physical, chemical, and biological water properties. However, it involves analysis, interpretation, and communication of these properties within the broader context of human activity, conservation of the natural environment, and varied needs at local, national, or international levels. The ultimate goal is to provide management-relevant information, which can vary based on institutions, resources, and priorities. Water quality monitoring and assessment serve diverse purposes and can be conducted from multiple perspectives.

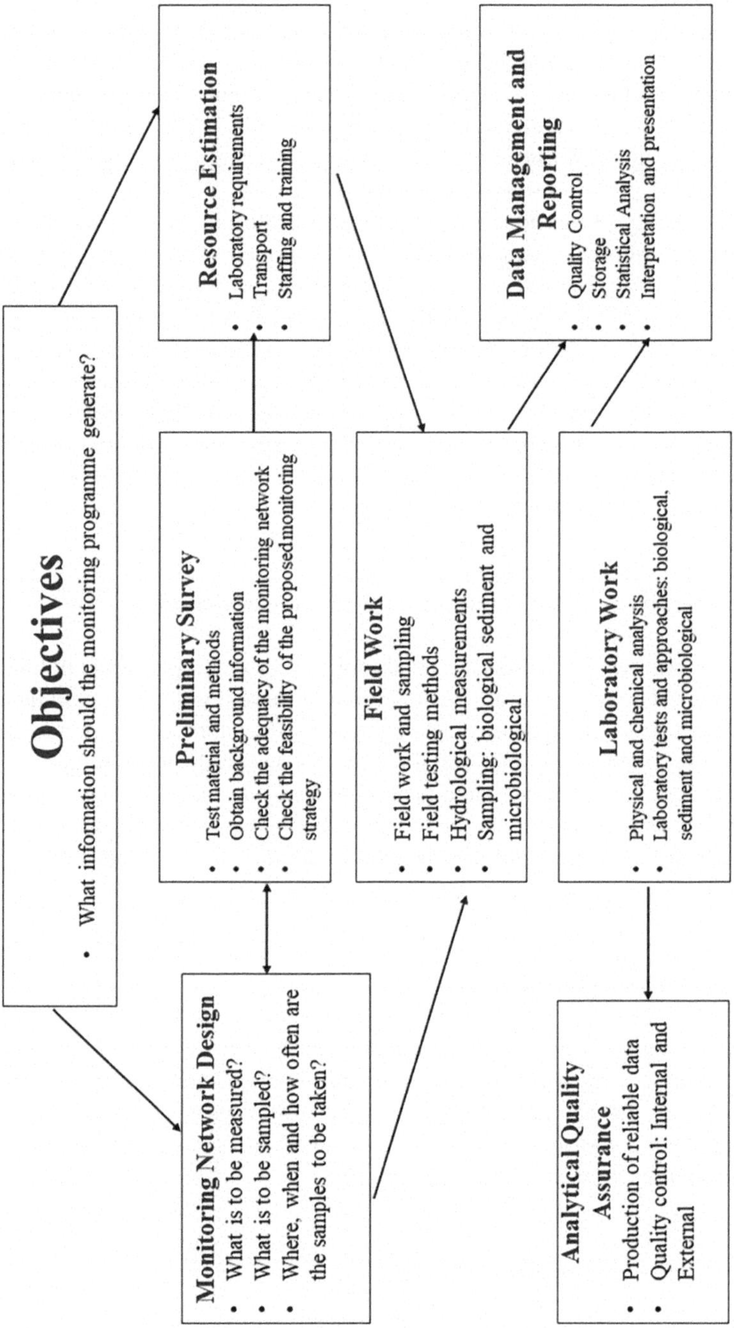

Fig. 1.3 Links between the critical elements of water quality monitoring

1.9.2.1 To Identify Whether Waters Are Meeting Designated Uses

To ascertain whether water bodies are meeting their designated uses, states have established specific criteria that dictate allowable concentrations of chemical pollutants. When these pollutants exceed the maximum or minimum allowable concentrations, the waters may no longer support beneficial uses like fishing, swimming, and drinking. These designated uses, coupled with specific criteria and antidegradation statements, collectively form water quality standards.

State water quality professionals evaluate water quality by comparing chemical pollutant concentrations in streams to the criteria outlined in the state's standards. However, water quality monitoring alone may not suffice to determine if the needs of aquatic life are being met in a stream. While certain constituents like dissolved oxygen and temperature are crucial for sustaining healthy fish and aquatic insect populations, other factors such as the physical structure and habitat condition of the stream also play a significant role, if not a greater one. Maintaining water quality standards is vital to ensure water bodies can support their designated uses. Through monitoring and assessing water quality, areas in need of improvement can be identified, prompting action to safeguard our invaluable water resources.

1.9.2.2 To Identify Specific Pollutants and Sources of Pollution

Water quality monitoring serves as a crucial tool for pinpointing specific pollutants and their sources. Through accurate identification of these problematic pollutants, a direct correlation with stream quality can be established, enabling appropriate action to be taken. It is worth noting that certain activities are predisposed to generating specific pollutants. For example, bacteria and nutrients are more likely to originate from an animal feedlot than an automotive repair shop. By establishing an initial link between the pollution source and the detected pollutants, the necessity for further investigation or ongoing monitoring can be justified.

1.9.2.3 To Determine Trends

Trend analysis is another critical application of water quality monitoring. By continuously monitoring chemical constituents at regular intervals and employing consistent methodologies, patterns and trends that emerge over time can be discerned. This analytical approach is indispensable in detecting and addressing potential pollution problems before they worsen.

1.9.2.4 To Screen for Impairment

Water quality monitoring serves as an invaluable early warning system for detecting impairments in our water bodies. When excessive levels of one or more chemical constituents are identified, it serves as a clear indication of potential pollution problems. By promptly identifying these issues, we can swiftly take action to prevent any further damage to our precious waterways.

1.9.3 Objectives of Water Quality Monitoring

Clear and concise objectives play a pivotal role in the success of any water quality monitoring program (Bartram & Ballance, 1996; Demetillo et al., 2019). By meticulously documenting these objectives, one can carefully evaluate the available options, leading to a more effective and efficient program. To facilitate the design of sampling networks, the following list presents a range of typical monitoring objectives. Although not exhaustive, the sample can serve as a valuable starting point:

- Identification of baseline conditions in the watercourse system
- Detection of any signs of deterioration in water quality
- Identification of any water bodies in the watercourse system that do not meet the desired water quality standards
- Identification of any contaminated areas
- Determination of the extent and effects of specific waste discharges
- Estimation of the pollution load carried by a watercourse system or subsystem
- Evaluation of the effectiveness of a water quality management intervention
- Development of water quality guidelines and/or standards for specific water use
- Development of regulations covering the quantity and quality of waste discharge
- Development of a water pollution control program.

There are two distinct types of monitoring programs: single objective programs, designed to address a specific issue, and multi-objective programs, covering a range of water uses, including drinking water supply, industrial manufacturing, fisheries, irrigation, and aquatic life. In practice, programs and projects typically combine multiple objectives, and data is used for various purposes.

Monitoring programs with multiple objectives are often the first major national programs established in a country. The design of such programs requires preliminary survey work to ensure that the selection of sampling sites takes into account various factors, such as actual and potential water uses, sources of pollution, pollution control operations, local geochemical conditions, and the type of water body. By implementing a multi-objective monitoring program, countries can effectively manage their water resources and ensure that they are being used sustainably. This type of program provides valuable data that can be used to inform policy decisions and improve water management practices. It is essential to establish a comprehensive monitoring program to ensure that water resources are protected and managed effectively.

1.10 Sampling Method for Water Quality Monitoring

Water sampling is a crucial process that involves taking a portion of water for analysis or testing, ensuring that water quality standards are met for various purposes such as drinking, bathing, or industrial use. The primary objective of water sampling is to provide accurate information on water quality and quantity, as well as the hydrodynamic properties of the systems under observation. To achieve this, certain conditions must be met:

- Problem-oriented sampling devices are required to safeguard the original qualitative and quantitative properties of subsurface waters.
- The water collected at the sampling point should reflect groundwater at the same depth in the aquifer system, and the flow should not be disturbed by well construction.
- In aquifer systems with strong time-dependent characteristics, sampling must be adapted to the actual representative information of the system by adequate pumping before sampling.
- By adhering to these conditions, water sampling can provide valuable insights into the quality and quantity of water resources, essential for effective management and conservation.

Sampling water for subsequent analysis is often considered easier than sampling soils for two main reasons. First, water tends to be more homogeneous than soil, meaning there is less variability between samples collected within the same vicinity. Second, it is often physically easier to collect water samples because it can be done with pumps and hose lines. Additionally, known volumes of water can be collected from known depths with relative ease. The amount of water collected depends on the sample being evaluated and can vary from 100 mL for potable drinking water collected from water utilities to 1000 L for groundwater samples. Sampling strategies are also less complicated for water samples since, in many cases, the water is mobile, and a set number of bulk samples are collected from the same point at various time intervals. For marine water, samples are often collected sequentially in time within the defined area of interest (Speight, 2019).

The objective of sampling is to collect a representative sample, in which the relative proportions or concentrations of all pertinent components are the same as in the material being sampled (CPCB, 2018; Eaton et al., 2000; Etikan & Bala, 2017; Keith, 1996). The sample volume should be small enough for transportation but large enough for analytical purposes. It is essential to ensure that the sample is representative to obtain accurate results.

1.10.1 General Requirements of Water and Wastewater Sampling

To ensure accurate analysis, it is crucial to obtain a sample that meets the requirements of the sampling program. Additionally, it is important to handle the sample with care to prevent deterioration, contamination, or compromise, prior to analysis. This can be achieved by following proper handling procedures and utilizing appropriate storage methods. By taking these precautions the integrity of the sample can be maintained, resulting in reliable and meaningful data. Before use, all sampling equipment should be thoroughly cleaned and quality-assured, to make them free of any contaminants as for organic analysis sampling it is recommended to bake the bottles at 450 °C.

It is important not to premix the sample, when filling the sample containers, as premixing can result in the loss of any pre-added preservative and may even bias the results if certain components are adhere to the container's sides. Depending on the determinations to be performed, the container should either be filled full (for most organic compound analysis) or left with space for aeration, mixing, etc. (for microbiological and inorganic analyses). If the bottle already contains preservative, care must be taken to avoid overfilling, as the preservative may be lost or diluted. For sampling volatile organic compounds, an air space equivalent to approximately 1% of the container volume should be left to allow for thermal expansion during shipment. Special precautions must be taken when sampling for organic compounds and trace metals, as they may be present in low concentrations, making them easily lost or contaminated if proper sampling and preservation procedures are not followed. By following these guidelines we can ensure accurate and reliable results for our samples.

Collecting over time, at various depths, or at various sampling points can yield composite samples. However, the details of collection may vary depending on local conditions, so specific recommendations cannot be universally applied. It is important to note that composite sampling may not be suitable for certain analytes where quantitative values are desired due to the inherent instability of certain properties and compounds. Examples of such analytes include residual, iodine, hexavalent chromium, nitrate, volatile organic compounds, radon-222, dissolved oxygen, ozone, temperature, and pH. However, regulatory agencies frequently use composite samples for BOD. It is recommended to refrigerate composite samples for BOD and nitrite.

Several factors can affect the accuracy of results, including the presence of suspended matter or turbidity, the method chosen for removing a sample from its container, and the physical and chemical changes brought about by storage or aeration. Therefore, detailed procedures are essential when processing samples to be analyzed for trace constituents, especially metals and organic compounds.

For metals, it is often appropriate to collect both a filtered and an unfiltered sample to differentiate between total and dissolved metals present in the matrix. It is important to add the same relative amount of acid to all samples and use an ultrapure acid preservative to prevent contamination. When filtered samples are to be collected, it is recommended to filter them in the field or at the point of collection before preserving them with acid. It is crucial to maintain a record of the sample collection process, including the date, time, location, and any relevant details. By following these guidelines, accurate and reliable results can be obtained for environmental monitoring and analysis.

General Information

- Sample identification number
- Location
- Sample collector
- Date and hour
- Sample type (Grab or composite).

Specific Information

- Water temperature
- Stream flow
- Water level
- Any other information.

The information may be attached using a tag, labeling or writing on container with water proof ink.

- Description of sampling points
- Using landmarks on map
- Using global positioning system.

- When collecting samples from a river or stream, it is important to consider various factors that may affect the observed results, such as depth, stream flow, and distance from each shore. The selection of the number and distribution of sampling sites should be based on study objectives, stream characteristics, available equipment, and other relevant factors.
- If equipment is available, it is recommended to take an integrated sample from top to bottom in the middle of the main channel of the stream or from side to side at mid-depth. However, if only grab or catch samples are collected, it is preferable to take them at various points of equal distance across the stream. If only one sample is collected, it should be taken in the middle of the main channel of the stream and at mid-depth.
- As rivers, streams, lakes, and reservoirs are subjected to considerable variations from normal causes such as seasonal stratification, diurnal variations, rainfall, runoff, and wind, the location, depth, and frequency of sampling should be chosen based on local conditions and the purpose of the investigation.
- During sample collection, it is important to avoid areas of excessive turbulence, as it may result in the loss of volatile constituents and the potential presence of denser-than-air-toxic vapors. Sampling at weirs should also be avoided if possible, as such locations tend to favor the retrieval of lighter-than-water, immiscible compounds.

- To ensure accurate results, samples should be collected beneath the surface in quiescent areas and the sampling container should be opened below the surface with the mouth directed toward the current to avoid collecting surface scum unless oils and greases are the constituents of interest. If composite samples are required, it is important to ensure that sample constituents are not lost during compositing due to improper handling of the portions being composited. If samples are to be analyzed for organic constituents, composited portions should be refrigerated.

1.10.2 Safety Considerations

During sampling, it is vital to prioritize safety by adhering to the following precautions:

1. **Avoid Consumption**: Eating, drinking, or smoking should be strictly prohibited in the laboratory or near samples and sampling locations to prevent contamination.
2. **Prevent Ignition Sources**: Keep sparks, flames, or excessive heat sources away from samples and sampling locations. If flammable compounds are suspected or known to be present, refrigerate the samples and use specially designed explosion-proof refrigerators.
3. **Proper Labeling**: Label any sample that is known or suspected to be hazardous due to flammability, corrosivity, toxicity, oxidizing chemicals, or radioactivity. This ensures that appropriate precautions are taken during sample handling, storage, and disposal.

By following these safety guidelines, the laboratory personnel can be protected, and the accuracy of the results can be maintained.

1.10.3 Collection of Water Samples

When conducting a physical examination of water, it is essential to use fully cleaned, ordinary buckets or plastic cans. However, for chemical tests, glass bottles with a capacity of more than two liters are recommended. These bottles should be thoroughly washed and cleaned, and containers for bacteriological tests must be handled by individuals free from any disease. Furthermore, containers and bottles should be cleaned with sulfuric acid, potassium dichromate, or alkaline permanganate, then rinsed with distilled water before sterilization. After collecting water samples, immediately close the bottles and cover them with a cloth to prevent dirt accumulation. It is advisable to test water samples as soon as possible for accurate results.

During water sample collection, the following things need to be considered:

- **Tap Water**: Allow a sufficient quantity of wastewater to pass through the tap before collecting the sample to eliminate stagnant water.

- **Surface Water**: Collect samples about 40–50 cm below the surface of streams or rivers to avoid surface impurities such as oils and tree leaves. Use strainers to remove impurities during intake collection.
- **Ground Sources**: Pump out a sufficient quantity of water from wells or tube wells before collecting samples. This ensures representative sampling and minimizes contamination.

1.10.4 Types of Samples

1.10.4.1 Grab Sample

Grab samples are a valuable tool in environmental sampling, providing a snapshot of a specific location at a particular moment in time. These samples can be collected as discrete grab samples and depth-integrated grab samples. Discrete grab samples are taken at a specific location, depth, and time, while depth-integrated grab samples are collected over a predetermined part of the entire depth of a water column at a selected location and time in a given body of water (Valenzuela et al., 2020).

It is important to note that a sample can only represent the composition of its source at the time and place of collection. However, if a source is known to be relatively constant in composition over an extended period or distance, then a single grab sample may represent a longer time period and/or a larger volume than the specific time and place at which it was collected. Examples of sources that can be adequately represented by single-grab samples include protected groundwater supplies, water supplies receiving conventional treatment, and some well-mixed surface waters. However, wastewater streams, rivers, large lakes, shorelines, estuaries, and groundwater plumes may require more frequent sampling due to their variability.

When a source is known to vary with time, grab samples collected at suitable intervals and analyzed separately can document the extent, frequency, and duration of these variations (Diwan et al., 2013; Valenzuela et al., 2020). Sampling intervals should be chosen based on the expected frequency of changes, which may vary from as little as 5 min to as long as 1 h or more. Seasonal variations in natural systems may necessitate sampling over months. When the source composition varies in space, samples should be collected from appropriate locations that meet the objectives of the study, such as upstream and downstream from a point source.

1.10.4.2 Grab Sampling Methods for Sewage Treatment Plants

At sewage treatment plants, grab samples are collected in situations where frequent changes in character and concentrations could affect the treatment process. These samples are taken when undesirable constituents are suspected, and the quality is not expected to remain constant, or when on-the-spot analysis is required for parameters such as dissolved oxygen (DO), pH, and residual chlorine. For instance, testing suspended solids (SS) in the clarifier overflow requires independent samples correlated with the time of sampling, as SS levels can vary between low flows, average flows, and peak flows. Typically, SS levels during peak flows, occurring for a few hours in the early morning, may be higher. Without timing information, it might wrongly suggest that the entire 24 h performance has a higher SS value.

To ensure accuracy, representative samples should be taken with good judgment and analyzed within 2–3 h of sampling. A well-washed, clean PVC or plastic bucket connected to a sturdy, long handle may be suitable for grab sampling (Kumar et al., 2011; Valenzuela et al., 2020). The quality of the sample collected is critical to the accuracy of the analysis. Therefore, it is essential to use appropriate equipment and follow proper sampling procedures to obtain reliable results. By doing so, we can ensure that the sewage treatment plant operates efficiently and effectively, protecting public health and the environment.

1.10.4.3 Composite Samples

Composite samples are mixtures of individual samples, typically taken from different locations, times, or sub-samples, and combined to form a single sample for analysis. When collecting samples for analysis, employing time-composite sampling is crucial for determining components that remain unchanged throughout the process of sample collection, preservation, and storage. However, this method should only be used if there is a high level of confidence that the components being tested will not undergo any change during the sampling process.

To collect individual portions, it is recommended to use a wide-mouth bottle and collect samples on an hourly basis. In some cases, it may be necessary to collect samples every half hour or even every 5 min. At the end of the sampling period, it is essential to mix the samples together in the bottle or combine them in a single bottle as they are being collected. If preservatives are used, they should be added to the sample bottle initially to ensure that all portions of the composite are preserved as soon as they are collected (Bhushan, 2019; Dodd et al., 2006).

Notably some automatic sampling devices are available; however, they should only be used if the sample is preserved as described above. Composite samplers running for extended periods, ranging from weeks to months, should undergo regular cleaning of containers and sample lines to minimize the growth of samples and the formation of deposits. By following these guidelines, samples can be collected and preserved in a manner that yields accurate and reliable results.

1.10.4.4 Composite Sampling Methods for Sewage Treatment Plants

To ensure accurate results in sewage treatment plant testing, continuous sample analysis is essential because sewage quality fluctuates throughout the day. Continuous analysis minimizes the time required for operators to focus on plant operation. However, for tests sensitive to rapid changes, such as dissolved oxygen and sulfide tests, samples can be taken at hourly or two-hourly intervals. During sample collection, immediate preservation is crucial using suitable ice boxes until transportation to a laboratory and storage in a refrigerator for analysis. Preventing anaerobic decomposition is important to maintain the composition and characteristics of organic portions like BOD. The required hourly flow and sampling volume for a 12 h composite sampling at a sewage treatment plant can be found in Table 1.10 (Valenzuela et al., 2020). After collecting samples for a 24 h period, they should be composited together to form a single 24 h composite sample, accurately representing sewage quality for the entire day. Following these procedures ensures accurate and reliable results for sewage treatment plant testing.

It is crucial to exclude large sewage solids from any sample, especially those exceeding 6 mm in diameter. To ensure accuracy, the resulting total volume of a composite sample should be taken as 1140 mL. The factor (*) mentioned in the table is a multiplying factor for flow, adjusted to arrive at a convenient sample volume. This factor remains constant for a particular sewage treatment plant and during the composite sampling period to maintain proportionality between flow and sample volumes. During composite sampling and at the exact moment of testing, it is essential to remix the collected samples to ensure they have the same composition. Failure to mix the samples can cause changes in results, particularly for samples of solids that settle out rapidly, such as those in activated sludge or raw sewage. Therefore, it is crucial to thoroughly mix the samples and pour them quickly before settling occurs. Failure to do so can result in errors of 25–50%, significantly impacting result accuracy.

Table 1.10 Hourly flow and sample volumes to be measured for a 12 h sampling during composite sampling

Time	Flow (MLD)	Factor (*)	Sample volume (mL)	Time	Flow (MLD)	Factor (*)	Sample volume (mL)
6 AM	0.2	100	20	12 N	1.5	100	150
7 AM	0.4	100	40	1 PM	1.2	100	120
8 AM	0.6	100	60	2 PM	1.0	100	100
9 AM	1.0	100	100	3 PM	1.0	100	100
10 AM	1.2	100	120	4 PM	1.0	100	100
11 AM	1.4	100	140	5 PM	0.9	100	90

1.10.4.5 Integrated (Discharge-Weighted) Samples

Integrated sampling, facilitated by discharge-weighted methods like equal width increment (EWI) or equal discharge increment (EDI) procedures, is a robust approach for obtaining accurate information about water composition. By combining grab samples collected from different points simultaneously, proportional to their relative flows, this method allows for a comprehensive evaluation of average composition or total loading in rivers, streams, lakes, and reservoirs. Integrated sampling is particularly valuable when assessing combined treatment for diverse wastewater streams, considering their interactions and overall treatment efficacy. However, in cases where local variations hold greater significance, analyzing samples separately is advisable. Effective implementation of integrated sampling requires specialized equipment capable of uniformly collecting water samples across the depth profile, along with a thorough understanding of the water body's volume, movement, and composition. To ensure accuracy and reliability, integrated sampling procedures should be meticulously outlined in a comprehensive sampling plan.

Integrated sampling is indeed a valuable approach for gaining precise insights into the composition of water sources. Through the utilization of discharge-weighted methods and specialized equipment, integrated sampling allows for the collection of a mixture of samples representing different points across the cross-section. This facilitates a more accurate assessment of average composition or total loading. However, in scenarios where local variations hold greater significance, analyzing samples separately is advisable. Irrespective of the chosen approach, having a well-defined sampling plan is crucial to guarantee accurate and dependable results.

1.11 Chain-of-Custody Procedures

Property designed and executed chain-of-custody forms are essential for maintaining the integrity of samples throughout the entire process, from collection to data reporting. This involves meticulously tracking the possession and handling of the sample from the moment it is collected until its final disposition. The chain of custody is particularly crucial when the data derived from the samples are to be used for regulatory or legal purposes. However, even in cases where litigation is not a concern, chain-of-custody procedures are valuable for the routine control of samples (Roy, 2019). Understanding what constitutes custody of a sample is fundamental to ensuring its integrity. A sample is considered to be under someone's custody if it is in their physical possession, within their sight, securely and tamper-proofed by them, or stored in an area restricted to authorized personnel (American Public Health Association (APHA), 2004; Yasuda & Bowen, 2006). The key aspects of chain-of-custody procedures include proper sample collection, labeling, and documentation. Additionally, maintaining a secure and controlled environment for the sample, including appropriate storage and transportation measures, is essential. By adhering to these procedures, we can guarantee the integrity of the samples and the accuracy of the resulting data.

Filling out a chain-of-custody record for each sample or group of samples is essential to maintain integrity throughout the process. This record includes crucial information such as:

- Sample number
- Collector's signature
- Date, time, and address of collection
- Sample type
- Sample preservation requirements
- Signatures of individuals involved in the chain of possession
- Inclusive dates and times of possession.

By documenting this information, the chain-of-custody record ensures that the sample's integrity is maintained from collection to analysis.

1.12 Sample Labels (Including Bar Code Labels)

To prevent confusion or misidentification of samples, the use of labels is essential. Gummed paper labels or tags are commonly employed for this purpose. These labels should contain specific information such as a unique sample number, sample type, collector's name, date and time of collection, collection location, and sample preservative. Furthermore, including the date and time of preservation allows for comparison with the collection details. For accuracy, it is advisable to attach tags or self-adhesive labels to sample containers either before or at the time of collection. Proper labeling is a foundational aspect of any scientific study or experiment, ensuring correct identification and tracking of samples throughout the process. By incorporating all necessary information on the label, researchers can readily access and analyze the data, saving time and preventing errors or discrepancies due to misidentification. Adhering to established labeling guidelines is essential, as it guarantees accurate and reliable results.

1.13 Sample Seals

To detect unauthorized tampering with samples prior to analysis, employing sample seals is essential. The self-adhesive paper seals contain the essential information including:

- Sample number (matching the number on the sample label)
- Collector's name
- Date and time of sampling.

These seals act as an additional layer of security, ensuring the integrity of the samples until analysis is performed.

1.14 Field Log-Books

During field surveys or sampling activities, maintaining a detailed record in a bound logbook is essential for accurate documentation. The logbook should contain pertinent information such as:

- Purpose of the sampling
- Location of sampling points
- Name and address of the field contact
- Producer of the material being sampled, including their address if different from the location
- Type of sample
- Method, date, and time of preservation.

For wastewater samples, it is important to identify the process producing the waste stream and provide a suspected sample composition, including concentrations. Additionally, the logbook should include:

- Number and volume of sample(s) taken
- Description of sampling points and methods
- Date and time of collection
- Collector's sample identification number(s)
- Sample distribution and method of transportation
- Relevant references such as maps or photographs of the sampling site
- Field observations and measurements
- Signatures of personnel responsible for observations.

Recording sufficient information ensures that the sampling event can be reconstructed accurately without relying solely on the collector's memory. Therefore, it is crucial to safeguard the logbook and store it in a secure location. By adhering to these guidelines, the accuracy and thoroughness of field surveys or sampling activities can be ensured.

1.15 Sample Analysis Request Sheet

The sample analysis request sheet serves as a crucial document accompanying samples to the laboratory. The collector fills out the field portion, including pertinent information from the log book. The laboratory personnel completes the laboratory portion, which includes the name of the recipient, laboratory sample number, date of receipt, sample condition, and determinations to be performed. This sheet ensures that the laboratory has all necessary information for accurate analysis.

1.16 Sample Delivery to the Laboratory

It is crucial to deliver the sample(s) to the laboratory promptly after collection, preferably within two days. For shorter holding times, arrangements should be made to ensure timely delivery. Ensure that each sample is accompanied by a sample analysis request sheet and a complete chain-of-custody record. To guarantee proper handling and storage, the sample must be delivered to the sample custodian as quickly as possible.

1.17 Receipt and Logging of Samples

Upon arrival at the laboratory, the sample custodian conducts a thorough inspection of the samples to ensure they are in good condition and the seals are intact. They reconcile the label information and seal with the chain-of-custody record before accepting the sample for analysis. Upon acceptance, the custodian assigns a laboratory number, logs the sample in the laboratory's records, and stores it securely in a designated storage area at the appropriate temperature. The sample remains stored until it is assigned to an analyst for analysis.

1.18 Assignment of Samples for Analysis

The laboratory supervisor oversees the assignment of samples for analysis to an analyst, who assumes responsibility for the sample's care and custody throughout the analytical process. This ensures meticulous handling of the sample, thereby enhancing the accuracy and reliability of the results obtained.

1.19 Disposal

Samples should be retained for the specified duration required by the project or until the data have been thoroughly reviewed and accepted. The disposal process must be meticulously documented to ensure compliance with local, state, and US EPA-approved disposal methods (US Environmental Protection Agency 1986). Proper disposal procedures are imperative for safeguarding the environment and maintaining the safety of laboratory personnel.

1.20 Methods of Sampling

There are several methods involved in the sampling of water and wastewater that include.

1.20.1 Manual Sampling

This method, although requiring minimal equipment, can be time-consuming, especially for large-scale or routine sampling programs. It necessitates trained field technicians and is often essential for regulatory and research investigations that demand a meticulous assessment of field conditions and intricate sample collection techniques. Manual sampling proves particularly valuable for collecting samples containing oil and grease.

1.20.2 Automatic Sampling

It is a method used to collect samples from various environments without manual intervention, often at pre-set intervals or triggered by specific conditions. It ensures consistent and representative sampling over time, improving accuracy in monitoring fluctuating variables like water quality, or industrial discharge. Automatic samplers can be programmed to take discrete or composite samples, providing flexibility depending on the research objectives. This method reduces human error, enhances safety, and increases efficiency, particularly in remote or hazardous environments. Commonly used in environmental science, industrial monitoring, and wastewater management, automatic sampling aids in real-time data collection.

1.20.3 Sorbent Sampling

The utilization of solid sorbents, especially membrane-type disks, is gaining popularity. This method offers rapid and cost-effective sampling if the analytes of interest can be efficiently adsorbed and desorbed, and if the water matrix is devoid of particulates that could clog the sorbent.

Sampling water and wastewater is a vital process that demands careful selection of the appropriate method. Whether employing manual sampling, automatic sampling, or sorbent sampling, it is essential to ensure that the chosen method aligns with the specific requirements of the sampling program and that the collected samples accurately represent the water or wastewater under examination.

1.21 Sampling Remediation Protocol and Sampling Points

Regulatory authorities play a crucial role in implementing pre-determined monitoring programs aimed at ensuring the quality of water from natural water systems. These programs typically involve routine sampling and analysis to assess the water's condition and detect any pollution. If pollution is identified, it is essential to implement an agreed-upon remediation plan and monitor its effectiveness in reducing pollution levels.

To establish effective remediation protocols, guidelines and standard test methods must be available to assess water quality accurately. When selecting sampling points, it is important to consider each locality individually and adhere to general criteria and guidelines:

1. **Representative Samples**: Sampling points should be chosen to ensure that the samples taken accurately represent the source from which water is obtained. This ensures that the data obtained from the samples are reflective of the overall water quality.
2. **Contamination Points**: Sampling points should include locations that may be susceptible to contamination. This allows for the identification of potential sources of pollution and enables targeted remediation efforts.
3. **Uniform Distribution**: Sampling points should be evenly distributed throughout the water system to capture variations in water quality across different areas. This ensures comprehensive monitoring and assessment of the entire system.
4. **System Representation**: The chosen sampling points should generally yield samples that are representative of the entire water system. This ensures that the data obtained accurately reflect the overall condition of the system and facilitate informed decision-making regarding remediation efforts.

By adhering to these guidelines and selecting appropriate sampling points, regulatory authorities can effectively monitor water quality, detect pollution, and implement remediation measures to safeguard public health and the environment.

Following these guidelines ensures effective monitoring of water systems and proper implementation of remediation plans, ultimately contributing to a safer and healthier environment for all. Variable or random sampling regimes are advantageous for detecting local issues but may not be ideal for tracking changes over time. However, the accuracy of any analysis depends on the proper collection, storage, and handling of the samples (Minamoto et al., 2021).

Sample bottles should be clean, though sterilization may not always be necessary, and preservatives may be required for certain analytes. Immediate testing of parameters like residual chlorine, pH, and turbidity is essential, as they can change during storage and transport. Minimizing the time between sampling and analysis is crucial, and storing samples in glass or polyethylene bottles at a lower temperature in the dark helps maintain their integrity (APHA, 2004).

Microbiological examination of water is vital for assessing water quality, requiring isolation and enumeration of organisms indicating microbial contamination. However, only reference laboratories should isolate specific pathogens from water for disease outbreak analysis and management.

In rivers, sampling points typically focus on the surface water layer, avoiding debris in the top layer. Additional sampling throughout the water column may be necessary for specific study objectives. For lakes and oceans, sampling points vary based on factors like geography, inflows, depth, tides, and currents (Roy, 2019).

When sampling underground water, careful site selection is crucial, considering factors like water flow, geological structure, and nearby activities. Avoiding bias in site selection ensures a comprehensive understanding of the underground water in the area, leading to more accurate and reliable results (Nielsen & Nielsen, 2006).

1.22 Sampling Location for Sewage Treatment Plants

In a sewage treatment plant (STP), the number of sampling stations is theoretically limitless. However, effective monitoring requires a minimum number of strategically placed sampling locations and tests to avoid unnecessary data accumulation and resulting confusion. Since sewage passes through the STP on a time-deferred scale, simultaneous sampling from inlet to outfall may not accurately represent performance.

Raw sewage samples should be collected post-screens or grit chambers, while samples from primary or secondary clarifiers should be taken from troughs or pipes before discharge weirs. Sampling points in aeration tanks of the activated sludge process should be areas of good mixing (APHA, 2004).

Samples of water entering septic tanks, Imhoff tanks, and other single-treatment units like waste stabilization ponds should be taken in leading chambers or channels, while outlet samples should be collected outside these units in receiving weirs, channels, or chambers. Specific sampling within tanks should be specified in terms of depth or distance.

Raw sludge samples should be taken from sludge sumps or delivery sides of sludge pumps. A return sludge sample from the activated sludge plant should be collected at the discharge point into the aeration tank, while mixed primary and secondary sludge samples should be collected upon delivery to the digester.

Digested sludge samples can be drawn from digester sampling points or the discharge end of delivery pipes leading to drying beds or mechanical dewatering units. Digester supernatant can be drawn from sampling cocks or ports on the digester dome. For SVI determination, samples can be collected between the aeration tank and secondary clarifier. Following these guidelines ensures accurate and representative data, facilitating better monitoring and management of the STP.

References

Aggarwal, S. C., & Kumar, S. (2011). Industrial water demand in India: Challenges and implications for water pricing. *India Infrastructure Report*, 274–280.

American Public Health Association (APHA). (2004). *Standard methods for the examination of water and wastewater* (22nd ed.). American Water Works Association.

Arbués, F., García-Valiñas, M. Á., & Martínez-Espiñeira, R. (2003). Estimation of residential water demand: A state-of-the-art review. *Journal of Socio-Economics, 32*(1), 81–102.

Arbués, F., García-Valiñas, M. Á., & Villanúa, I. (2020). Making decisions on industrial water sources: The case of Zaragoza, Spain. *Urban Water Journal, 17*(2), 122–135.

Bakker, K. (2012). Water security: Research challenges and opportunities. *Science, 337*, 914–915.

Bartram, J., & Ballance, R. (Eds.). (1996). *Water quality monitoring: A practical guide to the design and implementation of freshwater quality studies and monitoring programmes*. CRC Press.

Bhushan, B. (2019). Bioinspired water collection methods to supplement water supply. *Philosophical Transactions of the Royal Society A, 377*(2150), 20190119.

Central Pollution Control Board (CPCB). (2008). *Guidelines for water quality monitoring* (35 pp.). Parivesh Bhawan, East Arjun Nagar.

Central Pollution Control Board (CPCB). (2018). *Guide manual: Water and wastewater analysis* (181 pp.).

Chan, A., & Han, H. (2018). Water quality assessment and pollution sources in a river basin using multivariate statistical techniques. *Environmental Science and Pollution Research, 25*(24), 24107–24119.

Chandra, S., Singh, P. K., Tiwari, A. K., Panigrahy, B., & Kumar, A. (2015). Evaluation of hydrogeological factor and their relationship with seasonal water table fluctuation in Dhanbad district, Jharkhand, India. *ISH Journal of Hydraulic Engineering, 21*, 193–206.

Chilton, J. (1996). Groundwater pollution—Developments in water science. In *Water quality assessments—A guide to use of biota, sediments and water in environmental monitoring* (2nd ed., Vol. 5, p. 273). Chapman and Hall Ltd.

Cloete, N. A., Malekian, R., & Nair, L. (2016). Design of smart sensors for real-time water quality monitoring. *IEEE Access, 4*, 3975–3990.

Demetillo, A. T., Japitana, M. V., & Taboada, E. B. (2019). A system for monitoring water quality in a large aquatic area using wireless sensor network technology. *Sustainable Environment Research, 29*(1), 1–9.

Diwan, V., Lundborg, C. S., & Tamhankar, A. J. (2013). Seasonal and temporal variation in release of antibiotics in hospital wastewater: Estimation using continuous and grab sampling. *PLoS ONE, 8*(7), e68715.

Dodd, P. A., Price, M. R., Heywood, K. J., & Pebody, M. (2006). Collection of water samples from an autonomous underwater vehicle for tracer analysis. *Journal of Atmospheric and Oceanic Technology, 23*(12), 1759–1767.

Drexler, I. L. C., Prieto, A. L., & Yeh, D. (2014). Wastewater constituents. *Earth Systems and Environmental Sciences, Comprehensive Water Quality and Purification, 3*, 7–29.

Eaton, A. D., Clesceri, L. S., & Greenberg, A. E. (2000). *APHA, AWWA, WEF, standard methods for the examination of water and wastewater* (19th ed.).

Edokpayi, J. N., Enitan, A. M., Mutileni, N., & Odiyo, J. O. (2018). Evaluation of water quality and human risk assessment due to heavy metals in groundwater around Muledane area of Vhembe District, Limpopo Province, South Africa. *Chemistry Central Journal, 12*(2).

Edokpayi, J. N., Odiyo, J. O., & Durowoju, O. S. (2017). Impact of wastewater on surface water quality in developing countries: A case study of South Africa (Chap. 18). In T. Hlanganani (Ed.), *Water quality* (pp. 401–416). IntechOpen.

Environmental Protection Agency (EPA). (2017). *Water quality standards handbook: Chapter 3: Water quality criteria* (EPA-823-B-17-001). EPA Office of Water, Office of Science and Technology.

Etikan, I., & Bala, K. (2017). Sampling and sampling methods. *Biometrics & Biostatistics International Journal, 5*(6), 00149.

Fan, L., Liu, G., Wang, F., Ritsema, C. J., & Geissen, V. (2014). Domestic water consumption under intermittent and continuous modes of water supply. *Water Resources Management, 28*(3), 853–865.

Gary, G. (2005). From oasis to mirage: The aquifers that won't replenish. *World Watch, 8*, 30–37.

Gasim, M. B. (2015). Water quality degradation of Cempaka Lake, Bangi, Selangor, Malaysia as an impact of excessive *E. coli* and nutrient concentrations. *Malaysian Journal of Analytical Sciences, 19*(6), 1391–1404.

GESAMP. (1988). *Report of the eighteenth session, Paris 11–15 April 1988* (GESAMP reports and studies No. 33). United Nations Educational, Scientific and Cultural Organization, Paris.

Gleick, P. H. (2003). Water use. *Annual Review of Environment and Resources, 28*(1), 275–314.

Gleick, P. H., & Iwra, M. (1996). Basic water requirements for human activities: Meeting basic needs. *International Water, 21*(2), 83–92.

Goldar, B. N. (2003). Water use in Indian industry: Estimates of value and price elasticity of demand. In K. Chopra, C. H. Hanumantha Rao, & R. Sengupta (Eds.), *Resources, sustainable livelihoods and eco-system services*. Concept Publishing Company.

Hall, D. (2011, January). Public–public partnerships in water: An overview. In *Confluence of Ideas and Organisations. International Conference on Water Partnerships Towards Meeting the Climate Challenge*, Chennai, India (pp. 53–62).

Hoo, R. (2019). Managing water demand in Singapore through a systems perspective. *International Journal of Water Resources Development, 36*(6), 879–887.

Howard, G., Bartram, J., Brocklehurst, C., Colford Jr, J. M., Costa, F., Cunliffe, D., ... & Wright, C. Y. (2020). COVID-19: urgent actions, critical reflections and future relevance of 'WaSH': lessons for the current and future pandemics. *Journal of water and health, 18*(5), 613–630.

Keith, L. H. (1996). *Compilation of EPA's sampling and analysis methods* (2nd ed.). Lewis Publ./ CRC Press.

Kumar, V., Nakada, N., Yasojima, M., Yamashita, N., Johnson, A. C., & Tanaka, H. (2011). The arrival and discharge of conjugated estrogens from a range of different sewage treatment plants in the UK. *Chemosphere, 82*(8), 1124–1128.

Labaster, J. S., & Lloyd, R. S. (2013). *Water quality criteria for freshwater fish*. Elsevier.

Lee, D. D. (2004). *When smoke ran like water: Tales of environmental deception and the battle against pollution*. Basic Books.

Marcos, A. C., Al-Kassir, A., Cuadros, F., & Yusaf, T. (2017). Treatment of slaughterhouse waste water mixed with serum from lacteal industry of extremadura in Spain to produce clean energy. *Energies, 10*(6), 765.

McComas, S. (2003). *Lake and pond management*. Lewis Publishers. https://doi.org/10.1201/978 0203010082

Metcalf & Eddy, Inc., Tchobanoglous, G., Burton, F. L., & Stensel, H. D. (2007). *Wastewater engineering: Treatment and reuse* (4th ed.). McGraw-Hill Education.

Meybeck, M., & Helmer, R. (1996). An introduction to water quality. In D. Chapman (Ed.), *Water quality assessments: A guide to the use of biota, sediments, and water in environmental monitoring* (2nd ed., 648 pp.). Chapman and Hall.

Meybeck, M., Kimstach, V., & Helmer, R. (1992). *Water quality assessment: A guide to the use of biota, sediments, and water in environmental monitoring* (pp. 19–50). Chapman and Hall.

Min, K. J., Kim, J. H., & Park, K. Y. (2021). Characteristics of heavy metal separation and determination of limiting current density in a pilot-scale electrodialysis process for plating wastewater treatment. *Science of the Total Environment, 757*, 143762.

Minamoto, C., Fujiwara, N., Shigekawa, Y., Tada, K., Yano, J., Yokoyama, T., ... & Nakayama, S. (2021). Effect of acidic conditions on decomposition of methylene blue in aqueous solution by air microbubbles. *Chemosphere, 263*, 128141.

Mohamed, A. K., Liu, D., Mohamed, M. A. A., et al. (2018). Groundwater quality assessment of the quaternary unconsolidated sedimentary basin near the Pi river using fuzzy evaluation technique. *Applied Water Science, 8*, 65. https://doi.org/10.1007/s13201-018-0711-0

Mohandas, M. (2013). India's water crisis: A situation we can solve together. *Water Resource Research, 24*, 675–683.

Moparthi, N. R., Mukesh, C., & Sagar, P. V. (2018). Water quality monitoring system using IoT. In *2018 Fourth International Conference on Advances in Electrical, Electronics, Information, Communication, and Bio-Informatics (AEEICB)* (pp. 1–5). IEEE.

Nagarajan, R. (2005). *Water conservation*. Capital Publishing Company.

Nema, A. K. (2012). *Water: Sources and requirement: Water and wastewater treatment processes*. Department of Civil Engineering, I.I.T. Delhi.

Nielsen, D. M., & Nielsen, G. (2006). *The essential handbook of ground-water sampling*. CRC Press.

Pasika, S., & Gandla, S. T. (2020). Smart water quality monitoring system with cost-effective using IoT. *Heliyon, 6*(7), e04096.

Prüss-Ustün, A., Wolf, J., Bartram, J., Clasen, T., Cumming, O., Freeman, M. C., Gordon, B., Hunter, P. R., Medlicott, K., & Johnston, R. (2019). Burden of disease from inadequate water, sanitation, and hygiene for selected adverse health outcomes: An updated analysis with a focus on low- and middle-income countries. *International Journal of Hygiene and Environmental Health, 222*(5), 765–777. https://doi.org/10.1016/j.ijheh.2019.05.004

Ramulongo, L., Nethengwe, N. S., & Musyoki, A. (2017). The nature of urban household water demand and consumption in Makhado Local Municipality: A case study of Makhado Newtown. *Procedia Environmental Sciences, 37*, 182–194.

Roy, P.K., Majumder, A., Mazumdar, S., Majumder, M., & Roy, M.B. (2011). Impact of enhanced flow on the flow system and wastewater characteristics of sewage-fed fisheries in India. *African Journal of Environmental Science and Technology, 5*(7), 512–521.

Roy, R. (2019). An introduction to water quality analysis. *ESSENCE International Journal of Environmental Rehabilitation and Conservation, IX*(1), 94–100.

Roy, R., & Majumder, M. (2017). Comparison of surface water quality to land use: A case study from Tripura, India. *Desalination and Water Treatment, 85*, 147–153.

Safe Drinking Water Foundation. (2015). Retrieved September 2015, from www.safewater.org

Satapathy, I., & Syed, T. H. (2015). Characterization of groundwater potential and artificial recharge sites in Bokaro district, Jharkhand (India), using remote sensing and GIS-based techniques. *Environmental Earth Sciences, 74*, 4215–4232.

Sharma, S., Sarma, S. J., & Tay, J. H. (2019). Aerobic granulation in wastewater treatment: A general overview. *Microbial Wastewater Treatment*, 57–81.

Soltani, A. A., Oukil, A., Boutaghane, H., Bermad, A., & Boulassel, M. R. (2021). A new methodology for assessing water quality, based on data envelopment analysis: Application to Algerian dams. *Ecological Indicators, 121*, 106952.

Speight, J. G. (2019). *Natural water remediation: Chemistry and technology*. Butterworth-Heinemann.

Sugapriyan, K., Krishnamoorthy, R., & Radhakrishnan, S. (2018). Assessment of water quality index for the determination of water quality of drinking water sources in Tamil Nadu, India. *Environmental Monitoring and Assessment, 190*(7), 404.

Swistock, B., & Clark, J. (2015). Pre-gas drilling drinking water testing—An educational opportunity for extension. *The Journal of Extension, 53*(1), 22.

Tyagi, S., Sharma, B., Singh, P., & Dobhal, R. (2013). Water quality assessment in terms of water quality index. *American Journal of Water Resources, 1*(3), 34–38.

US Environmental Protection Agency. (1986). *Test methods for evaluating solid waste: Physical/chemical methods* (3rd ed.) (Publication No. SW-846). Office of Solid Waste and Emergency Response, Washington, D.C.

USEPA (2017) National Primary Drinking Water Regulations. U.S. Environmental Protection Agency, Washington DC. https://www.epa.gov/ground-water-and-drinking-water/national-primary-drinking-water-regulations

U.S. Geological Survey. (2015). *Mineral commodity summaries 2015*. U.S. Geological Survey. https://minerals.usgs.gov/minerals/pubs/mcs/2015/mcs2015.pdf

Valenzuela, E. F., Menezes, H. C., & Cardeal, Z. L. (2020). Passive and grab sampling methods to assess pesticide residues in water: A review. *Environmental Chemistry Letters, 18*(4), 1019–1048.

Van Wyk, P., & Scarpa, J. (1999). Water quality requirements and management. In *Farming marine shrimp in recirculating freshwater systems* (pp. 141–162).

Verma, P., Singh, P. K., Sinha, R. R., et al. (2020). Assessment of groundwater quality status by using water quality index (WQI) and geographic information system (GIS) approaches: A case study of the Bokaro district, India. *Applied Water Science, 10*(1), 27. https://doi.org/10.1007/s13201-019-09792-6

Wang, Y., Wan, T., & Tortajada, C. (2018). Water demand framework and water development: The case of China. *Water, 10*(1). https://doi.org/10.3390/w10010010

Winter, M., & Kickuth, R. (1989). Elimination of sulphur compounds from wastewater by the root zone process—I. Performance of a large-scale purification plant at a textile finishing industry. *Water Research, 23*(5), 535–546.

World Health Organization. (2004). *Guidelines for drinking-water quality* (3rd ed.). WHO. http://www.who.int/water_sanitation_health/dwq/gdwq3/en/index.html

World Health Organization (WHO). (2004). *Guidelines for drinking water quality: Training pack*. WHO.

World Health Organization (WHO). (2017). *Guidelines for drinking-water quality* (4th ed.). World Health Organization.

WWAP (UNESCO World Water Assessment Programme). (2019). *The United Nations world water development report 2019: Leaving no one behind*. UNESCO.

Yasuda, T., & Bowen, R. E. (2006). Chain of custody as an organizing framework in seafood risk reduction. *Marine Pollution Bulletin, 53*(10–12), 640–649. https://doi.org/10.1016/j.marpolbul.2006.10.012

Yotova, G., Lazarova, S., Kudłak, B., Zlateva, B., Mihaylova, V., Wieczerzak, M., Venelinov, T., & Tsakovski, S. (2019). Assessment of the Bulgarian wastewater treatment plants' impact on the receiving water bodies. *Molecules, 24*(12), 2274. https://doi.org/10.3390/molecules24122274

Chapter 2
Water and Wastewater Treatment Technologies

2.1 Water Treatment

Water treatment is defined as any unit process that alters the chemical, physical, and bacteriological characteristics of water to render it safe for human consumption and appealing to consumers. Additionally, treatment serves to safeguard water distribution system components from corrosion (Wacławek et al., 2017). Water treatment encompasses a series of unit processes, such as screening, sedimentation, flotation, coagulation, filtration, adsorption, ion exchange, gas transfer, oxidation, biological reactions, and disinfection. When implemented continuously with all necessary engineering components and accessories, this assembly of processes constitutes a water treatment plant (WTP). The primary objective of a WTP is to bring about the desired change in water quality, irrespective of the source of the water under treatment.

The choice of treatment process employed in a WTP hinges on various factors, including the source of water, capital and operating expenses, reliability, ease of operation, traditions, and prevailing practices (Droste & Gehr, 2018). The relative significance of these factors fluctuates depending on the specific project context. Water and wastewater treatment systems are indispensable facilities designed to remove harmful substances that can cause illness or nuisance. At its core, the fundamental aim of water treatment operations is to safeguard public health, with the overarching objective of delivering potable and aesthetically pleasing water. Ultimately, well-executed water treatment protocols yield water that is safe for consumption and aesthetically pleasing in appearance, taste, and odor.

S. A. Bandh and B. Mushtaq, *Wastewater Treatment Technology*, Springer Water,
https://doi.org/10.1007/978-3-031-86684-5_2

2.2 Historical Background of Wastewater Treatment

Water and wastewater treatment constitute a crucial practice that has undergone significant evolution. While drainage systems date back to before the nineteenth century, waste disposal posed considerable challenges. The common practice involved depositing "night soil" in buckets along streets and subsequently emptying them into "honeywagon" tanks, with the waste then transported to rural areas and spread over agricultural lands. However, the introduction of flush toilets in the nineteenth century resulted in a surge in waste volume, posing transportation difficulties. Consequently, cities began directing water and wastewater through drainage and storm sewers into water bodies, despite Chadwick's (1842) recommendation to segregate rainwater and sewage, directing the latter to the soil. Unfortunately, this approach led to severe pollution and health issues downstream.

In 1842, English engineer Lindley constructed the first "modern" sewerage system for water and wastewater conveyance in Hamburg, Germany. Although the Lindley system has undergone enhancements with improved materials, inclusion of manholes, and sewer appurtenances, its underlying principles persist. The imperative for water and wastewater treatment arose when water bodies exceeded their assimilative capacity, leading to intolerable health concerns. Numerous methods were experimented with between the late 1800s and early 1900s until the development of contemporary treatment processes around 1920. However, the design remained largely empirical until the mid-twentieth century.

Centralized water and wastewater systems were designed and promoted, with the expenses of treatment borne by communities discharging into the plant. Over time, the evolution of modern sewerage systems and treatment processes has substantially enhanced the quality of life for people globally. Nevertheless, continuous improvement and innovation in this domain are indispensable to ensure the provision of clean and safe water for future generations.

2.3 Objectives of Water Treatment

The evolution of wastewater treatment methods initially stemmed from concerns regarding public health and the adverse impacts of wastewater discharge on the environment. As urban areas expanded, the land available for wastewater disposal via irrigation and intermittent filtration became inadequate, necessitating alternative treatment approaches capable of expediting natural processes under controlled conditions in smaller facilities.

From about 1900 to early 1970s the primary objectives of wastewater treatment centered on the removal of suspended and floatable materials, treatment of biodegradable organics, and elimination of pathogenic organisms. However, these objectives were inconsistently met across the United States, evidenced by numerous plants discharging partially treated wastewater into wells since the 1960s.

Subsequently, the objectives shifted to prioritize aesthetic and environmental concerns. While earlier goals of reducing biological oxygen demand, suspended solids, and pathogenic organisms persisted, they were addressed at heightened levels. Moreover, the removal of nutrients such as nitrogen and phosphorus gained attention, particularly in inland streams and lakes.

State and national agencies embarked on a significant endeavor to achieve more effective and widespread wastewater treatment to enhance surface water quality. This initiative was propelled by a deeper understanding of the environmental repercussions of wastewater discharges, the long-term adverse effects of specific wastewater constituents, and a national commitment to environmental preservation.

Since 1980, the landscape of wastewater treatment has undergone substantial transformation owing to heightened scientific knowledge and an expanded information repository. The focus shifted towards health concerns associated with toxic and potentially toxic chemicals released into the environment. While water quality improvement objectives from the 1970s remain pertinent, emphasis is now pivoted towards defining and eliminating toxic and trace compounds that may induce long-term health repercussions. Consequently, the requisite degree of treatment significantly escalated, with additional treatment objectives and goals being incorporated. Notably, the removal of toxic compounds like refractory organics and heavy metals exemplifies additional treatment objectives under consideration.

In the contemporary technological era, wastewater treatment objectives garnered significant attention. Globally, the prime objectives of wastewater treatment are being redefined to ensure efficacy in removing toxic and trace compounds posing risks to human health and the environment. Addressing biodegradable organic substance reduction is a critical imperative, necessitating the breakdown of organic matter containing carbon, nitrogen, phosphorus, and sulfur via oxidation into gases released or remaining in solution. Similarly, curbing nutrient concentration in the environment is vital, as wastewater laden with nitrogen and phosphorus can fuel eutrophication and proliferation of algae and aquatic plants, subsequently depleting oxygen levels and endangering aquatic life. Pathogen elimination is paramount in wastewater treatment, considering the potential disease transmission from bacteria, viruses, fungi, protozoa, and helminths present in fecal matter.

Moreover, water recycling and reuse emerge as essential facets. Water scarcity, exacerbated by population growth and urbanization, underscores the need for sustainable water resource management. Agriculture, a significant water consumer, further emphasizes the importance of water recycling and reuse. By addressing biodegradable organic substances, nutrient concentration, eliminating pathogens, and prioritizing water recycling and reuse, we can safeguard water resources and foster a healthy and sustainable environment for future generations.

2.4 Purpose of Water Treatment

Water treatment stands as a pivotal process aimed at conditioning, altering, or eliminating undesirable impurities from water sources, ultimately striving to deliver safe, palatable, and satisfactory water to consumers. While this serves as the primary objective of water treatment, its necessity is also mandated by various regulations. For instance, some regulations stipulate water treatment if contaminants listed under them surpass maximum contaminant levels (MCLs). Moreover, treatment becomes imperative for surface-influenced well or spring sources, irrespective of the actual presence of contamination.

Certain impurities impact the aesthetic characteristics of water, necessitating treatment if they exceed secondary MCLs set by the United States Environmental Protection Agency (USEPA). Assuming groundwater as the typical water supply system source, several prevalent groundwater issues may necessitate treatment. Water requiring treatment for one problem may concurrently exhibit other issues, including bacteriological contamination, hydrogen sulfide odor, hard water, corrosive tendencies, iron, and manganese.

Ensuring compliance with regulations and addressing prevalent groundwater issues constitutes integral facets of water treatment. Through the implementation of effective water treatment measures, we can ascertain the safety, palatability, and acceptability of our water sources to consumers.

2.5 Stages of Water Treatment

- Before raw water can be provided to the public for various purposes, including domestic and industrial use, it must undergo thorough treatment and purification. The extent of treatment hinges on the characteristics and quality of the available water, alongside the quality requirements stipulated for its intended application. Raw water frequently harbors suspended, colloidal, and dissolved impurities, which can be objectionable not only in terms of taste and odor but also from a public health standpoint. The overarching goal of water treatment is to eliminate these impurities and ensure that the water is safe and suitable for consumption.

A conventional water treatment plant typically adheres to a specific layout, as depicted in Fig. 2.1.

By adhering to this process, raw water undergoes a transformative journey, emerging as clean, safe, and high-quality water tailored to meet the public's needs.

In the treatment of surface waters, several processes are typically employed to safeguard water for consumption. These processes encompass screening, aeration, sedimentation, filtration, disinfection, and various ancillary procedures.

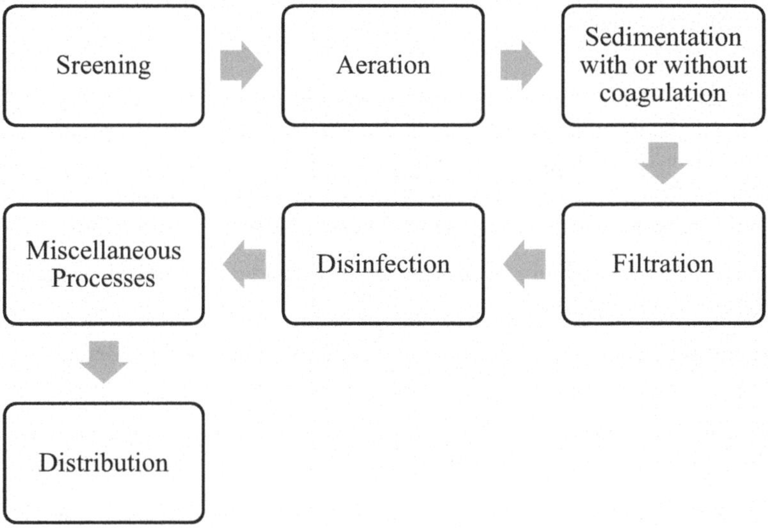

Fig. 2.1 Stages of water treatment

Screening: This initial step involves removing floating matter from surface waters, typically executed at the intake point to ensure the water is devoid of visible impurities.

Aeration: Employed to address objectionable taste and color, aeration aids in eliminating dissolved gases like carbon dioxide and hydrogen sulfide while oxidizing iron and manganese. Although optional, it is essential if the water exhibits undesirable taste or odor.

Sedimentation: Utilized to extract suspended impurities, plain sedimentation targets silt and sand removal, while coagulant-assisted sedimentation tackles very fine suspended particles and certain bacteria.

Filtration: A pivotal purification stage, filtration eradicates minute suspended and colloidal impurities that may have evaded sedimentation tanks. It also significantly reduces microorganism content in the water.

Disinfection: Implemented to eradicate or diminish residual microorganisms and prevent water contamination during transit from the treatment plant to consumption sites.

Miscellaneous Processes: Supplementary methods such as water softening, desalination, and the removal of iron, manganese, and other harmful constituents may also be employed to ensure water safety.

By embracing these treatment processes, surface waters can undergo purification, ensuring their safety for consumption and upholding the health and well-being of the population.

2.6 Unit Processes

A water treatment unit process represents a meticulously engineered system crafted to induce specific alterations in the state of water. This comprehensive process encompasses an array of techniques, including screening, gravity settling, coagulation, flocculation, filtration, gas transfer, ion exchange, adsorption, membrane separations, biological treatment, disinfection, oxidation, and chemical precipitation.

Screening: Involves retaining a substance by a screen with a mesh size smaller than the substance to be retained.

Gravity Settling: Entails sedimentation, where particles descend under the influence of gravity.

Coagulation: Encompasses the charge neutralization of a negatively charged colloid, typically achieved through chemical means like alum or a ferric compound.

Flocculation: Represents a unit process fostering collisions between particles, causing them to adhere upon contact and grow in size, thereby enhancing settling velocity.

Filtration: Involves passing a water stream through porous media to retain suspended particles within the media.

Gas Transfer: Refers to the exchange of gas between the dissolved phase in water and a gas phase.

Ion Exchange: Entails the exchange of ions (e.g., Na^+) bonded to sites within an ion-exchange material (e.g., zeolite mineral or synthetic resin) to be displaced by an ion targeted for removal (e.g., Ca^{2+}) with a stronger bonding force.

Adsorption: Involves the attachment of a molecule to an adsorption site provided by the internal surface of an adsorbent material. Activated carbon stands as the most renowned adsorbent for engineered systems, although virtually any solid material can furnish adsorption sites.

Membrane Separation: Membrane Separation processes are of following four types

(1) Microfiltration (removes colloids and bacteria)
(2) Ultrafiltration (removes viruses)
(3) Nano-filtration (removes large molecular weight organic molecules and some ions)
(4) Hyper-filtration, more commonly called "reverse osmosis" (removes molecules and ions).

2.7 Wastewater Treatment Technologies

In the domain of unit processes, technology assumes the role of a facilitator, providing the means for implementation. For instance, a rapid filter serves as the vehicle for executing a "deep-bed" filtration process. This technological solution encompasses various components that enable its operation, including the filter box depth and area, under-drain system, and backwash system. The sizing and specifications of each component are determined through a blend of rules of thumb, established practices, and manufacturer's standards (Neoh et al., 2016).

Essentially, technology embodies a comprehensive array of design principles and components that coalesce into a system supporting functional processes. Manufacturers may offer a complete package or a selection of ancillary components to fulfill specific requirements.

2.8 Breadth of Unit Processes and Technologies

To provide a more comprehensive understanding of the concept, Table 2.1 showcases 15 unit processes along with examples of their corresponding treatment technologies (Muga & Mihelcic, 2008). The list of associated technologies is not exhaustive and only represents a generic overview. It serves to highlight the vast array of technologies that have been developed (Cheremisinoff, 2019).

2.9 Proprietary Technologies

Proprietary innovation stands as a pivotal component of modern technologies, with manufacturers offering their distinct "packages" of technology for various processes, such as deep-bed filtration. These proprietary innovations play a significant role in the success of a given process.

For example, the Parkson Dynasand filter represents a moving bed filter that removes floc-saturated media from the bottom, replenishes the filter bed from the top, and eliminates the need for backwashing. Similarly, the Culligan Multi-Tech filter serves as a comprehensive package plant facilitating flocculation within coarse media before the main filter, designed for complete automation of the filtration process.

Furthermore, the Infilco Dregemont ABW filtration system showcases a shallow-bed filter comprising transverse cells, each 305 mm (1.0 ft.) wide. This system features a traveling bridge equipped with a hood to collect backwash flow and a backwash pump with a "shoe" that directs the backwash plumbing over the under-drain opening in the finished water channel. This design allows for short-duration backwash of a single cell, ensuring continuous operation of the filter as a whole (Tong & Elimelech, 2016).

Table 2.1 Unit processes and associated technologies

S. No.	Unit process	Principal	Treatment technologies
1	Screening	Retention of objects or particles larger than screen opening	Trash rack Bar screen Fine screen Comminutor Micro screen
2	Chemical precipitation	Solubility product of reaction "product" is very low, such as 10^{-10}–10^{-30} precipitate settle	Lime softening Metal remover
3	Coagulation	Negatively charged colloids are charged neutralized by cation cloud	Rapid mix In-line mixer Static mixer Submerged jet
4	Flocculation	Micro-flocs contact with each other to form settleable flocs. Contacts are induced by controlled turbulence or advection through a sludge blanket	Paddle-wheel flocculator Porous media Solids contact flocculation Turbine flocculator
5	Settling	Gravity causes fall of particles heavier than water and rise of particles lighter than water	Horizontal flow Up-flow Tube Plate American Petroleum Institute (API) Oil separator
6	Flotation	Gas bubbles attach to particles to create buoyant force causing rise	Dissolved air Diffused air
7	Filtration-deep bed	Large neutralized micro-flocs are advected to "collectors" (usually sand and anthracite) where they "attach" (by van der Waal's forces)	Conventional 760 mm (30 in.) dual media Mixed media Mono media, 1–3 m (3–10 ft.) deep
8	Cake filtration	Particles are retained by smaller pores as media is added concurrently (so that the media hydraulic conductivity does not change)	Slow sand Proprietary variations Diatomaceous earth Shapes of septum Plate and frame Candle
9	Membrane filtration	Particles, e.g., ions, organic molecules, viruses, bacteria, cysts, mineral matter, etc., that are larger than the membrane pore size are retained while water and matter smaller than the pore are transported, by advection, under a pressure gradient	Microfiltration Ultrafiltration Nano-filtration Hyper-filtration Shapes of membrane Spiral-wound sheets Hollow fiber electrolysis

(continued)

Table 2.1 (continued)

S. No.	Unit process	Principal	Treatment technologies
10	Adsorption	Molecules and particles adsorb on an "adsorbent," a material that provides "sites" for attachment by van der Waal's forces ion-exchange is the same except ions attach to sites and force is electrostatic	Granular activated carbon Powdered activated carbon Activated alumina Ion-exchange Zeolite Resi
11	Gas transfer	Concentration gradient for dissolved gas is induced to cause mass transfer rate of mass transfer is proportional to gradient	Turbine aeration Diffused air packed towers Steam stripping
12	Oxidation	Electrons are removed from outer shell of substance being oxidized	Chlorine Chlorine dioxide Ozone Ozone-hydrogen peroxide Potassium permanganate Ultraviolet radiation
13	Supercritical oxidation	Pressure and temperature are raised to create supercritical conditions	Wet air oxidation High pressure high temperature Oxidation
14	Aerobic biological treatment	Bacterial enzymes permit metabolism of organic molecules	Activated sludge Complete mix Plug flow Aerated lagoon Facultative pond Fixed film reactor Trickling filters (traditional) Bio-filters (forced air) Rotating biological contactor
15	Anaerobic biological treatment	Two stage reaction: acid formers metabolize organic molecules with organic acids as products and methane formers metabolized organic acids to produce methane and carbon dioxide	Anaerobic biological reactor Digester Anaerobic pond Anaerobic filter

The exemplified filtration process underscores the critical role of proprietary companies in achieving process success. An array of proprietary technologies has been developed for each unit process, either to supplement a generic design with ancillary equipment or to offer a complete operational package technology. Notably, several unit processes have significantly evolved through proprietary research efforts, such as membrane processes, which have garnered widespread commercial adoption since the 1960s. This highlights how proprietary innovation serves as a driving force behind the development and success of modern technologies, enabling manufacturers to offer unique and effective solutions for various processes.

2.10 Management of Wastewater: Conventional and Advances Technologies for Treatment

Traditional methods of wastewater treatment are encountering escalating challenges due to the identification of an expanding array of contaminants, rapid population growth, increasing industrial activities, and diminishing fresh water sources. While conventional processes have proven effective in eliminating numerous chemical and microbial contaminants from wastewater, their efficacy has waned in recent decades in light of new challenges. Heightened awareness regarding water pollution consequences and public demands for higher water quality have spurred the implementation of stricter regulations, broadening the scope of regulated contaminants and reducing maximum contaminant levels for wastewater discharge (Rout et al., 2021). Notably, the removal of nutrients (nitrogen and phosphorus) and synthetic organic compounds (SOCs) stands out as major challenges due to their significant impacts on public health and the environment.

The second factor exacerbating these challenges is the dwindling water resources amidst rapid population and industrial growth. The imperative to reuse municipal and industrial wastewaters and recover potential pollutants used in industrial processes has intensified, particularly in arid regions where transporting drinking water entails substantial costs. Reclamation has gained prominence amid mounting concerns over water resource contamination from toxic compounds discharged by industries (Bairagi & Ali, 2020). Advanced treatment technologies are indispensable for eliminating various potentially harmful compounds that elude effective removal by conventional treatment processes.

The third challenge stems from advancements in the manufacturing industry and the burgeoning market of advanced treatment processes, leading to substantial enhancements in industrial-scale treatment processes. Consequently, adopting advanced treatment technologies is imperative to address these challenges and ensure the provision of safe and clean water for public consumption. This proactive approach not only safeguards public health and the environment but also promotes sustainable development.

In response to emerging challenges, a diverse range of wastewater treatment technologies have been proposed, tested, and implemented. These include membrane filtration systems, automatic variable filtration (AVF), advanced oxidation processes (AOP), and UV irradiation. These innovative technologies have been tailored to meet both existing and anticipated treatment requirements.

Through rigorous testing, these new treatment technologies have showcased their ability to effectively remove a broad spectrum of challenging contaminants from wastewater. Consequently, they have gained traction among wastewater treatment facilities seeking to enhance their treatment processes and comply with stringent regulatory mandates. The adoption of these cutting-edge wastewater treatment technologies represents a significant stride forward in wastewater management. By harnessing the latest technological advancements, we can safeguard and preserve our water resources for future generations.

2.11 Water Stress

The dynamics of our global environment are undergoing rapid transformation. With a burgeoning world population, escalating standards of living, and heightened consumption patterns, including water, we are confronting a resource scarcity. The demand for water is escalating at an unprecedented pace, surpassing the current resources available on the Earth's surface. Projections indicate that with the projected population growth to 9 billion by 2050 and increasing standards of living, the required water volume will triple compared to current availability. Compounding this challenge is the fact that over 80% of used water is discharged into water bodies without prior treatment. Consequently, urgent and essential actions are imperative.

Recycling and reusing all resources, particularly water, are vital steps in addressing this pressing issue. Water stress exacerbates when water extraction exceeds replenishment, particularly when it returns laden with harmful pollutants (Osakabe et al., 2014). Global climatic changes further compound the strain on water resources.

It is evident that we necessitate more efficient treatment technologies and water management practices to meet global needs. Immediate action is imperative to address this issue and ensure adequate water availability for our expanding population. By implementing sustainable water management practices, we can guarantee sufficient water for everyone while safeguarding our environment. It is therefore time for collective action to secure a sustainable future.

2.12 Role of Technology in Making Wastewater as a Resource

Innovations in wastewater treatment underscore the notion that wastewater can be a valuable resource. These advancements often integrate novel technologies and concepts, driven by the principle that reusing wastewater minimizes water extraction, thereby conserving our dwindling natural resource. This approach holds promise in addressing escalating water demands and fostering exploration into novel wastewater treatment methodologies.

However, the success of wastewater reuse and resource recovery hinges on several critical factors, including operational costs, potential revenues, resource value, public acceptance, and requisite engineering (Hoek et al., 2016). The challenge lies in developing and implementing innovative wastewater treatment and resource management strategies, bolstered by cutting-edge technologies, to establish sustainable systems for the future.

Human activities generate wastewater that can exert detrimental effects on the environment, including exacerbating water scarcity (Goh et al., 2022). Contamination of rivers and groundwater by untreated wastewater renders these water resources unusable. Therefore, treating wastewater before its release into the environment, and ideally, treating it to potable standards, is imperative. Wastewater encompasses all water utilized in households that flows into drains or sewage systems, with industries and businesses often contributing significant volumes of wastewater to sewage collection systems.

Proper disposal of wastewater generated from industrial or domestic activities is vital due to stringent disposal regulations. Consequently, selecting appropriate wastewater treatment technology emerges as a paramount concern during the planning stages of any wastewater treatment plant, whether for industrial or domestic purposes. Irrespective of whether conventional or advanced treatment systems are employed, wastewater follows a predefined treatment trajectory to achieve water quality standards (Goh et al., 2022; Sutton et al., 2011). As it traverses the wastewater treatment facility, wastewater is commonly referred to as influent. Wastewater treatment plants play a pivotal role in safeguarding water from excessive pollution, with the nature of treatment and the engineering scale of the plant contingent upon the degree and type of wastewater. Most wastewater treatment plants comprise primary and secondary treatment components, with tertiary treatment facilities also available at various sewage treatment plants.

2.13 Processes/Technologies Available for Wastewater Treatment

The treatment of wastewater typically progresses through three main stages: Primary Treatment, Secondary Treatment, and Tertiary Treatment (Table 2.2). However, in instances where wastewater presents extreme characteristics, specialized treatments such as thermal destruction (e.g., evaporation or incineration) may be warranted. Additionally, preliminary treatments like grit removal and oil and grease removal may be employed based on the source, transportation, and quality of the incoming wastewater (Moharir et al., 2020).

2.13.1 Primary Treatment

Primary Treatment also referred to as physico-chemical treatment, encompasses both physical and chemical treatment methods. It involves several key processes, including flow equalization, pH correction, flocculation, and sedimentation. The treatment of wastewater is a multifaceted and crucial process that demands meticulous attention to detail. By deploying effective treatment methods, we can ensure the safe and responsible management of our wastewater resources.

Flow Equalization and pH Correction

- Wastewater from different section of plant have different characteristic, so, in the first instance it needs to equalize effluents characteristics (quality and quantity) in Equalization tanks with large capacities.
- Acid or Alkalis are used to neutralize wastewater.

Coagulation and Flocculation

- Coagulation agents such as Lime, $FeCl_3$, $FeSO_4$, Poly aluminum chloride (PAC) are used to enhance coagulation.
- Afterwards, polyelectrolytes are added to create large flocs of coagulated mass to facilitate better settlement and removal during sedimentation.

Sedimentation

It is generally, the last stage of primary treatment section. Here the coagulated mass present in the wastewater is allowed to settle and is removed as sludge from bottom of sedimentation tanks, generally called clarifiers.

Table 2.2 Various stages of wastewater treatment

Treatment section	Brief description of treatment technology
Preliminary treatment	
Screening	• It is adopted to remove floating matter and shall be provided at the intake point
Grit removal	• Used when WWTP has to deal with rainwater which normally entrains a considerable amount of sand
Oil and grease removal	• Oil and grease are skimmed-off by passing the waste water through skimming tank. This process can be rendered more efficient by dissolved air flotation or vacuum flotation
Primary treatment	
Equalization	• Applicable for wastewaters having different characteristics at different intervals of time and where uniform treatment is required • Each unit volume of waste is mixed thoroughly with other unit volumes of other wastes to produce homogeneous and equalized effluent • Gives better mixing of different unit volumes of effluents
Neutralization	• Applicable for highly acidic and highly alkaline effluents • Acidic effluents may be neutralized by treatment with lime or lime slurry or caustic soda • Alkaline waste may be neutralized by treatment with acids
Sedimentation	• Separation of suspended particles by gravitational settling and floating material • Clarifies collected rainwater from solid content (sand or dust) • Clarifies wastewater from inert contents (sand or comparable particles) • Clarifies wastewater from reaction material (emulsified metal compounds, polymers and their monomers) • Separates heavy metals or other dissolved components after preceding flocculation process • Removes suspended solids in the primary clarifier • Removes biological sludge in secondary clarifier of a biological WWTP
Secondary treatment	
Aerobic treatment	
Activated sludge process	• Applicable to all biodegradable industrial wastewater streams • The effluent from primary treatment processes are collected in aeration tank and are aerated with mechanical devices such as fixed/floating/diffused aeration/oxygen injection etc.
Aerated lagoons	• The effluent from primary treatment processes are collected in lagoons and are aerated with mechanical devices such as floating/fixed aerators

(continued)

Table 2.2 (continued)

Treatment section	Brief description of treatment technology
Trickling filters/bio filters	• This process the microorganisms are attached to a highly permeable medium through which the wastewater is trickled or percolated • Trickling filters are effectively used for the treatment of industrial waste water • Used to treat urban and industrial wastewater • Used when effluent is highly loaded with COD • Used to upgrade an existing activated sludge plant
Sequential batch reactors (SBR)	• The operation is in sequence of fill, aerate, settle and waste sludge and draw treated waste water but not with secondary clarifier
Sub-merged aerobic fixed film reactor	• This technology utilizes an aerobic fixed film process that is a combination of submerged attached growth and activated sludge process • This system is designed to be installed into a two compartment, where the first compartment provides majority of BOD removal, and the second compartment polishes the BOD • Rigid block-type media is submerged within the treatment module, providing surface area for microbial growth
Membrane bioreactor	• Treats municipal and industrial waste water • It is particularly suitable for effluents with high COD and/or ammoniacal nitrogen loads; envisaging recycling of waste water, stringent discharge regulations, sensitive receiving water bodies, sludges which are hard to settle, upgrading existing plants, compact installations
Anaerobic treatment	
Anaerobic treatment	• Used only as pretreatment for waste water which is characterized by high organic load (> 2 g/L) • Applicable mostly for effluents of high BOD loads
Anaerobic contact reactor	• In anaerobic contact process (ACP) waste water is mixed with recycled sludge and digested in a sealed reactor, the waste water/sludge mixture externally separated (sedimentation, or vacuum fine screening flotation) and the supernatant discharged for further downstream treatment
Up flow anaerobic sludge blanket (UASB)	• In the UASB process, waste water is introduced at the bottom of the reactor, from where it flows upward through a sludge blanket composed of biologically formed granules or particles
Fixed-bed reactor	• In the fixed-bed or anaerobic filter process, waste water flows upward or downward (depending on the solids content of the influent) through a column with various types of solid media on which anaerobic micro-organisms grow and are retained
Biological removal of sulphur compounds/heavy metals	• Much lower solubility of metal sulphides compared to their hydroxides • Applicable to all waste water streams that contain a considerable amount of sulphate

(continued)

Table 2.2 (continued)

Treatment section	Brief description of treatment technology
Tertiary treatment	
Sand filters	• Removes undissolved pollutants such as suspended solids, undissolved phosphate and attached organics • Flexible for modifications in basic design structure to accommodate site specific criteria
Carbon filters	• Activated carbon adsorbs organics • Flexible for modifications in basic design structure to accommodate site specific criteria
Micro filtration	• Applied when a solid free waste water for downstream facilities is desired such as reverse osmosis or complete removal of hazardous contaminants • Used in metal particle recovery • Used in metal plating waste water treatment • Used in sludge separation after activated sludge process in a central biological WWTP, replacing secondary clarifier
Ultra filtration	• It is a pre-treatment step prior to reverse osmosis or ion exchange • Removes SS along with attached COD as a polishing step and avoiding secondary clarification • Removes pollutants such as proteins and other macromolecular compounds and toxic non-degradable components • Separates heavy metals after complexation or precipitation • Separates components not readily degradable in sewage treatment effluents which are subsequently recycled to the biological stage
Retention ponds	• Used to avoid hydraulic overload of downstream facilities • Separates solid pollutants (such as sediment, organic matter, dissolved metal compounds and nutrients) from rainwater • Applied to industrial sites with highly contaminated surfaces
Nano filtration	• Applied to remove larger organic molecules and multivalent ions in order to recycle and reuse the waste water or reduce its volume • Increase the concentration of contaminants to such an extent that subsequent destruction processes are feasible
Reverse osmosis (RO)	• Separates water and dissolved constituents down to ionic species • It is applied when a high purity water is required • The segregated water phase is recycled and reused such as desalination, final removal of degradable components if biological treatment is not available, heavy metals, toxic components and segregation of pollutants with the aim of concentrating or further processing • It is often used in combination with post treatment techniques for the permeate

(continued)

Table 2.2 (continued)

Treatment section	Brief description of treatment technology
Ion exchange	• Applied to remove unwanted ionic and ionizable species from wastewater • Its greatest value lies in recovery potential • It recovers rinse water and process chemicals **Examples** • Heavy metal ions—cationic or anionic, e.g. Cr^{3+} or cadmium and its compounds, with low feed concentrations Ionizable inorganic compounds, such as H_3BO_3 Soluble, ionic or ionizable organic compounds, e.g. carboxylic acids, sulphonic acids, some phenols, amines as acid salt, quaternary amines, alkyl sulphates and organic mercury can be removed • Ion exchange is the removal of undesired or hazardous ionic constituents of waste water and their replacement by more acceptable ions from an ion exchange resin, where they are temporarily retained and afterwards released into a regeneration or backwashing liquid
Evaporation	• It is applicable to remove or concentrate inorganics

Source Technical EIA Guidance Manual for Common Effluent Treatment Plant, Ministry of Environment Forests and Climate Change, New Delhi

2.13.2 Secondary Treatment

Secondary treatment, also known as biological treatment, constitutes a pivotal phase in industrial wastewater and raw sewage treatment. Following the removal of suspended particles in the primary treatment stage, the wastewater undergoes biological treatment, wherein microbes degrade the organic material present in the incoming wastewater, thus mitigating the pollution load.

This process plays a critical role in ensuring that the wastewater meets environmental discharge standards. By decomposing organic material, biological treatment aids in diminishing the environmental impact of industrial activities and sewage disposal. Overall, secondary treatment stands as an indispensable step in the wastewater treatment process, necessitating vigilant monitoring and management to ensure its efficacy in reducing pollution levels. With proper implementation, biological treatment fosters environmental protection and facilitates the promotion of sustainable industrial practices.

Biological treatment can be broadly categorized into two primary classifications.

2.13.2.1 Aerobic Biological Treatment

- Here, additional oxygen is provided to microbes by means of blowers, surface aerator or any other suitable methods.

2.13.2.2 Anaerobic Biological Treatment

- Here, no oxygen is provided and degradation of organic mass is done in absence of oxygen.

2.13.3 Tertiary Treatment

Tertiary treatment, also known as polishing treatment, serves as an important stage in wastewater treatment, primarily aimed at eliminating trace amounts or residual pollution loads present in the wastewater. This phase encompasses a range of techniques, including Advanced Oxidation Processes and adsorption over charcoal. In scenarios necessitating wastewater recycling, membrane separation technology is commonly employed during tertiary treatment, typically followed by the thermal evaporation of the concentrated fraction (Collins et al., 2005).

Tertiary treatment holds paramount importance in the wastewater treatment process, ensuring that the water is rendered safe for reuse or discharge into the environment. The Advanced Oxidation Process and adsorption over charcoal stand out as highly efficient methods for eliminating any lingering pollutants in the water. Moreover, membrane separation technology plays a pivotal role in wastewater recycling, a cornerstone of sustainable water management practices.

Overall, tertiary treatment significantly contributes to the maintenance of water resource quality. Through the utilization of these advanced techniques, we can guarantee the safety of our water for both human consumption and environmental preservation.

2.14 Selection of Wastewater Treatment Technology

Analytical parameters, including chemical oxygen demand (COD), biochemical oxygen demand (BOD), and total dissolved solids (TDS), serve as critical indicators in determining appropriate treatment technologies and sequences. Table 2.3 elucidates the significance of these parameters in the selection of treatment technologies. It is imperative to incorporate these analytical parameters into the selection process, as they furnish invaluable insights into wastewater composition.

COD quantifies the oxygen required for oxidizing organic matter, while BOD quantifies the oxygen consumed by microorganisms during organic matter breakdown. Conversely, TDS gauges the concentration of inorganic and organic substances dissolved in water. By evaluating these parameters, treatment technologies can be meticulously chosen to effectively address wastewater treatment requirements, ensuring that the treated water adheres to requisite standards for discharge or reuse. Thus, it is paramount to judiciously consider these analytical parameters during treatment technology selection (Table 2.3).

Table 2.3 Consideration of analytical parameters during treatment technology selection

Combination	Quality of effluent	Treatment options
High TDS, high COD and equivalently high BOD	Waste is not easily biodegradable but toxic	• Thermal decomposition (based on calorific value) • Chemical oxidation by hydrogen peroxide, ozone etc. • Evaporation + secured landfill
High TDS, high COD and high difference between COD and BOD	May be toxic; not suitable for biological treatment; mostly inorganic salts	• Chemical treatment (recovery, precipitation etc.) • Evaporation + secured landfill of evaporated residue
High TDS, high BOD and low difference between COD and BOD	Highly organic effluent fully biodegradable	• Anaerobic + aerobic treatment • If quantity is less, incineration (based on calorific value) + secure landfill of incineration ash
High TDS, low BOD and low BOD and COD difference	Only inorganic salts, no need for biological treatment	• Solar evaporation • Forced evaporation (after separation of volatile organic matter) • Membrane technologies
Low TDS, and high COD and equivalently high BOD	Highly organic effluent, may not be easily biodegradable	• Thermal decomposition • Chemical oxidation by hydrogen peroxide or ozone or sodium hypochlorite etc. • Chemical + biological treatment
Low TDS, high COD and high difference between COD and BOD	Highly inorganic effluent, not suitable for biological treatment	• Chemical recovery • Chemical oxidation + biological treatment
Low TDS, high BOD and low difference between COD and BOD	Organic effluent, fully biodegradable	• Anaerobic + aerobic treatment
Low TDS, low BOD and low BOD and COD difference	Low organic and low inorganic effluent	• Recycle and reuse (after preliminary treatment)

Source Technical EIA Guidance Manual for Common Effluent Treatment Plant, Ministry of Environment Forests and Climate Change, New Delhi

2.15 Sequence Batch Reactor (SBR) Technology

As environmental discharge standards become increasingly stringent, traditional continuous flow-based biological wastewater treatment processes encounter significant challenges. However, the sequencing batch reactor (SBR) technology presents a promising solution to this dilemma. SBR, a modification of the widely utilized activated sludge process (ASP), transforms the continuous nature of ASP-based treatment into a batch process. This conversion introduces various process flexibilities and alternatives in process controls and design, facilitating enhanced compliance with the latest effluent discharge standards.

The term "SBR" was initially coined by RL. Irvine in 1971, and since then, it has emerged as one of the most advanced treatment technologies for wastewater. SBR reactors treat wastewater, such as sewage or output from anaerobic digesters or mechanical biological treatment facilities, in batches. Oxygen is introduced into the wastewater to diminish biochemical oxygen demand (BOD) and chemical oxygen demand (COD), rendering it suitable for discharge into sewers or for land application.

Though there are several configurations of SBRs, the fundamental process remains akin. The setup typically comprises at least two identically equipped tanks with a shared inlet, which can be switched between them. These tanks employ a "flow through" system, where raw wastewater (influent) enters at one end and treated water (effluent) exits the other. At the inlet, a section of the tank known as the bio-selector directs the flow, aiding in mixing the incoming influent and the returned activated sludge, initiating the biological digestion process before entering the main part of the tank. SBR technology presents a flexible and effective solution to the challenges encountered by traditional continuous flow-based biological wastewater treatment processes.

2.15.1 What Is SBR Technology?

The Sequencing Batch Reactor (SBR) stands as a state-of-the-art technology employing a fill and draw activated sludge system for wastewater treatment, effectively targeting both industrial and municipal wastes. Distinguished by its utilization of a single batch reactor or tank for equalization, aeration, and clarification processes, the SBR diverges from other technologies, which often necessitate multiple batch reactors for diverse processes (Jafarinejad, 2017).

SBR technology garners widespread acclaim as one of the most efficient solutions in wastewater treatment. Its operational advantages include minimal maintenance requirements, low operational costs, and seamless management of continuous batch operations. Key features distinguishing SBR technology include:

- High removal capacity
- Versatility
- Compliance with stringent discharge standards
- Production of high-quality end products.

These attributes collectively underscore the significance of SBR technology in addressing contemporary wastewater treatment challenges while meeting regulatory standards and ensuring superior treatment outcomes.

2.15.2 Background

In the early 1900s, Ardern and Lockett (1914) established the fundamental principles of biological degradation processes using activated sludge. These researchers conducted "fill-and-draw" processes on crude sewage in Manchester, UK, and introduced the concept of sequencing batch reactors (SBRs). SBRs operate a single biological reactor basin using repetitive cycles of aeration, settlement, and discharge of the treated effluent. The original fill-and-draw variable-volume SBR systems were capable of achieving excellent treated effluent quality but suffered from many operational difficulties. This led to the development of fixed volume continuous-flow activated sludge processes, which incorporate two separate units, one for aeration and one for settlement. Further developments of the SBR process did not occur until the 1950s when Pasveer (1955) incorporated interrupted and continuously fed batch treatment principles in their variable volume activated sludge system. The 1970s saw further development, mainly in Australia and the United States, with grant aid from the EPA and publication of the EPA's SBR Design Manuals in 1986 and 1992, leading to the widespread application of the technology worldwide.

Technological improvements, particularly reliable microprocessor control systems, aeration equipment, and mechanically actuated valves, have resolved the earlier operational difficulties. Through process performance monitoring and variations/modifications of the original SBR process, the modern generation of SBRs have found application in large scale municipalities (up to 1 million population equivalent), as well as the modular expansion and up-rating of existing wastewater treatment facilities.

2.15.3 SBR Operating Principles: Basic Treatment Process

The SBR system, in its most fundamental form, comprises a series of tanks that operate on a fill-and-draw basis. Each tank is filled during a specific period and then operated as a batch reactor. Once the desired treatment is achieved, the mixed liquor is allowed to settle, and the clarified supernatant is drawn from the tank. The cycle for each tank in a typical SBR is divided into five distinct periods: Fill, React, Settle, Draw, and Idle, as illustrated in Fig. 2.1. There are various types of Fill and React periods, which differ based on aeration and mixing procedures. Sludge wasting may occur near the end of React, or during Settle, Draw, or Idle (Piotrowski et al., 2019). The central aspect of SBR design is the use of a single tank for multiple aspects of wastewater treatment, and the full SBR cycle is completed in four stages, as shown in Fig. 2.2 (Filling, Aeration, Settling, and Decanting).

The SBR system is a highly effective and efficient method of wastewater treatment, offering numerous benefits over traditional treatment methods (Fig. 2.3). Its modular design allows for easy expansion and modification, making it an ideal solution for facilities with changing wastewater treatment needs. Additionally, the SBR

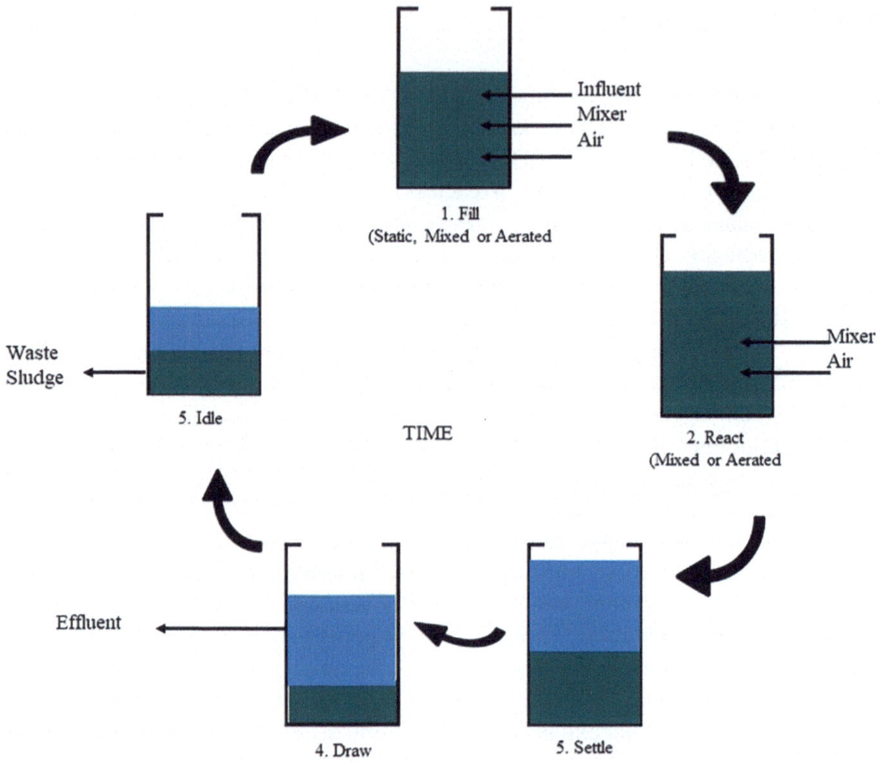

Fig. 2.2 SBR operation

system's ability to operate on a fill-and-draw basis allows for greater flexibility in treatment schedules, reducing energy consumption and operating costs. It is a versatile and reliable method of wastewater treatment, offering numerous advantages over traditional treatment methods. Its modular design, flexibility, and efficiency make it an ideal solution for facilities of all sizes and types.

2.15.3.1 Filling

Water purification is indeed a critical process, and the Sequential Batch Reactor (SBR) plays a significant role in achieving it. The first step in the purification process involves filling the influent wastewater into the SBR tank. This step is essential to maintain an appropriate food to microorganisms' ratio, ensuring efficient degradation of contaminants (Fig. 2.4).

The influent to the tank may consist of raw wastewater that has been previously screened and degritted, or it may be primary effluent. The volume of feed is determined based on various factors, including desired loading and detention time, as well as the expected settling characteristics of the microorganisms present in the reactor.

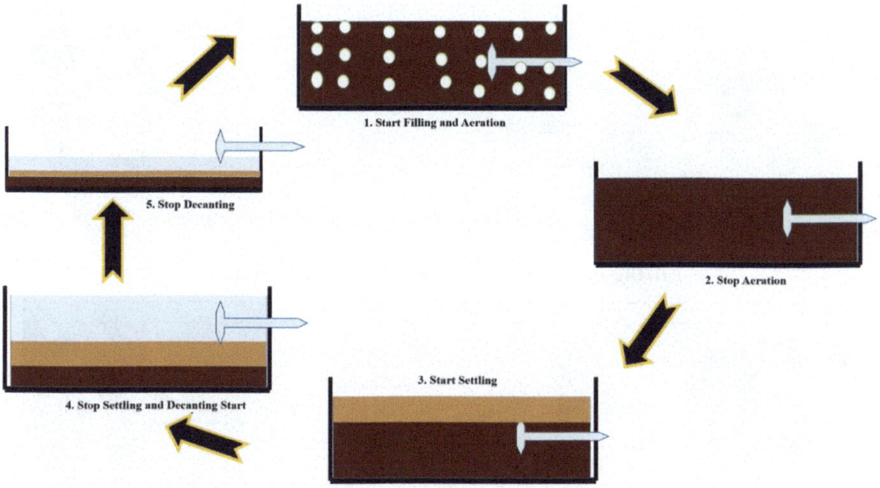

Fig. 2.3 Flow diagram of SBR process for one cycle

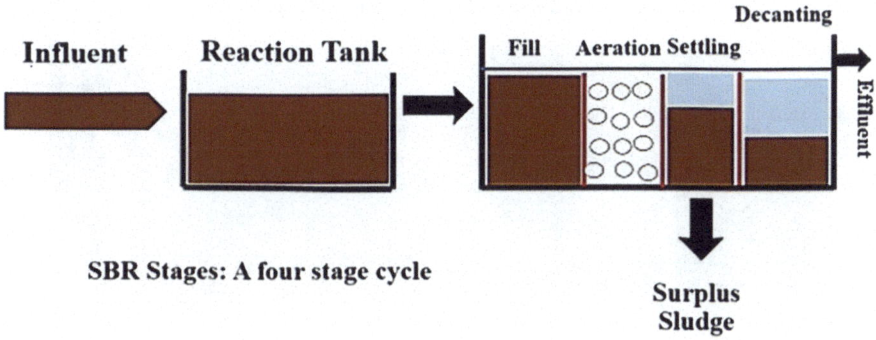

Fig. 2.4 Main stages of SBR cycle

During the filling process, aeration is a critical aspect. Various aeration systems can be employed, such as diffused aeration, floating mechanical aerators, or jet aerators. The ideal aeration system should provide a range of mixing intensities, allowing for both gentle agitation and complete mixing as needed. Additionally, the flexibility to mix without aeration is important. Level sensing devices, timers, or in-tank probes can be utilized to control the operation of aerators and mixers, ensuring optimal conditions for wastewater treatment (Fig. 2.5).

Overall, the filling and aeration process in the SBR tank are crucial aspects of water purification. By carefully considering various factors and employing appropriate techniques, contaminants and impurities can be efficiently removed from wastewater, contributing to the production of clean and safe water for various applications.

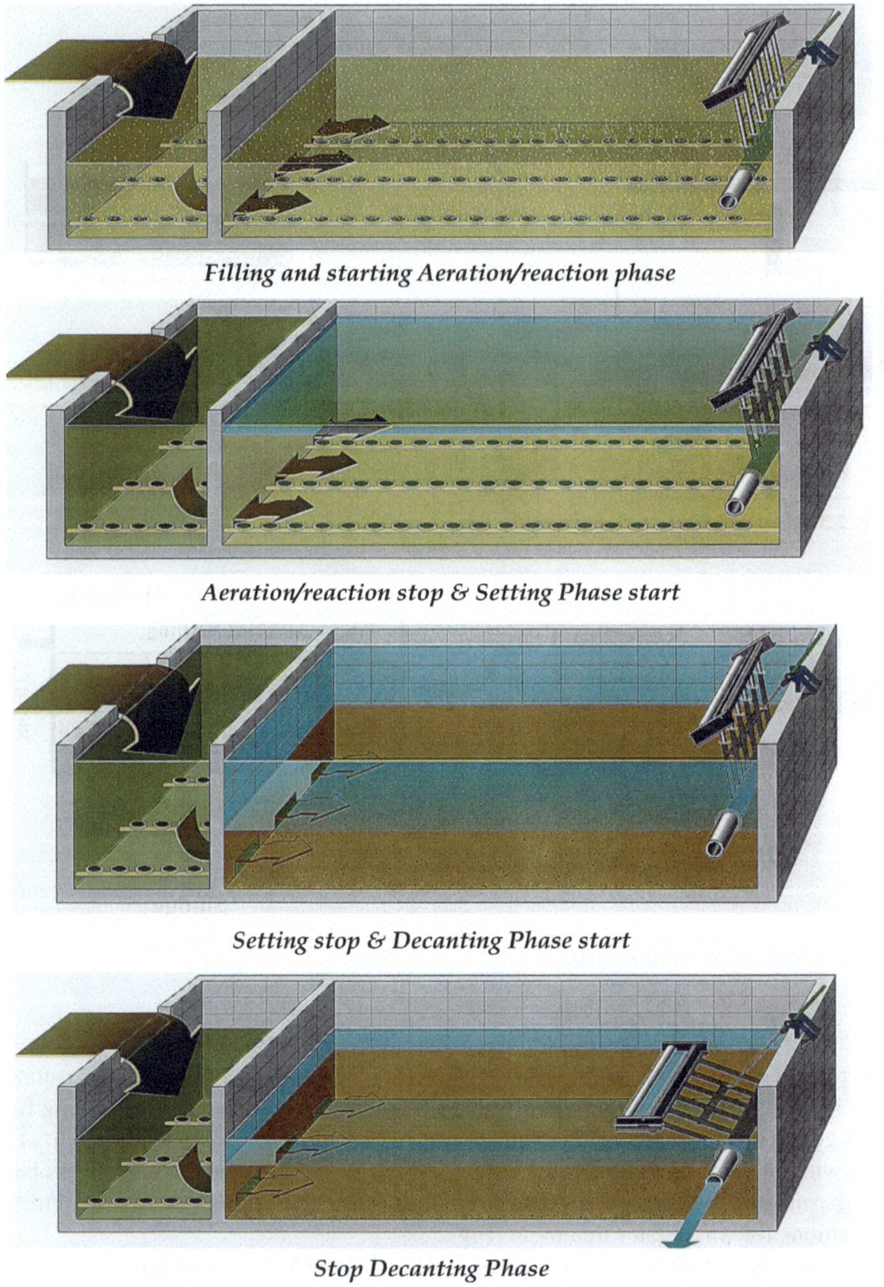

Fig. 2.5 Pictorial examples of SBR operation

2.15.3.2 Reaction

The React stage is a critical step in the wastewater treatment process, where microorganisms utilize ammonia, nitrogen, and Biochemical Oxygen Demand (BOD) to further break down organic matter. The efficiency of this stage depends on factors such as sludge mass, aeration time, impurities, and the degree of nitrification. Alternating conditions of low and high dissolved oxygen concentrations may be required to optimize the process. Sludge wasting during this stage helps control the sludge age and decrease solids volume, contributing to process efficiency.

2.15.3.3 Settling

Settling follows the React stage and typically lasts between 60 and 90 min. This step involves the cessation of aeration and the settling of sludge. In the SBR system, separation of solids occurs in a tank under quiescent conditions, without inflow or outflow. Unlike conventional clarifiers, there's no need for underflow hardware in SBR systems, as all biomass remains in the tank until some fraction is wasted.

2.15.3.4 Draw/Decant

The Draw/Decant stage involves removing effluent from the tank without disturbing settled sludge. This is achieved using a withdrawal mechanism, such as a pipe with flow regulated by an automatic valve or a pump, or an adjustable or floating weir. Proper design and operation of the withdrawal mechanism prevent the discharge of floating matter. It is important to monitor the duration of the Draw phase to prevent problems with rising sludge.

2.15.3.5 Idle

The Idle stage refers to the time between two filling cycles. Despite its name, it can be effectively used to dispose of settled sludge. It is recommended to implement sludge wasting programs more frequently to maintain process efficiency and sludge settling, thereby improving overall effectiveness. These stages collectively contribute to the successful operation of the SBR system in wastewater treatment.

2.15.3.6 Sludge Wasting

Sludge wasting is a crucial aspect of wastewater treatment to ensure the proper functioning of the system and maintain optimal conditions for microbial activity. Two key parameters play a significant role in controlling the quality of sludge wasting:

1. **Mean Cell Residence Time (MCRT) or Solids Retention Time (SRT)**: MCRT/SRT is the duration for which solids or bacteria are retained within the activated sludge system. It is typically measured in days and is essential for maintaining stable microbial populations and achieving effective treatment. A longer MCRT/SRT allows for better microbial acclimatization to varying influent conditions, leading to improved treatment efficiency and stability.
2. **Mixed Liquor Suspended Solids (MLSS)**: MLSS represents the concentration of suspended solids in the mixed liquor of the activated sludge system, typically measured in grams per liter. This parameter provides insight into the biomass concentration and the overall efficiency of the treatment process. Controlling MLSS levels ensures proper microbial activity and nutrient removal while preventing issues such as sludge settling or bulking.

By monitoring and controlling MCRT/SRT and MLSS levels within optimal ranges, wastewater treatment plants can effectively manage sludge wasting and ensure consistent treatment performance. Adjusting these parameters as needed helps maintain the system's stability and efficiency, contributing to the overall success of the wastewater treatment process.

2.15.4 SBR Process Configuration

The essential components of SBR configuration are:

- Reactor basin
- Waste sludge draw-off mechanism
- Aeration equipment
- Effluent decanter
- Process control system.

The Sequencing Batch Reactor (SBR) system offers a versatile and efficient approach to managing the continuous inflow of wastewater. Unlike conventional treatment systems, SBRs typically require fewer structures and can be easily adapted to varying loading conditions. The key features and benefits of the SBR system are:

1. **Storage/Equalization Tank**: The SBR system may include a storage or equalization tank to handle fluctuations in wastewater flow rates and provide preliminary treatment. This tank helps to balance the influent wastewater flow and minimize the impact of peak flows on the treatment process.
2. **Reactor Design**: SBR reactors are typically constructed as circular, square, or rectangular tanks made from materials like concrete, steel, or lagoon structures. These reactors serve dual purposes, acting as both aeration tanks and final clarifiers, which reduces the need for additional structures in the treatment plant layout.
3. **Compact Layout**: The integrated design of SBR reactors allows for a more compact layout of the treatment plant site compared to conventional systems.

This streamlined layout minimizes space requirements and simplifies facility construction and expansion.

4. **Modular Design**: SBR systems can easily accommodate future loading conditions by adding modular basins using common wall construction. This modular design approach offers flexibility and scalability, allowing treatment plants to adapt to changing operational needs and increased wastewater volumes over time.
5. **Hydraulic Volume**: The volume treated per batch, or "hydraulic" volume, is typically up to 30% of the designated top water level volume in SBR reactors. This volume is carefully designed to provide sufficient mass of activated sludge for effective biological treatment processes.

Overall, the SBR system's combination of efficient reactor design, compact layout, and modular construction makes it a reliable and sustainable solution for wastewater treatment. By effectively managing continuous inflow and minimizing the need for additional structures, SBR systems help treatment plants optimize performance and meet regulatory requirements.

2.15.5 Waste Activated Sludge

In the realm of wastewater treatment, withdrawing waste activated sludge (WAS) from Sequencing Batch Reactor (SBR) systems is a crucial step typically accomplished during the decant or idle phase of the cycle. WAS concentrations typically fall within the range of 8000–15,000 mg/L, and after withdrawal, it can undergo further treatment through aerobic or anaerobic processes or be thickened and dewatered before disposal.

In North America, SBR processes commonly utilize air blowers for aeration to facilitate the biological degradation of organic components in wastewater. Various aeration systems are employed, including low maintenance coarse bubble diffused aeration, jet aeration systems (particularly in industrial facilities), and surface aeration equipment. Flexible membrane, self-sealing, fine bubble diffusers have gained widespread adoption in SBR plants due to their high efficiency in oxygen transfer per unit of energy.

During SBR operation, air can only be supplied during the fill, aerate, and react periods. For a cycle with 50% aeration, air must be supplied to the SBR tank in a 12-h period every day. In a two-tank system, continuous blower operation is necessary, with each tank receiving airflow directed by motorized valves controlled by the process control center. Varying aeration intensity and duration offers turndown capability to prevent over-aeration.

Nitrification and denitrification can be achieved by controlling aeration during fill and react phases to create macro anoxic conditions within activated sludge flocs. Biological phosphorus removal can be accomplished by incorporating an anaerobic phase during filling. Treated effluent is removed after the settle period, with sufficient time allowed for solids interface to reach a low enough position to avoid entrainment during decanting.

Effluent removal systems for SBRs include fixed decanters with submerged outlet pipes and moving devices like weir troughs and floating weirs. Design considerations for decanters include uniform withdrawal of effluent, prevention of scum and floating materials compromising effluent quality, and flexibility to accommodate process changes. Prioritizing these considerations ensures optimal performance and reliability of decanters in SBR systems.

2.15.6 Biological Nitrogen Removal Process

In Sequencing Batch Reactor (SBR) systems, the cycle operates in time rather than space, allowing for flexibility in nutrient removal through the alteration of aeration and mixing regimes. This creates alternating aerobic and anoxic environments during the Bio-React phase, facilitating biological nutrient removal.

Nitrification is the first step in the biological nutrient removal process, involving the conversion of organic nitrogen into ammonia–nitrogen, and then to nitrate–nitrogen by chemoautotrophic bacteria. This process, catalyzed by Ammonia oxidizing bacteria (AOB) and Nitrite oxidizing bacteria (NOB), requires a strictly aerobic environment achieved through aeration in the SBR tank.

Denitrification, another crucial process in the nitrogen cycle, occurs under anaerobic conditions with the utilization of complex organic compounds by anaerobic heterotrophic bacteria. Nitrate serves as the electron acceptor, resulting in the formation of nitrogen gas, which exits the aqueous phase.

Several parameters such as temperature, pH, dissolved oxygen (DO) level, and solids retention time (SRT) influence the kinetics of nitrification. Optimal temperature for nitrification falls between 25 and 35 °C, and pH between 7.5 and 9.8. A DO level greater than 2.0 mg/L maximizes the rate of nitrification, with increased nitrifier concentration boosting the process. Denitrification is effective under anoxic conditions when DO levels are below 0.5 mg/L, with an optimum pH between 6.5 and 9. Inhibition of denitrification occurs when DO levels exceed 1.0 mg/L. Optimizing these parameters ensures efficient nitrification and denitrification in SBR systems (Wu et al., 2016).

2.15.7 Applicability

Sequential Batch Reactors (SBRs) are indeed a preferred option for wastewater treatment facilities dealing with flow rates of 5 MGD or lower. Larger SBR plants can become operationally complex, which often deters their use for higher flow rates. Nevertheless, their compact footprint renders them particularly suitable for regions with land constraints.

Table 2.4 Key design parameter of a wastewater system

Factors	Municipal	Industrial
Food to mass (F:M)	0.15–0.4/day	0.15–0.6/day
Treatment cycle duration (h)	4.0	4.0–24
Typically low water level mixed liquor suspended solids (mg/L)	2000–2500	2000–4000
Hydraulic retention time	6–14 h	Varies

The paramount advantage of SBRs lies in their flexibility. These systems boast easily modifiable cycles, enabling seamless adjustment to meet evolving nutrient removal demands. This adaptability proves invaluable in navigating changes in regulatory requirements for effluent parameters, ensuring consistent compliance with environmental standards. Furthermore, SBRs offer a cost-effective solution for facilities necessitating treatment beyond mere biological processes, such as filtration. Their demonstrated efficiency and effectiveness have cemented their status as a favored choice among wastewater treatment facilities worldwide.

2.15.8 Design Criteria

After assessing the influent and effluent characteristics of the system, engineers often consult SBR manufacturers for guidance on designing the system. Considering site-specific factors like temperature, they select key design parameters tailored to the system's requirements. Table 2.4 provides an illustrative example of these parameters for a wastewater system loading.

After establishing the key design parameters, such as the number of cycles per day, number of basins, decant volume, reactor size, and detention times, engineers can proceed to calculate the suitable sizing for aeration equipment, decanters, and associated piping (Passoni et al., 2021). However, it is crucial to acknowledge that additional site-specific details are necessary to accurately size the aeration equipment. These include the site's elevation above mean sea level, wastewater temperature, and total dissolved solids concentration. By integrating these factors into the design process, optimization can be achieved, ensuring maximum efficiency and effectiveness.

2.15.9 Construction

The construction of Sequencing Batch Reactor (SBR) systems is indeed recognized for requiring a smaller footprint compared to conventional activated sludge systems, primarily because the SBR eliminates the need for primary clarifiers and never requires secondary clarifiers. The size of the SBR tanks is contingent upon

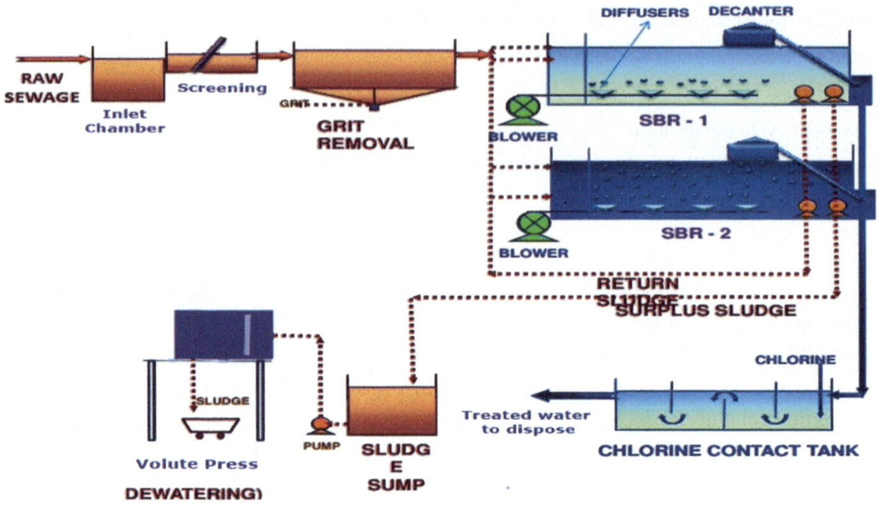

Fig. 2.6 Design layout of SBR technology based sewage treatment plant

the specific site, but the SBR system proves particularly advantageous in scenarios where space is limited (Fig. 2.6).

2.15.10 Tank and Equipment Description

The SBR system, a sophisticated wastewater treatment system, comprises various components, including a tank, aeration and mixing equipment, a decanter, and a control system. Central to the SBR system are the control unit and automatic switches and valves, which sequence and time the different operations. It is advisable to seek recommendations from SBR manufacturers for tanks and equipment selection, with a preference for a complete SBR system recommended and supplied by a single manufacturer.

SBR tanks are typically constructed using steel or concrete. For industrial applications, steel tanks coated for corrosion control are common, while concrete tanks are prevalent in municipal treatment of domestic wastewater (Shammas & Wang, 2009). Jet aeration systems are commonly used for mixing and aeration due to their capability to mix either with or without aeration, although other aeration and mixing systems are also utilized. Positive displacement blowers are typically employed for SBR design to manage wastewater level variations in the reactor.

The decanter is a critical component that distinguishes different SBR manufacturers. There are two main types of decanters: floating and fixed. Floating decanters offer the advantage of maintaining the inlet orifice slightly below the water surface, minimizing the removal of solids in the effluent during the DRAW step. They also provide operating flexibility to vary fill-and-draw volumes. On the other hand, fixed decanters are integrated into the side of the basin and may be used if the Settle step is extended to reduce the chance of solids in the wastewater floating over the fixed decanter. While fixed decanters may be less expensive and can be designed to allow adjustment of the decanter level, they lack the operational flexibility of floating decanters.

2.15.11 Operation and Maintenance

The operation and maintenance of SBR technology offer notable ease and simplicity compared to other wastewater treatment technologies. One of its significant advantages is the elimination of separate primary and secondary clarifiers in most municipal systems, reducing operational and maintenance requirements. In contrast, conventional biological nutrient removal systems may necessitate various additional components, such as anoxic basins and toxic basin aeration equipment, which increase maintenance needs. However, in SBR systems, these functions are consolidated in one reactor, minimizing the maintenance required for clarifiers and pumps.

The core of the SBR system lies in its controls, automatic valves, and switches, which may require more maintenance compared to a conventional activated sludge system. The higher level of sophistication in SBR technology can lead to more potential failure points or maintenance needs, especially in larger wastewater treatment plants where the system complexity is advanced.

A key advantage of SBR systems is their significant operating flexibility. They can be configured to replicate various conventional activated sludge processes, including Biological Nitrogen Removal (BNR) systems. For instance, adjustments to holding times in the Aerated React mode of an SBR can mimic different treatment systems, such as contact stabilization or extended aeration treatment systems.

In BNR plants, alternating between aerated react mode (oxic conditions) and mixed react mode (anoxic conditions) facilitates nitrification and denitrification, respectively. Additionally, mixed fill mode and mixed react mode can be utilized for denitrification under anoxic conditions, and ultimately for achieving anaerobic conditions for phosphorus removal. SBRs operate in time rather than space, allowing for variations in the number of cycles per day to control effluent limits, providing additional flexibility compared to conventional activated sludge systems.

2.15.12 *Modifications of SBR: Different Variants of SBR Technology*

2.15.12.1 Cyclic Activated Sludge System

The Cyclic Activated Sludge System (CASS) represents an advanced wastewater treatment process characterized by a single basin with variable volume, operating in an alternating mode. This system offers a unique combination of plug flow in the initial zone followed by a completely mixed reactor basin with secondary and main aeration zones. The activated sludge from the main aeration zone is recirculated into a selector zone ahead of the complete-mix unit, where it mixes with the incoming raw wastewater. A well-designed high rate plug-flow selector ensures stable and uniform metabolic activity levels of the sludge in the complete-mix aeration tank, leading to accelerated digestion of organic matter and enhanced settleability of flocs. Consequently, the system operates effectively regardless of variations in influent flow rate and organic concentration.

In addition to these advantages, the CASS system achieves superior simultaneous nitrification and denitrification, as well as biological phosphorus removal, compared to conventional SBR processes. This versatility makes it suitable for a wide range of applications in both industrial and municipal wastewater treatment.

Overall, the Cyclic Activated Sludge System stands out as a highly efficient and effective solution for wastewater treatment, offering numerous advantages over traditional methods. Its innovative design and advanced capabilities make it a valuable asset for organizations seeking to enhance their wastewater treatment processes.

2.15.12.2 UNITANK Technology

The UNITANK Technology represents a highly effective wastewater treatment system that integrates the advantages of SBR, three-ditch oxidation treatment, and a conventional aeration tank. This innovative system is designed with a single tank divided into three hydraulically connected compartments in series, each equipped with its own aeration system and without external sludge recirculation provision.

During a single operational cycle, the UNITANK system undergoes two main stages, each comprising three basic steps performed symmetrically, starting from either of the outer compartments. These outer compartments alternate between acting as aeration and sludge settling tanks, while the middle compartment functions solely as an aeration unit. Unlike traditional sedimentation tanks, the outer compartments of the UNITANK system feature sludge slots and fixed effluent weirs.

For enhanced nitrogen and phosphorus removal, an advanced variant of the UNITANK system is employed. This configuration includes additional anaerobic/anoxic compartments with internal recirculation of mixed liquor (Khan et al., 2022; Sun et al., 2015).

The UNITANK system is particularly suitable for small- to middle-sized wastewater treatment plants, offering a range of benefits such as simple structure, minimal land occupation, cost-efficiency, and reliable operation. Its successful implementation in various countries, including China, Mexico, Argentina, Brazil, Vietnam, among others, underscores its versatility and effectiveness in wastewater treatment applications.

2.15.12.3 Intermediate Cycle Extended Aeration System

The Intermediate Cycle Extended Aeration System (ICEAS) represents an advanced replication of the standard SBR system, enabling continuous inflow of wastewater through a distributor box that evenly allocates variable inflows among all tanks, to prevent overloading in any single tank. The pre-react zone, featuring high F/M (food to microorganisms ratio), serves as a selector, facilitating enhanced sludge settling and inhibition of filamentous growth.

The main-react zone, following the pre-react zone, operates in three basic modes: Aeration, Settle, and Draw. Equal loading of all basins during continuous inflow simplifies operation and process control, making maintenance more manageable. Additionally, ICEAS requires only a single set of tanks, leading to significant capital cost reduction compared to conventional SBR processes. Complex process control associated with conventional SBR processes is also mitigated, as all basins receive equal loading and flow at any given time.

ICEAS is gaining traction in various countries, including China, the US, the UK, Peru, and Qatar, where it replaces old STPs or fulfills the needs of new plants with limited space or a demand for enhanced effluent quality. Consequently, ICEAS stands as a highly effective and efficient wastewater treatment system, offering numerous advantages over conventional SBR processes.

Large-scale applications of Sequential Batch Reactors (SBRs) commonly rely on two systems: the Cyclic Activated Sludge System (CASSTM) and the Intermittent Cycle Aerated Extended System (ICEAS®). These systems have been successfully implemented in municipal wastewater treatment plants serving populations exceeding one million. For example, the Dublin Bay WWTP in Ireland utilizes the ICEAS® process in 24 SBR basins stacked in two levels, treating an average daily flow of 320,000 m^3/d and a wet weather peak flow of 960,000 m^3/d (Cao et al., 2023).

Another illustration of the compact design of the SBR process can be found in Bangkok, Thailand. Two separate CASSTM SBR facilities, each with an average daily flow of 200,000 m^3/d and peak flow of 500,000 m^3/d, employ tanks stacked on four levels, achieving a treatment plant footprint of 6000 m^2 (Cao et al., 2023). Both facilities operate as Biological Nutrient Removal (BNR) systems, where equivalent conventional activated sludge systems would have either required more footprint (not feasible due to existing buildings) or additional levels (resulting in higher capital and operating costs).

2.15.13 Advantages and Disadvantages

Some advantages and disadvantages of SBRs are:

2.15.13.1 Advantages

- Equalization, primary clarification (in most cases), biological treatment, and secondary clarification can be achieved in a single reactor vessel.
- Operating flexibility and control.
- Minimal footprint.
- Potential capital cost savings by eliminating clarifiers and other equipment.

2.15.13.2 Disadvantages

- A higher level of sophistication is required (compared to conventional systems), especially for larger systems, of timing units and controls.
- Higher level of maintenance (compared to conventional systems) associated with more sophisticated controls, automated switches and valves.
- Potential of discharging floating or settled sludge during the DRAW or decant phase with some SBR configurations.
- Potential plugging of aeration devices during selected operating cycles, depending on the aeration system used by the manufacturer.
- Potential requirement for equalization after the SBR, depending on the downstream processes.

2.16 Activated Sludge Process (ASP)

The activated sludge process is a highly effective biological method for treating industrial wastewater. This process involves the use of air or oxygen to break down solids and create a biological environment that digests organic matter and pollutants. Unlike many three-stage sewage treatment plants, activated sludge plants do not require a primary settlement chamber, which needs regular emptying. Instead, excess sewage liquor is discharged into a clarification chamber where live bacteria settle to the bottom, dead bacteria rise to the top, and a clear liquid forms in the middle. This clear water is then discharged into either a watercourse or a soakway. The live bacteria, known as "activated bacterial sludge," are returned to the digestion chamber to re-seed the incoming raw sewage, while the dead bacterial crust is manually removed. This process is highly effective in treating industrial wastewater and is widely used across various industries. Proper maintenance and monitoring are essential to ensure the continued effectiveness of the activated sludge process.

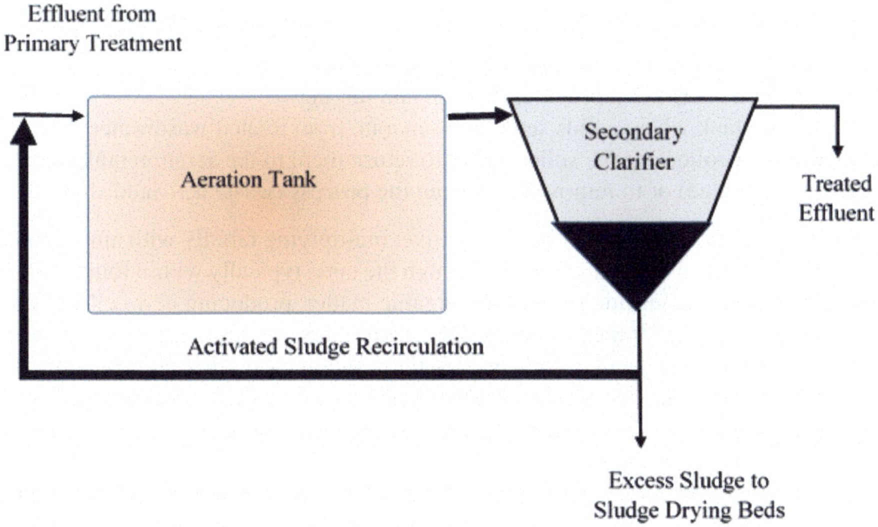

Fig. 2.7 Activated sludge process flow diagram

By implementing this process, industries can effectively treat their wastewater and reduce their environmental impact (Fig. 2.7).

The fundamental principle of the activated sludge process is the growth of microorganisms that form flocs, which settle at the bottom of the tank, leaving a partially clear liquid free of organic material and suspended solids. The process involves mixing screened wastewater with recycled liquid containing a high proportion of organisms from a secondary clarifying tank, resulting in a product called mixed liquor. This mixture is continuously stirred and fed with large quantities of air to maintain oxygen levels and keep solids in suspension. After some time, the mixed liquor flows to a clarifier where it is allowed to settle, and a portion of the bacteria is removed as activated sludge, while the partially cleaned water is taken for further treatment. The settled solids or activated sludge are returned to the aeration tank to restart the process.

Developed in England in the early 1900s, the activated sludge process has undergone several variations, including extended aeration, sequencing batch reactors (SBRs), and oxidation ditches. The process is a popular biological treatment method for larger and smaller plants due to its capability of producing high-quality effluent at a lower cost. Additional advantages include low construction costs and relatively small land requirements, making it widely required by large cities and communities where large volumes of wastewater need highly efficient and economical treatment. The process is also suitable for isolated facilities, such as hospitals or hotels, cluster situations, subdivisions, and small communities.

A basic activated sludge process consists of several interrelated components:

- An aeration tank where the biological reactions occur
- An aeration source that provides oxygen and mixing
- A clarifier tank where solids settle and separate from treated wastewater
- A means of collecting the solids either to return them to the aeration tank (return activated sludge) or to remove them from the process (waste activated sludge).

In the aeration tank, aerobic bacteria thrive, multiplying rapidly with ample food and oxygen. As the wastewater moves through the tank, typically within four to eight hours, the bacteria consume most of the organic matter, producing new cells. These microorganisms settle at the bottom of the clarifier tank, separating from the clearer water. This sludge is either pumped back to the aeration tank, where it is mixed with incoming wastewater, or removed from the system as excess, known as wasting. The relatively clear liquid above the sludge, known as the supernatant, is then sent for further treatment as required.

The efficient functioning of this process is crucial for proper wastewater treatment. The proliferation of aerobic bacteria is essential for breaking down organic matter, and the separation of sludge from the clearer water ensures the treatment process's effectiveness. By understanding the intricacies of this process, we can ensure that wastewater is treated efficiently and effectively, protecting both the environment and public health.

2.16.1 Principles of Operation

The process of biodegradation in the activated sludge system involves the conversion of coagulated organic matter into harmless products and new cell tissue through the action of microorganisms. This system relies on a mixture of raw wastewater and returned activated sludge, referred to as mixed liquor, which contains a high concentration of microorganisms. These microorganisms live by using oxygen and work by adsorbing and breaking down organic matter present in the wastewater. Oxygen is essential for providing the energy needed for cell growth and producing carbon dioxide (CO_2) and water (H_2O) as final by-products. The cell growth ensures the production of new organisms, maintaining the continuity of the process.

Microorganisms in the activated sludge system consist of two main types: 'floc formers' and 'filament formers.' The floc formers are beneficial as they tend to settle easily in the clarifiers, creating a dense, settleable floc that can be easily separated from the treated water. In contrast, filament formers do not settle as effectively and tend to float on the surface of the tank, potentially being carried out with the final effluent. Proper management of these microorganisms is crucial for maintaining the system's efficiency.

Key factors influencing the efficiency of the activated sludge system include:

1. **pH**: The optimal pH range for the activated sludge process is typically between 6.5 and 8.5. Deviations from this range can affect microbial activity and the overall efficiency of the treatment process.
2. **Temperature**: Microbial activity in the activated sludge process is temperature-dependent. The optimal temperature range is generally between 20 and 35 °C. Lower temperatures can slow down microbial metabolism, while higher temperatures may cause thermal stress to the microorganisms.
3. **Air Contact**: Adequate aeration is essential to provide the necessary oxygen for microbial respiration. Insufficient aeration can lead to reduced treatment efficiency and the formation of anaerobic zones, which may cause undesirable by-products such as hydrogen sulfide.
4. **Suspended Solids**: The concentration of suspended solids in the mixed liquor affects the system's performance. High concentrations can lead to poor settling characteristics, while low concentrations may result in insufficient microbial biomass for effective treatment.
5. **Culture Growth**: The health and growth rate of the microbial culture are vital for the process. Factors such as nutrient availability and the presence of toxic substances can impact microbial growth and activity.
6. **Concentration of Floc**: Maintaining an appropriate concentration of floc in the mixed liquor is crucial for effective sedimentation and the production of clear effluent. Regular monitoring and adjustment of the sludge return rate help maintain this balance.

Excess sludge produced in the system must be regularly removed to prevent overloading and maintain the normal balance of the process. This waste activated sludge (WAS) is typically treated separately through processes such as thickening, digestion, and dewatering before final disposal or reuse.

By understanding and managing these factors, the activated sludge process can effectively treat wastewater, producing high-quality effluent and minimizing environmental impact. Proper maintenance and monitoring are essential to ensure the continued effectiveness and efficiency of this biological treatment method.

2.16.2 Activated Sludge Process Description

The wastewater treatment process involves two primary components: the Aeration Tank and the Secondary Clarifier. Each plays a crucial role in the biological treatment and clarification of wastewater.

2.16.2.1 Aeration Tank

1. **Microbial Degradation**: In the Aeration Tank, organic waste is broken down by a cultivated microbial culture. This culture, known as mixed liquor, contains a variety of microorganisms that metabolize organic pollutants, converting them into carbon dioxide, water, and new microbial biomass.
2. **Oxygen Supply**: Oxygen is essential for the aerobic bacteria in the mixed liquor to thrive and perform the biodegradation process effectively. This is supplied through an Air Blower system, which pushes air through diffusers placed strategically within the tank. These diffusers ensure even distribution of oxygen throughout the tank, promoting optimal microbial activity.
3. **Nutrient Addition**: To support microbial growth and activity, essential nutrients such as nitrogen and phosphorus are added to the Aeration Tank. These nutrients are necessary for the synthesis of new cellular material and the overall metabolic processes of the bacteria.
4. **Mixed Liquor Suspended Solids (MLSS)**: Maintaining a constant MLSS level is critical for the efficiency of the process. The MLSS is a measure of the concentration of microbial biomass in the aeration tank. To achieve this, activated sludge is continuously recycled using a non-clog, centrifugal, semi-open type Sludge Recirculation pump. This pump ensures a consistent return of activated sludge to the aeration tank, maintaining the desired biomass concentration and enhancing the treatment process.

2.16.2.2 Secondary Clarifier

1. **Sedimentation**: The effluent from the Aeration Tank flows into the Secondary Clarifier. Here, the microbial flocs formed during the aeration process settle to the bottom of the tank due to gravity. This settling process separates the biomass (activated sludge) from the treated water.
2. **Sludge Recycling and Disposal**: The settled activated sludge at the bottom of the Secondary Clarifier is partially recycled back to the Aeration Tank to maintain the MLSS. This recycling is crucial for maintaining the population of microorganisms necessary for continuous wastewater treatment. The excess sludge, which accumulates over time, is withdrawn and transferred to sludge drying beds.
3. **Sludge Drying Beds**: In the sludge drying beds, the excess sludge is dried through natural evaporation and drainage. The dried sludge is then converted into sludge cakes, which can be used as high-quality manure in agricultural applications due to their nutrient content or for landfilling purposes.

2.16.2.3 Process Benefits

This advanced wastewater treatment process, leveraging microbial cultures and efficient aeration systems, ensures the effective breakdown of organic waste. The result is not only the production of treated effluent that meets environmental safety standards but also the generation of valuable by-products like nutrient-rich manure. This integration of advanced technology and biological treatment methods promotes sustainable wastewater management and resource recovery, contributing to environmental protection and agricultural productivity.

2.16.2.4 Seeding in Activated Sludge Plants

The wastewater treatment process using the activated sludge method relies heavily on the effective separation of flocculent microbial material, known as sludge, from the treated effluent in a secondary settler or clarifier. This process is crucial for ensuring that the activated sludge, which is rich in beneficial microorganisms, is maintained without contact with raw organic matter. The activated sludge is then partially recycled back to the aeration tank to serve as a seed for new microbial growth, while the excess sludge is wasted.

Microbial Seeding

The initiation of an activated sludge plant often involves seeding with microorganisms sourced from nearby operating plants or specialized microbial populations. The seeding material may include sludge underflow from the clarifier, mixed liquor from the aeration basin, or mixed liquor from an aerobic digester.

Preferred Seed Sources

High-quality seed material is crucial for a successful start-up. Ideal sources include activated sludge from:

- Nearby plants with similar influent characteristics.
- Plants with a proven track record of effective operation.
- Commercially available microbial cultures tested and proven effective for similar applications.

High-quality seed material must contain a significant number of microorganisms that form an active biological population with good flocculation properties. These bacteria play a vital role in determining the direction, rate, and effectiveness of pollutant degradation.

2.16.2.5 Benefits of Seeding

1. **Reduction in Start-Up Time**: Effective seeding can significantly reduce the start-up time by quickly establishing a robust microbial community.

2. **Process Efficiency**: High-quality seed material can decrease the critical sludge age, preventing sludge bulking and enhancing pollutant removal.
3. **Adaptation to Influent Changes**: Seeding helps maintain the required MLSS levels, particularly during variations in influent load, which can cause a drop in MLSS.

2.16.2.6 Maintenance and Monitoring

- **Consistent Recycling**: During plant start-up and operation, consistent recycling of activated sludge is essential. This helps acclimatize the bacterial culture, ensuring a steady MLSS level.
- **Re-seeding**: In cases of drastic changes in influent load or when MLSS levels drop, re-seeding with high-quality bacterial cultures is necessary to maintain process efficiency.

2.16.2.7 Key Elements of Activated Sludge Process

Separation and Recycling

- **Secondary Clarifier**: The effluent from the aeration tank enters the secondary clarifier, where gravity allows the sludge to settle at the bottom. The clear supernatant, free of organic material, is then discharged, while the settled sludge becomes "activated."
- **Sludge Recycling**: A portion of this activated sludge is recycled back to the aeration tank to maintain a high concentration of microorganisms, essential for the breakdown of organic pollutants. The remaining sludge is wasted to keep the biomass concentration in balance.

Bioaugmentation in Activated Sludge Systems

Bioaugmentation: This involves adding specialized microbial cultures to enhance the degradation of specific pollutants. However, bioaugmentation is not widely adopted due to its unpredictability and control issues, despite commercial reports suggesting potential benefits (Jauffur et al., 2014).

2.16.3 Modifications in Activated Sludge Process

The management of a specific type of wastewater entering activated sludge plants has necessitated modifications in the systems, leading to various variations in the activated sludge process.

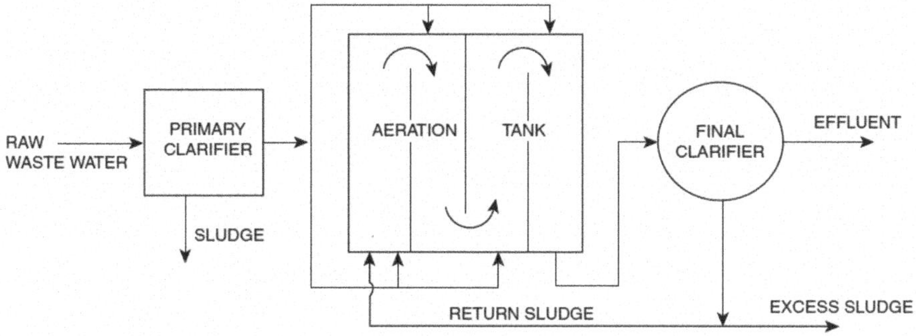

Fig. 2.8 Step aeration plant

2.16.3.1 Step Aeration

Step Aeration is a wastewater treatment process that involves introducing influent wastewater at various points along the length of the aeration tank. Sludge return rates typically range between 25 and 50%. During the 3–7 h aeration period, the oxygen requirement is approximately half that of the conventional process (Wang et al., 2009) due to more effective biomass utilization in the aeration basin, which allows for organic loadings of 30–50 pounds of biochemical oxygen demand per 1000 cubic feet per day. This is a significant improvement compared to the conventional systems, which only permit loadings of 30–40 pounds of biochemical oxygen demand per 1000 cubic feet per day (Hu et al., 2022).

Step Aeration is a highly efficient and effective method for treating wastewater. By introducing influent wastewater at various points along the aeration tank, the process allows for more effective biomass utilization, resulting in a reduced oxygen requirement during the aeration period. This not only saves energy but also allows for higher organic loadings, making it a more cost-effective solution compared to conventional systems. With sludge return rates ranging between 25 and 50%, Step Aeration is a sustainable and environmentally friendly option for wastewater treatment (Fig. 2.8).

2.16.3.2 Contact Stabilization

The contact stabilization activated sludge process is a highly effective wastewater treatment method employing a two-step aeration system. Initially, raw or primary-settled wastewater is mixed with activated sludge in the contact tank, where it undergoes short-duration aeration for 30 min–2 h. Following this, the effluent is directed to a final settling tank for separation. The settled activated sludge from the final clarifier is then routed to a separate re-aeration basin for 3–8 h of aeration before being returned to the contact aeration basin to mix with incoming raw wastewater or primary settled effluent.

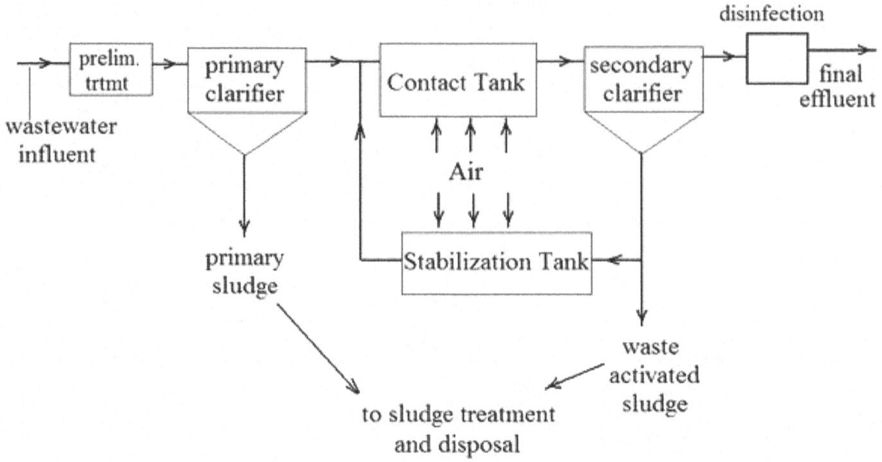

Fig. 2.9 Contact stabilization activated sludge process

A significant advantage of the contact stabilization process is its enhanced ability to manage greater shock and toxic loadings compared to conventional systems, attributed to the buffering capacity of the biomass in the stabilization tank. During periods of abnormal loadings, the majority of the activated sludge is isolated from the main stream of plant flow, thereby maintaining system stability and efficacy.

However, contact stabilization plants are not recommended for situations with daily hydraulic or organic loading variations routinely exceeding a 3:1 ratio on consecutive days, nor for plants with average flows of less than 0.1 million gallons per day, unless prior approval from the owner is obtained.

Overall, the contact stabilization activated sludge process is a robust and reliable wastewater treatment method, offering numerous advantages over conventional systems. Its two-step aeration process, coupled with the buffering capacity provided by the biomass in the stabilization tank, makes it an ideal choice for handling shock and toxic loadings, ensuring efficient and effective wastewater treatment (Fig. 2.9).

2.16.3.3 Tapered Aeration

Tapered aeration is a strategic approach to balancing oxygen supply and demand within an aeration tank, optimizing the efficiency of wastewater treatment systems. This process involves modifying the conventional method to decrease oxygen demand progressively along the tank's length. Consequently, in a tapered aeration system, oxygen levels are highest at the influent end and gradually decrease towards the effluent end. By allocating 55–75% of the total air supply to the first

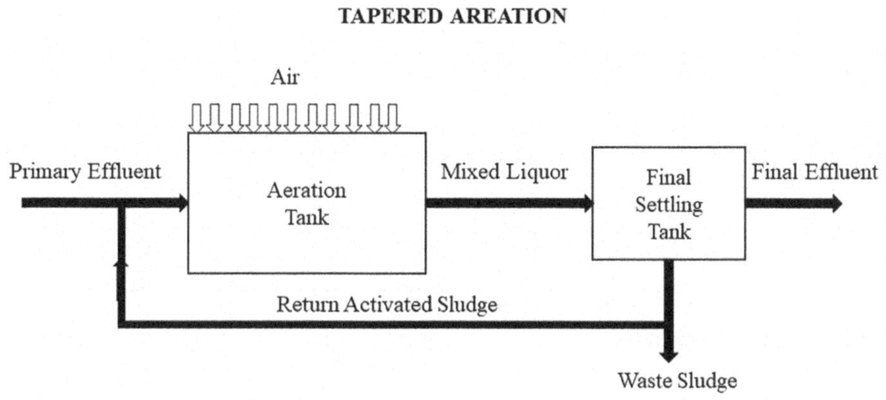

Fig. 2.10 Tapered aeration process

half of the tank, optimal treatment outcomes are achieved. This technique is particularly beneficial for large-scale treatment plants, ensuring uniform oxygen distribution throughout the tank and enhancing treatment efficacy. Moreover, tapered aeration reduces energy consumption and operational expenses, offering a cost-effective solution for wastewater treatment (Fig. 2.10).

2.16.3.4 High Rate Activated Sludge (HRAS)

The High Rate Activated Sludge (HRAS) Process is a wastewater treatment method characterized by lower concentrations of biological floc and shorter aeration periods within the aeration tank. This approach aims to prevent microorganisms from transitioning into the endogenous phase, which is marked by decay, and instead maintains them in the lag and declining phases. Widely employed across various industries, this method proves highly effective in wastewater treatment by ensuring the continuous efficiency of the treatment process.

The HRAS Process stands as a dependable and efficient solution for wastewater treatment as its capacity to sustain microorganisms in the lag phase guarantees the effectiveness and efficiency of the treatment process, rendering it a favored choice across diverse industrial applications.

2.16.3.5 Complete Mix Process

The complete mix process stands as a highly efficient system renowned for its rapid and thorough distribution of influent flow and return sludge across the reaction zone. This meticulous distribution fosters uniform mixing and aeration, thereby ensuring a consistent concentration throughout the aeration tank. Moreover, the dual functionality of the aeration tank as an equalization tank optimizes the utilization of

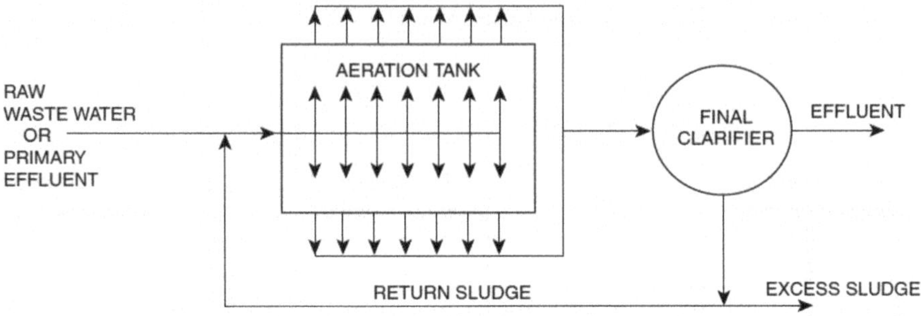

Fig. 2.11 Complete mix process

microorganisms within the activated sludge. Nonetheless, it is worth noting that this approach may incur higher aeration costs. Despite this consideration, the complete mix process remains a dependable and effective method for wastewater treatment, offering reliability and efficacy in achieving treatment objectives (Fig. 2.11).

2.16.3.6 Extended Aeration

Extended aeration is a wastewater treatment process characterized by providing ample aeration time to facilitate the metabolism of organics through cell growth and the oxidation of cell mass. This method involves an aeration tank with a prolonged detention time compared to conventional processes, followed by a settling tank with an extended detention period to achieve complete stabilization of organic matter under aerobic conditions.

2.16.3.7 Kraus Process

The Kraus process represents an enhancement of the conventional sludge process. Kraus achieved this by aerating a blend of anaerobic digester supernatant and a portion of the return sludge in a separate tank. This blend underwent significant nitrification and exhibited favorable settling properties (Metcalf & Eddy, 2014; Tchobanoglus et al., 2003). Subsequently, the nitrified mixture was introduced to the raw wastewater in the aeration tank, resulting in approximately 90% BOD removal. Kraus employed a dual aeration system, utilizing fine bubble aeration at the bottom and coarse aeration at the top (Fig. 2.12).

These two wastewater treatment processes play critical roles in ensuring the efficient removal of organic matter from wastewater. The extended aeration system, with its prolonged detention time, enables complete stabilization of organic matter under aerobic conditions. Conversely, the Kraus process enhances the performance of the conventional sludge process by aerating a mixture of anaerobic digester supernatant and return sludge, yielding a highly nitrified and well-settling mixture.

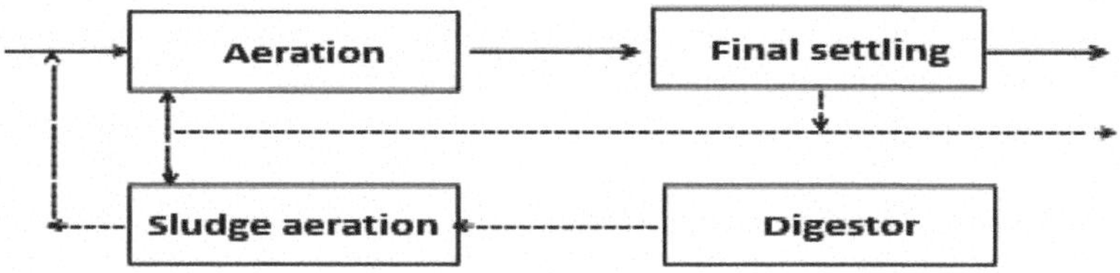

Fig. 2.12 Kraus process

2.16.4 Some Important Parameters in Activated Sludge Process

2.16.4.1 Mixed Liquor Suspended Solids (MLSS)

The Mixed Liquor Suspended Solids (MLSS) concentration is a pivotal parameter in both the design and operation of wastewater treatment systems. It denotes the quantity of solid particulate matter within the mixture of Activated Sludge and Raw Wastewater. The MLSS concentration holds significant importance in determining the efficacy of the aeration system, a critical component of the Extended Aeration System's functionality.

Maintaining the MLSS level within specified parameters is essential for ensuring optimal treatment of the incoming effluent stream. A high MLSS concentration, coupled with an adequate aeration period, is crucial for the system to operate effectively. Adjustments to the sludge wasting rate may be necessary to regulate the MLSS concentration within the desired range. If the MLSS concentration decreases, reducing sludge wasting can elevate the MLSS concentration in the aeration lagoon. Conversely, an increase in MLSS concentration may necessitate an increase in sludge wasting to restore the MLSS balance in the tank.

2.16.4.2 Sludge Volume Index

The Sludge Volume Index (SVI) serves as a critical metric in wastewater treatment plants, quantifying the ratio of the percent settled volume of sludge to the percent weight of Mixed Liquor Suspended Solids (MLSS) in grams. This index offers valuable insights into the settleability and physical state of the activated sludge.

A healthy operational state of the plant is typically denoted by an SVI falling within the range of 50–100, indicating proper settling of the sludge. However, when the SVI surpasses 150, it signifies inadequate sludge settlement, with values reaching 200 indicating the onset of "Sludge Bulking" issues.

To manage SVI values effectively, it is imperative to regulate both the MLSS concentration and the flow of effluent/return sludge in the aeration lagoon. When the SVI exceeds 100, adjustments such as increasing the amount of sludge waste and reducing the quantity of 'Return Active Sludge' from the clarifier/settling tank become necessary. By maintaining the SVI within the optimal range, wastewater treatment plants can uphold efficient and effective operations.

2.16.4.3 Food to Microorganism Ratio

The Food to Microorganism ratio (F/M), a pivotal parameter in wastewater treatment, denotes the proportion of incoming food or waste to microorganisms within the mixed liquor. As underscored by Fazli et al. (2020), the F/M ratio significantly influences the performance of the treatment process.

Maintaining an optimal F/M ratio is imperative for achieving efficient wastewater treatment. Elevated F/M ratios can prompt excessive microbial growth, resulting in poor settling and clarification of treated water. Conversely, low F/M ratios may curtail microbial activity, leading to incomplete degradation of organic matter in the wastewater.

Hence, meticulous monitoring and adjustment of the F/M ratio are paramount to ensuring optimal treatment performance. Through such measures, high-quality treated water meeting requisite environmental standards can be consistently attained.

2.16.4.4 Mixed Liquor Volatile Suspended Solids (MLVSS)

MLVSS, or Mixed Liquor Volatile Suspended Solids, is a vital indicator of the concentration of volatile suspended solids within the aeration basin of an activated sludge system. This parameter holds significant importance in assessing the efficiency of the treatment process and ensuring its optimal performance. Precise measurement of MLVSS allows wastewater treatment plants to diligently monitor and fine-tune their operations, thereby upholding the production of high-quality effluent.

Maintaining a professional tone when discussing technical terms like MLVSS is essential for fostering clear communication and understanding among industry professionals. This ensures that crucial concepts are effectively conveyed and comprehended, facilitating informed decision-making and proficient operation of wastewater treatment facilities.

2.16.4.5 Cell Residence Time (CRT)

The Cell Residence Time (CRT) refers to the average duration that an organism spends in the secondary treatment process. This metric is measured in days and calculated by dividing the total Mixed Liquor Volatile Suspended Solids (MLVSS) in kilograms by the total waste of MLVSS in kilograms per day. Regulating the CRT for a facility involves appropriately managing excess biomass disposal. Maintaining the desired CRT is essential to ensure optimal treatment efficiency.

To manage excess biomass effectively and maintain the desired CRT, proper waste disposal techniques must be implemented. This includes strategies for handling and disposing of excess biomass in a timely and efficient manner. By adopting these practices, wastewater treatment facilities can uphold the optimal duration of organism residence in the secondary treatment process. This, in turn, facilitates efficient and effective treatment processes, contributing to the overall performance and sustainability of the facility.

2.16.5 Features of Activated Sludge Process

The activated sludge system represents a highly efficient wastewater treatment method, offering distinct advantages over lagoon systems. One key benefit is its compact footprint, making it particularly suitable for sites with limited land availability. Unlike lagoon systems, the activated sludge system can achieve effluent standards without necessitating significant storage capacity. This streamlined approach allows for simultaneous wastewater treatment and processing, reducing retention time and resource consumption while optimizing treatment outcomes.

Moreover, the treated effluent generated by the activated sludge system holds potential for various applications, including replacing fresh water in production processes or fulfilling irrigation needs, provided the topography of the plant site allows for such practices. Integral to the system's operation are electro-mechanical components like aerators, crucial for maintaining optimal oxygen levels within the wastewater. While these aerators entail additional costs and require regular maintenance, they are indispensable for achieving desired treatment results in the activated sludge system.

2.17 Membrane Bio Reactor Technology (MBR)

MBR (Membrane Bioreactor) technology integrates conventional activated sludge treatment with membrane filtration, offering a highly efficient approach to wastewater treatment. The membrane employed in MBR systems typically features a pore size ranging from 10 nm to 0.4 μm, facilitating effective separation of sludge. Acting as a barrier, this membrane retains particles, colloids, bacteria, and viruses, ensuring thorough disinfection of treated water.

A notable advantage of MBR technology lies in its capability to operate at elevated sludge concentrations, reaching up to 12 g/L compared to the typical 4 g/L in conventional systems. This leads to a reduction in reactor volume and sludge production, enhancing cost-effectiveness. Among the configurations, the immersed setup is widely preferred in municipal wastewater treatment due to its lower operational costs. In this arrangement, the module is directly immersed into the process tank, minimizing energy consumption.

Commercial membrane modules for the immersed configuration are available in two main types: flat sheet and hollow fiber. While hollow fiber modules offer higher packing density, they are more prone to membrane clogging and sludging. Commercial membrane MBR products predominantly utilize fluorinated and sulphonated polymers such as polyvinylidene difluoride and polyethersulfone.

Understanding membrane fouling mechanisms is crucial in MBR operation, with factors categorized into four groups: sludge nature, operating parameters, membrane/module characteristics, and feed wastewater composition. By addressing these factors, MBR systems can maintain optimal performance and prolong membrane lifespan.

MBR technology represents a highly efficient and cost-effective solution for wastewater treatment. Its ability to operate at elevated sludge concentrations while ensuring complete disinfection underscores its significance in wastewater treatment processes.

2.17.1 Membrane Filtration

Membrane filtration stands as a highly efficient method for eliminating waterborne pollutants, offering a process where contaminated water flows across a specialized membrane. This membrane facilitates the permeation of water through a distinct channel for recovery, while the cross-flow movement ensures that retained materials do not accumulate on the membrane surface but are instead flushed out of the system for later retrieval or disposal (Fig. 2.13a).

The water that successfully passes through the membrane is termed permeate, while the portion containing more concentrated pollutants is referred to as concentrate or retentate. Membranes are commonly composed of cellulose or other polymer materials, with the maximum pore size determined during manufacturing to prevent

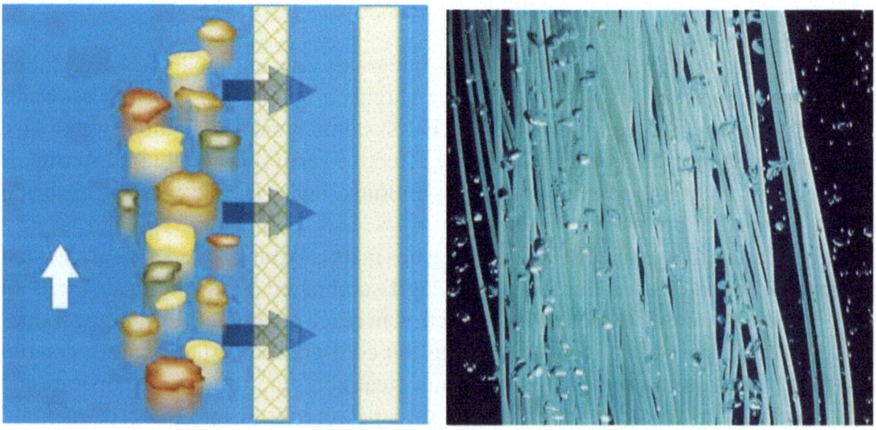

a b

Fig. 2.13 **a** Membrane filtration process. **b** Hollow-fibre membranes

the passage of particles the size of microorganisms, which typically measure approximately 1 μm (0.001 mm) (Hube et al., 2020). Although MBR systems excel at removing solid materials, supplementary treatment steps are often required to effectively eliminate dissolved wastewater constituents, ensuring the water's safety for reuse or discharge into the environment.

By harnessing membrane filtration, we can efficiently extract pollutants from water sources, safeguarding both human health and environmental integrity.

Membranes can be configured in diverse ways to cater to different applications. In the context of MBR (Membrane Bioreactor), two configurations are frequently employed: hollow fibers grouped in bundles, as illustrated in Fig. 2.13b, or flat plates (Konvensional, 2017). The hollow fiber configuration entails linking the hollow fiber bundles via manifolds, engineered for easy maintenance and replacement. This guarantees the MBR system operates at peak efficiency, delivering dependable and uniform outcomes.

2.17.2 Design and Operation Consideration

The IMBR (Integrated Membrane Bioreactor) configuration is extensively applied in large-scale applications (Zhang et al., 2016). During design and operation, several crucial factors are considered, including pre-treatment, design flux, hybrid systems, equalization tanks, membrane fouling control and cleaning, sludge retention time, biomass concentration, and membrane life.

2.17.2.1 Pre-treatment

Pre-treatment in the IMBR process is critical for removing suspended solids, organic matter, and other impurities from the influent. Coarse solids like plastics and fine particles such as hair found in wastewater can damage the membranes if not adequately removed. Proper pre-treatment and screening, typically with screens ranging from 1 to 3 mm openings, are essential to protect the membranes and ensure efficient operation of MBR's.

2.17.2.2 Design Flux, Hybrid Systems, and Equalization Tanks

Membrane permeate flux is a crucial parameter affecting both capital and operational expenditures in IMBR's for municipal wastewater treatment. Full-scale IMBR's typically operate at flux rates exceeding 10–30 L per square meter per hour (LMH). Considering membrane permeate flux during design and operation is vital for ensuring efficiency and cost-effectiveness. High flux rates in full-scale IMBR's indicate their high-performance capabilities, contributing to effective municipal wastewater treatment and environmental protection.

2.17.2.3 Membrane Fouling Control and Cleaning

Effective membrane fouling control is essential for optimal MBR operation, as fouling can increase energy demand and operational costs. Control methods include air sparging, relaxation, back flushing, and frequent maintenance cleanings. Air sparging, relaxation, and back flushing are performed at specific intervals to maintain membrane permeability, while frequent maintenance cleanings are necessary to prevent fouling buildup.

2.17.2.4 SRT and Biomass Concentration

Sludge Retention Time (SRT) and biomass concentration significantly influence treatment performance, membrane filtration, and system economics. Higher SRT results in increased sludge solids concentration, reducing bioreactor volume requirements and improving treatment performance. However, high solids concentration can lead to decreased air sparging efficiency, increased energy demand, membrane fouling, and risk of clogging. IMBRS are typically designed for an MLSS range of 8–12 g/L and an SRT range of 10–20 days.

2.17.2.5 Membrane Life

The lifespan of membranes in MBR systems can exceed 10 years, with studies showing a correlation between permeability loss and operation time. After approximately seven years of operation, membrane permeability may decline to a non-operative level. Regular maintenance and monitoring are essential to prolong membrane life and ensure optimal system performance.

2.17.3 Advantages

1. **Efficient Wastewater Treatment**: Membrane bioreactors (MBRs) are highly efficient in treating wastewater by effectively separating suspended and colloidal matter, resulting in high-quality effluents suitable for reuse.
2. **Complete Biomass Retention**: MBR systems retain biomass within the aerobic reactor, decoupling sludge retention time (SRT) from hydraulic retention time (HRT). This allows for increased biomass concentrations in the reaction basin, enhancing treatment efficiency.
3. **Compact Design**: Compared to conventional activated sludge processes (CAS), MBR systems are more compact. They integrate multiple processes into a single unit, saving space and facilitating easier installation.
4. **Reduced Pre-treatment Requirements**: MBR systems require only screening (1–3 mm) of feed wastewater just before the removal of larger solids that could

damage the membranes. This simplifies pre-treatment processes and reduces operational complexity.

The widespread adoption of membrane bioreactors (MBRs) is impeded by significant capital and operating expenditures (CAPEX and OPEX), primarily attributed to membrane installation, replacement, and high energy demand. Compared to conventional activated sludge (CAS) systems, MBRs require more energy due to the need to prevent and address membrane fouling, a complex phenomenon involving the accumulation of solid particles and macromolecules on the membrane surface.

Fouling mechanisms are heterogeneous and involve varying sizes of particles and macromolecules in microbial suspensions. Dynamic filtration processes further complicate fouling. Operational issues include membrane process complexity, specific cleaning requirements, foam formation from excessive aeration, limited sludge dewatering capacity, and sensitivity to shock loads.

High-pressure systems in MBRs increase electricity consumption, necessitating pre-treatment to prevent fouling, potentially increasing space requirements and costs. Additionally, concentrate disposal and residual handling may be needed. Membrane replacement every 3–5 years, scale formation, gradual flux rate decline, and suboptimal recovery rates present challenges.

Despite challenges, MBR technology remains popular. However, careful consideration of costs, product economic value, and potential performance loss from fouling is essential. High flux requirements in MBRs result in significant energy penalties for reducing membrane fouling (~ 10 kWh/m^3 product). While offering advantages, MBR implementation requires careful evaluation of cost and operational challenges, particularly related to membrane fouling.

2.18 Moving Bed Biological Reactor (MBBR)

The Moving Bed Biological Reactor (MBBR) is an advanced biological treatment process derived from the conventional activated sludge and biofilter processes. Operating as a completely mixed and continuously operated Biological Reactor, the MBBR employs small carrier elements with a density lighter than water, which are kept in motion within the reactor along with the water stream. This movement can be induced by aeration in aerobic reactors or by mechanical stirrers in anaerobic or anoxic reactors (Fig. 2.14).

The versatility of the MBBR makes it highly effective in treating various industrial effluents, including those from the pulp and paper industry, poultry processing, cheese factories, refineries, slaughterhouses, phenolic sources, dairy facilities, and municipal wastewater (Azizi et al., 2021). The selection of the carrier material is crucial for optimizing system performance, as the efficiency of the MBBR greatly depends on it.

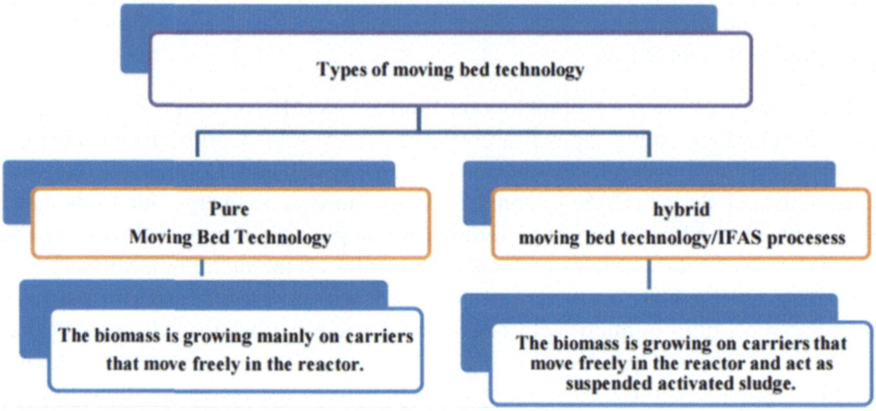

Fig. 2.14 Moving bed biological reactor

2.18.1 *Working of MBBR*

The Moving Bed Biofilm Reactor (MBBR) process is a highly efficient wastewater treatment method that utilizes floating plastic carriers, or media, within the aeration tank. These carriers significantly increase the available surface area for microorganisms to treat the wastewater, enhancing the treatment process compared to conventional secondary treatment methods. The microorganisms within the MBBR process consume organic material, and the media provides an ideal environment for their attachment and growth within the aeration tanks.

The increased surface area offered by the media allows for a reduction in the footprint of the tanks required for wastewater treatment, making the MBBR process a space-efficient solution. As depicted in Fig. 2.15, the media (Table 2.5) is continuously agitated by bubbles from the aeration system, which introduce oxygen into the first compartment of the aeration tank. In the middle compartment, a stone bed filters the wastewater to some extent before it enters the last compartment through provided openings.

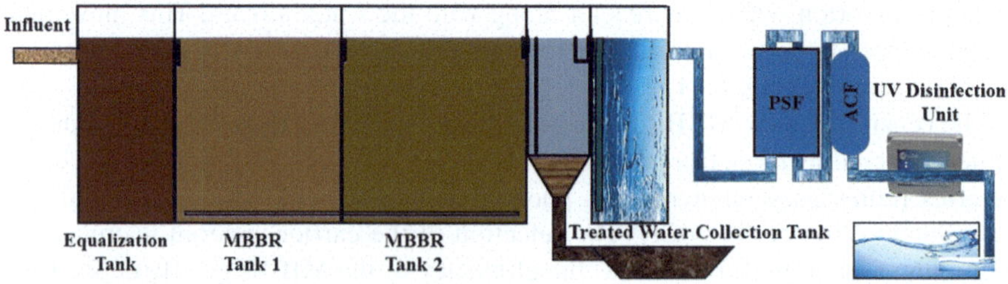

Fig. 2.15 Moving bed biofilm reactor experimental setup

Table 2.5 Characteristics of the biomedia

Material	Polypropylene, plastic, ceramic, porous
Shape	Corrugated cylinder, chips, hollow, curved density
Density	0.95 g cm^{-3}
Dimensions	10 × 15 mm
Specific surface	260 m^2 m^{-3}

In the final portion of the tank, bio carriers are filled, and turbulence is induced in the wastewater with the assistance of rotors. This ensures thorough treatment before the final effluent is discharged through the outlet. Overall, the MBBR process stands as a highly effective and sustainable method of wastewater treatment, offering a more environmentally friendly solution for various treatment applications.

2.18.2 Operating Principle

The Moving Bed Biofilm Reactor (MBBR) process is a highly effective technique in wastewater treatment, facilitating the growth of biomass on plastic supports that move within a biological reactor. This movement is enabled by either aeration systems in aerobic reactors or mechanical systems in anoxic or anaerobic reactors. The MBBR process has gained popularity due to its ability to offer increased specific surface area in the reactor for biomass growth, resulting in significant reductions in biological reactor volume. Moreover, the process can be operated at very high loads and is insensitive to load variations and other disturbances.

To grasp the MBBR process better, it is essential to examine its various components.

The Moving Bed Biofilm Reactor (MBBR) process relies on several key components to effectively treat wastewater.

1. **Basin or Reactor**

    - The basin, also known as the reactor or aeration tank, is the primary vessel where the MBBR process occurs.
    - Its size is determined based on the specific filtration requirements of the plant.
    - The influent, or wastewater to be treated, is introduced into the basin for treatment.
    - Depending on the treatment needs, the effluent may be directed to a secondary basin for additional processing or treatment.

2. **Media or Carriers**

    - The basin contains a significant quantity of small plastic chips or media, also known as carriers, which occupy a substantial portion (typically 50–70%) of the tank volume.

- These carriers are designed to optimize the surface area available for biofilm growth.
- The carriers provide a substrate for the attachment and growth of microorganisms that break down organic matter in the wastewater.
- The biofilm that develops on the carriers plays a crucial role in the treatment process by hosting microorganisms that degrade pollutants.

3. **Aeration Grid**

- The aeration grid is located at the bottom of the tank and serves to facilitate the movement of media within the reactor.
- It functions similarly to a fan, ensuring that the carriers remain in motion and come into contact with the wastewater.
- The aeration grid introduces oxygen into the tank, which is essential for the activity of aerobic microorganisms.

4. **Sieve**

- To prevent the media from escaping the tank, a sieve is affixed to the tank's outlet.
- The sieve allows water to pass through while retaining the plastic carriers within the basin.

5. **Design Considerations**

- When designing an MBBR system, it is essential to consider achieving treatment through a series of reactors, each designated for specific treatment functions.
- This compartmentalized approach allows for the development of specialized biofilms tailored to achieve specific treatment goals.
- By optimizing each reactor's design and function, the overall treatment scheme becomes more efficient and effective, resulting in higher-quality treated wastewater.

6. **Bio-carriers**

- Bio-carriers are crucial components of MBBR systems, serving as substrates for biofilm growth (Fig. 2.16).
- They come in various shapes and surface areas, with each factor influencing the carrier's performance.

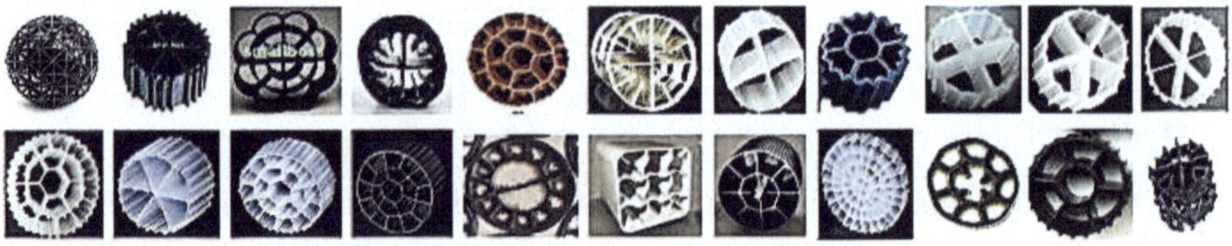

Fig. 2.16 Different shapes of bio carriers used in MBBR technology

- The carrier's shape, density, and surface area affect its strength, shearing, and colliding conditions, which impact biofilm growth.
- Common materials for bio-carriers include high-density polyethylene (HDPE), with different shapes and protected surface areas.

Overall, the integration of these components within an MBBR system facilitates efficient wastewater treatment by providing ample surface area for biofilm growth, ensuring proper mixing and oxygenation, and preventing media escape from the reactor. This results in a cost-effective and sustainable solution for wastewater treatment.

- Carriers occupy 25.70% of tank Volume.
- Specific surface area of carrier is about 500 m^2/m^3.
- The carriers are slightly buoyant and have a specific gravity between 0.94 and 0.96 g/cm^3.
- Media carrier life is around 15–20 years.

SSA (Specific Surface area of the reactor) = Surface area of the biofilm on the carrier + Reactor Volume.

2.18.3 Designs

2.18.3.1 High Rate Designs

When a compact high-rate system is required to meet basic secondary treatment standards, the moving-bed reactor is a viable option. In high-rate applications, the primary objective is to eliminate the soluble and easily degradable BOD from the influent stream. However, under high loading conditions, the settling character of the biofilm slough diminishes (Safwat, 2019). To address this issue, the high-rate MBBR can be combined with chemical coagulation and flocculation of the treated effluent, flotation, or a solids contact process step to remove the solids. Despite these additional steps, the overall result is a compact system that can meet basic secondary treatment standards in a short hydraulic retention time (HRT). The moving-bed reactor is an excellent option for those seeking a high-rate system that can meet basic secondary treatment standards. By combining the MBBR with other processes, such as chemical coagulation and flocculation, flotation, or a solids contact process step, the system can effectively remove solids and achieve the desired level of treatment.

2.18.3.2 Normal-Rate Designs

When considering conventional basic secondary treatment approaches, the moving-bed reactor offers a viable technological option. Typically, two reactors in series are utilized to achieve the desired level of treatment. To provide a summary of the 7-day

BOD (BOD7) removal performance at four normal-rate MBBR WWTPs designed for BOD removal followed by chemical removal of phosphorus, it is important to note that these MBBRs were designed with an organic loading rate of 7–10 g BOD m^2/d at 10 °C. Additionally, they are combined with follow-up chemical addition and flocculation for phosphorus removal and enhanced solids separation.

This information, as reported by Safwat (2019), highlights the effectiveness of MBBRS in achieving the desired level of treatment. By utilizing this technology, wastewater treatment plants can effectively remove BOD and phosphorus, resulting in improved water quality and environmental sustainability.

2.18.3.3 Low-Rate Designs

When designing reactors for the removal of organic matter upstream of a nitrification reactor. it is crucial to consider a low loading-rate design. This approach ensures that a high nitrification rate can be achieved in the downstream nitrifying moving-bed reactor, resulting in the most economical design. However, if the BOD load to the nitrification MBBR is not sufficiently reduced, the nitrification rate is significantly reduced, rendering the reactor ineffective.

This scenario can occur when a too-high of BOD load is selected for the prior organic matter removal stage of treatment, resulting in carryover to the nitrification stage of treatment. For instance, a nitrification rate of 0.8 $g/m^2/d$ can be achieved at a BOD load of 2 $g/m^2/d$ and a bulk dissolved oxygen of 6 mg/L. However, this rate would be reduced by approximately 50% if the BOD load increased to 3 $g/m^2/d$ (Safwat, 2019).

To address this issue, the operator could adjust by using a higher bulk dissolved oxygen level to compensate for the impeded condition, or the fill fraction could be increased to reduce the surface area loading rate. However, neither of these options results in an economical and effective design approach. Instead, the preferred design approach would be to size the MBBRS designated for BOD removal by using a conservative low loading rate, ensuring maximum efficiency in the downstream nitrifying MBBR.

2.18.4 Advantages

The MBBR is a complete mix, continuous flow process which combines the advantage of fixed film and suspended growth processes including:

1. Compact units with small size
2. Increased treatment capacity
3. Complete solids removal
4. Improved settling characteristics
5. Operation at higher suspended biomass

6. Concentrations resulting in long sludge retention times
7. Enhanced process stability
8. Low head loss
9. No filter channelling
10. No need of periodic backwashing
11. Reduced sludge production and no problems with Sludge bulking.

2.18.5 Application

1. Handle high sewage loads
2. Improve treatment quality standards
3. Treated water can be reused
4. River water treatment: Denitrification, Dephosphorization
5. Remove ammonia–nitrogen for water purification.

2.19 Nanotechnology

The field of wastewater treatment has undergone a revolutionary transformation with the emergence of nanotechnology and the integration of living microorganisms in biomicroelectronic devices. Nanotechnology has the unique ability to seamlessly merge with other technologies, modify existing concepts, and offer innovative approaches to develop and exploit these processes in completely new ways.

One of the most promising applications of nanotechnology in wastewater treatment is the development of higher-performing membranes with fewer fouling characteristics and improved hydraulic conductivity. Researchers are currently investigating the use of nanomaterials for the fabrication of membranes that can effectively decompose toxic compounds during the treatment process. This enables the effective segregation of metals, bimetallic nanoparticles, mixed oxides, zeolites, and carbon compounds from wastewater resources. With improved membranes and configurations, more efficient pumping and energy recovery systems become possible.

Various types of nanomaterials can be used in wastewater treatment, including polymeric nanoparticles, metal nanoparticles, carbon-based nanomaterials, zeolites, self-assembled monolayers on mesoporous supports, biopolymers, and many more. Nanotechnology-based pathways employed for wastewater remediation include adsorption and biosorption, nanofiltration, photocatalysis, disinfection and pathological control, sensing, and monitoring. The integration of nanotechnology in wastewater treatment has opened up new possibilities for more efficient and effective treatment processes. As research in this field continues to advance, we can expect to see even more innovative solutions that will help us better manage our precious water resources.

2.19.1 Adsorption and Biosorption

Adsorption and biosorption are pivotal processes in the realm of wastewater treatment. Adsorption is a surface phenomenon that involves the transfer of a phase, such as a molecule or ion, from a liquid or gaseous bulk onto a solid surface called an adsorbent. This process occurs through physicochemical or chemical interactions under specific conditions.

Biosorption, on the other hand, is a specialized form of adsorption that employs biological materials, such as bacteria, algae, or fungi, as adsorbents due to their inherent ability to bind and accumulate heavy metals. The biosorption process encompasses mechanisms like microprecipitation, ion exchange, and cell-surface complexation. This can occur via metabolically mediated pathways, which utilize ATP, or through spontaneous physicochemical pathways of uptake. Biosorption is particularly effective for removing heavy metals from dilute aqueous solutions.

Several advanced techniques for adsorption and biosorption hold significant applicability in wastewater treatment.

Each of these techniques comes with its own set of advantages and disadvantages, and the choice of technique is determined by the specific requirements of the wastewater treatment process, including the type of contaminants present, the concentration of pollutants, and the operational conditions of the treatment facility.

2.19.2 Carbon-Based Nano-adsorbents

Carbon has been a focal point of extensive research across various scientific and technological disciplines for decades. Studies on carbon nanostructures have revealed a variety of low-dimensional allotropes, including activated carbon, carbon nanotubes (CNTs), the C60 family of buckyballs, graphite, and graphene (Nasir et al., 2018). These carbon nanostructures are extensively utilized as nanoadsorbents in wastewater treatment due to their abundant availability, cost-effectiveness, high chemical and thermal stability, large active surface areas, excellent adsorption capacities, and environmentally friendly properties (Sayed et al., 2021).

Activated carbon has historically been the most widely used adsorbent due to its high porosity and extensive surface area. However, its high production cost limits its broader application. Consequently, researchers are investigating various carbon allotropes and functionalized carbons as potential nanoadsorbents.

The use of carbon nanostructures in wastewater treatment holds significant promise and potential for transformative impacts on the field. The unique properties of these materials make them ideal for adsorption applications, and their cost-effectiveness and environmental sustainability, position them as attractive alternatives to traditional adsorbents. Continued advancements in this area of research are likely to yield innovative applications of carbon nanostructures in wastewater treatment.

2.19.2.1 Carbon Nanotubes (CNTs)

CNTs are cylindrical molecules composed of hybridized carbon atoms arranged in a hexagonal lattice. They can be synthesized by rolling single or multiple graphene sheets, resulting in single-walled CNTs (SWCNTs) or multi-walled CNTs (MWCNTs), respectively (Ray et al., 2021). Researchers are increasingly interested in CNTs due to their exceptional properties, including mechanical flexibility, high specific surface areas, and large pore volumes, making them highly effective in wastewater treatment.

CNTs have shown superior performance as adsorbents compared to other materials due to their tunable surface chemistry, chemical inertness, hollow structure, high specific surface area, light mass density, high porosity, and strong interactions with pollutants (Verma et al., 2020). These properties make them particularly suitable for treating wastewater contaminated with heavy metals and ions, which pose significant environmental and health risks.

Researchers investigated the adsorption and desorption of Mn^{7+} ions by CNTs and found that CNTs effectively reduced Mn^{7+} ion concentrations from 150 to 3 ppm, as determined by UV–visible spectrophotometric analysis (Verma et al., 2020). Overall, CNTs represent a promising solution for wastewater treatment due to their exceptional properties and superior adsorbent performance. Continued research in this field is expected to further elucidate and enhance the role of CNTs in mitigating heavy metal and ion pollution in wastewater.

2.19.2.2 Graphene and Graphene Oxide (GO)

Graphene is an extraordinary two-dimensional material composed of sp^2 hybridized carbon atoms arranged in a hexagonal pattern (Fig. 2.17). It exhibits unique properties, including ambipolar electric field effects, classical thermal conductivity, and quantum Hall effects at room temperature. Additionally, graphene possesses a high surface area and porosity, making it an excellent candidate for adsorbing gases such as methane, hydrogen, and carbon dioxide. Its high surface area, numerous active sites, extensive delocalized π-electron systems, and good chemical stability make it highly suitable for wastewater treatment applications.

The properties of graphene can be modulated by altering the number of layers and their stacking. Graphene oxide (GO), a single monomolecular layer of graphite, features various oxygen-containing functionalities such as hydroxyl, carboxyl, carbonyl, and epoxide groups. These functionalities endow GO with versatility and a wide range of potential applications in wastewater treatment.

Fig. 2.17 Graphical representation of structure and absorption sites of graphene sheets

2.19.3 Metal-Based Nano-adsorbents

Metal-based nanomaterials have garnered significant research interest in recent years. Nanometals and their oxides, such as Fe_3O_4, TiO_2, MnO_2, MgO, ZnO, and CdO, have shown considerable potential in removing heavy metals, ions, and dyes from wastewater (Sing et al., 2023). These nanometal oxides are considered superior to activated carbon as adsorbents for heavy and radioactive metals due to their smaller size and larger surface area, which facilitate shorter intraparticle diffusion distances without compromising surface area.

The sorption process in these nanomaterials is primarily driven by the complexation between dissolved metals and the oxygen in metal oxides. For instance, wastewater from oil refineries often contains various ions and metals, such as calcium (Ca^{2+}) and copper (Cu^{2+}). To tackle this issue, He et al. (2021) developed reusable nanoadsorbents based on Fe_3O_4/GO-COOH through the magnetization and carboxylation of graphene oxide (GO). These nanoadsorbents achieved remarkable removal rates of 78.4% for Ca^{2+} and 51% for Cu^{2+} within 60 min. Even after five adsorption–desorption cycles, the nanoadsorbents maintained high recovery rates (82.1% for Ca^{2+} and 91.8% for Cu^{2+}) and removal percentages (72.3% for Ca^{2+} and 49.33% for Cu^{2+}) (He et al., 2021).

The development of reusable nanoadsorbents, such as Fe_3O_4/GO-COOH, underscores the significant potential of metal-based nanomaterials in effectively treating wastewater from various industries. In another study, Peralta et al. (2021) developed silica-based magnetic nanoadsorbents to evaluate their potential in eliminating organic pollutants, including contaminants of emerging concern (CECs), polyaromatic hydrocarbons (PAHs), and aliphatic hydrocarbons (AHs). These nanoadsorbents, characterized by a silica shell and 3-(trimethoxysilyl) propyloctadecyldimethyl-ammonium chloride (3-TPODAC) as a structure-directing agent, were further modified with trimethoxyphenylsilane to obtain phenyl-functionalized nanoadsorbents (MMSSNPs-ph).

Characterization and batch sorption tests with single and mixed contaminants showed that MMSSNPs-ph were more effective in adsorbing PAHs and AHs. The presence of phenyl and 3-TPODAC moieties on the mesostructured silica scaffolds played a crucial role in achieving high PAH uptake from aqueous media. Additionally, MMSSNPs-ph were tested against CECs such as carbamazepine, diclofenac, and ibuprofen, demonstrating that although MMSSNPs had better adsorption capacities for CECs overall, MMSSNPs-ph achieved high uptakes for ibuprofen and diclofenac.

Overall, studies have highlighted the promising potential of MMSSNPs-ph as effective adsorbents for removing organic pollutants from aqueous media and these findings have significant implications for the development of efficient and sustainable water treatment technologies.

2.19.4 Polymer-Based Nano-adsorbents

In recent years, extensive research has focused on polymer nanocomposites to enhance environmental sustainability and improve wastewater treatment. These composites offer several advantages, including a high surface area for rapid decontamination, remarkable stability, improved processability, and cost-effectiveness. They also exhibit selectivity in eliminating various pollutants from wastewater. Polysaccharides such as chitosan (CS), cyclodextrin, nano-magnetic polymers, covalent organic polymers, and extracellular polymeric substances are commonly used as cost-effective polymeric adsorbents. Nanocellulosics, derived from cellulose, are non-toxic, ubiquitous, excellent adsorbents, and easily modifiable, making them ideal for wastewater remediation. Additionally, lignin-derived nanomaterials have shown significant potential for water and wastewater treatment, effectively catalyzing the degradation of dyes, nitroarenes, and the removal of heavy metals (Sajjadi et al., 2021).

Some researchers developed a novel approach to capturing both Methyl Orange (MO) and Pb^{2+} from wastewater using bi-functionalized β-cyclodextrin (β-CD) and polyethyleneimine (PEI) magnetic nanoadsorbents (β-CD/PEI-Fe_3O_4) (Chen et al., 2019). This system features spatially separated sorption sites, where the N-bearing groups of PEI and the cavities of β-CD play key roles in MO removal via electrostatic attraction and host–guest inclusion, respectively. The edges of β-CD with oxygen-bearing groups and free amino moieties on PEI serve as active sites for efficient Pb^{2+} uptake from wastewater.

Similarly, Moharrami and Motamedi (2020) developed a nanocomposite hydrogel of biological origin using starch-grafted copolymers of 2-acrylamido-2-methylpropane sulfonate and acrylic acid (S-g-(AMPS-co-AA)) and cellulose nanocrystals functionalized with magnetite (M-CNCs) for the adsorption of cationic dyes such as crystal violet (CV) and methylene blue (MB). This system demonstrated impressive adsorption capacities of 2500.0 mg/g and 1428.6 mg/g for CV and MB, respectively.

These innovative approaches to wastewater treatment underscore the potential of nanoadsorbents and nanocomposite hydrogels in addressing environmental challenges. By leveraging the unique properties of these materials, researchers are developing efficient and effective solutions for capturing harmful pollutants and improving water quality.

2.19.5 Zeolites

Zeolites are highly structured, three-dimensional microporous materials characterized by well-defined voids and conduits of distinct size. Composed of aluminum, silicon, and oxygen, they occur naturally as silicate minerals or can be synthesized as magnetically modified zeolite or bio-zeolite. In recent years, zeolitic nanomaterials have garnered significant attention in environmental applications due to their stability in water, low production cost, high surface area, selectivity, and compatibility with the natural environment. Extensive studies have confirmed their exceptional performance in removing metal cations from wastewater. Zeolites find diverse research applications across various industries, with their adsorbent property being extensively explored. Their capability for regeneration and reuse makes them suitable candidates for wastewater treatment.

For instance, Zhou et al. (2016) prepared cubic NaA zeolite, a microporous crystalline aluminosilicate zeolite composed of Na_2O/Al_2O_3, from natural halloysite minerals with nanotubular structures as source materials for adsorbing ammonium ions (NH^{4+}) from wastewaters. The maximum adsorption capacity of the prepared NaA zeolite for NH^{4+} ions was determined to be 44.3 mg/g. The constructed adsorbent system exhibited reusability and demonstrated potential application in eliminating NH^{4+} pollutants from wastewater.

Overall, zeolites offer a promising solution for wastewater treatment and hold the potential to revolutionize the field of environmental remediation.

2.19.6 Nanofilters

Water filtration plays a pivotal role in eliminating or reducing particulate matter, including suspended particles, microorganisms, and various harmful biological and chemical contaminants from contaminated water sources. This process is imperative to guarantee the safety and purity of water for consumption, pharmaceutical, and medical purposes.

In recent years, membrane technology has emerged as a prominent solution, with nanofiltration (NF) membranes representing a significant advancement in this field. NF membranes, characterized by a molecular weight cut-off (MWCO) for uncharged particles in the nanometer range, have gained prominence as a preferred method for drinking water and wastewater treatment.

NF operates as a pressure-driven membrane process situated between ultrafiltration (UF) and reverse osmosis (RO), typically operating at pressures between 5 and 20 bars and featuring pore sizes ranging from 0.5 to 2.0 nm. It boasts high rejection of divalent or higher-valent ions, low rejection of monovalent ions, high flux, low energy consumption compared to RO, and high rejection compared to UF.

As a significant and widely utilized technique in wastewater treatment, NF offers exclusive filtration mechanisms and a range of membrane options. It effectively filters out organic and inorganic contaminants, including harmful microbes, from wastewater. NF membranes are characterized by their flexibility, cost-effectiveness, and ease of production, with polymeric and ceramic membranes being the two most common types.

Polymeric membranes, while cost-effective, exhibit a shorter lifespan due to inferior chemical resistance and high fouling rates, whereas ceramic membranes offer higher mechanical, chemical, and thermal stabilities.

Researchers (Khan et al., 2019) have explored innovative NF systems utilizing carbon nanotubes (CNTs) for virus and yeast filtration, demonstrating impressive filtration efficiency and removal of heavy metal ions. These studies underscore the potential of CNT-based materials for water purification applications.

Efficient filtration systems are essential for ensuring access to clean and safe drinking water. Advances in nanotechnology, particularly in the development of NF membranes, offer promising solutions to address this critical need.

2.20 Photocatalysis

The term "photocatalysis" originates from the Greek words "photo," meaning "light," and "catalysis," meaning the substances that accelerate chemical reactions without undergoing changes themselves. Essentially, photocatalysis involves a process where a catalyst, when activated by light, drives and speeds up a chemical reaction. In this process, a solid material known as a photocatalyst absorbs light (photons), initiating a chemical reaction. Photocatalysis is categorized as one of the Advanced Oxidation Processes (AOPs), which involve the in-situ generation of highly reactive chemical oxidants using agents like Fenton's reagent, hydrogen peroxide (H_2O_2), UV light, ozone (O_3), or a catalyst. The resulting hydroxyl radicals (OH) are potent enough to oxidize even stubborn organic compounds, making AOPs effective methods for removing Contaminants of Emerging Concern (CECs) from wastewater effluent (Fig. 2.18).

Photocatalysis has garnered significant attention from the scientific community for wastewater treatment due to its efficacy in breaking down a wide array of organic materials, including estrogens, dyes, organic acids, pesticides, crude oil, microbes (including viruses and chlorine-resistant organisms), and some inorganic molecules

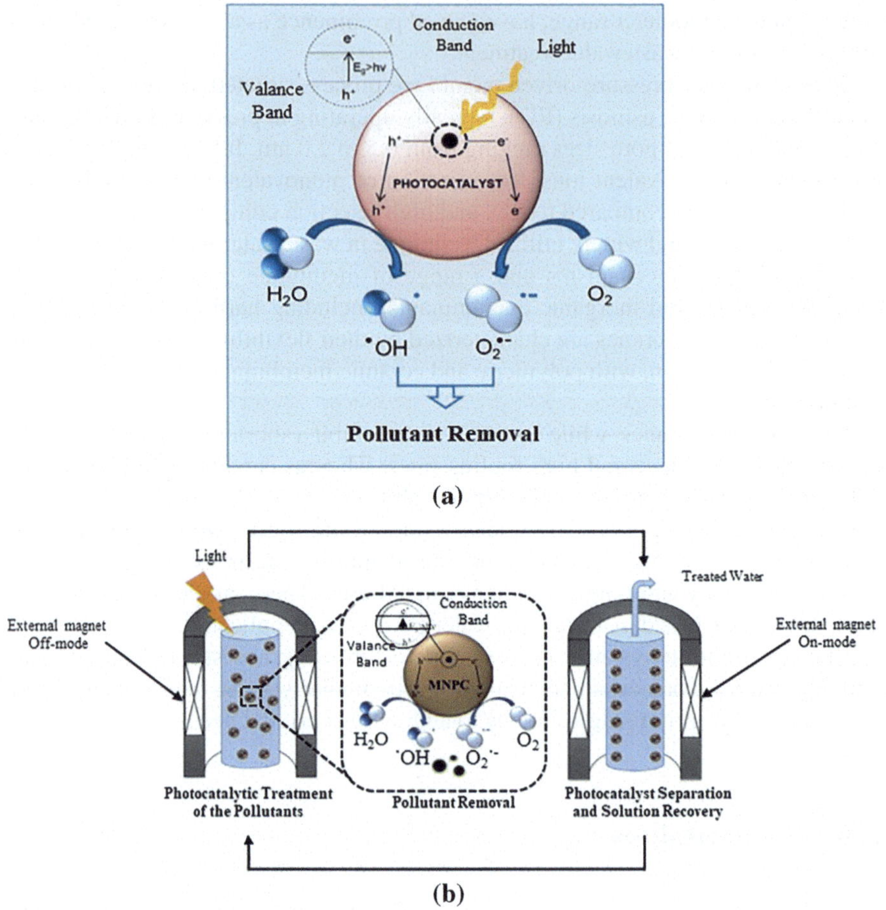

Fig. 2.18 Graphical representation of **a** excitation of a nanophotocatalyst during the photocatalytic process; **b** photocatalytic treatment of polluted water and recovery of nanophotocatalyst

such as nitrous oxides. When coupled with filtration or precipitation, photocatalysis can also eliminate metals (such as mercury) present in wastewater.

Photocatalysts have become a subject of intense research interest among researchers as they represent a potent approach for wastewater treatment, leveraging light-induced reactions to degrade a diverse range of organic and inorganic compounds.

Photocatalysis is a surface phenomenon and it involves five basic steps:

- Diffusion of reactants/pollutants to the surface of photocatalyst
- Adsorption of reactants/pollutants on the surface of photocatalyst
- Reaction of adsorbed reactants/pollutants
- Desorption of products from the surface
- Removal/diffusion of products from interface.

2.21 Disinfection and Pathological Control

According to the World Health Organization (WHO), contaminated water consumption is attributed to 80% of diseases in developing nations. Pathogens including bacteria, viruses, prions, fungi, protozoa, and rickettsia present a significant risk to human health, leading to millions of fatalities annually worldwide (Rikta et al., 2019). Disinfection, the process of reducing microbial counts on surfaces or within materials to acceptable levels, is achieved through physical or chemical means. Traditional disinfection methods in wastewater treatment encompass free chlorine, hypochlorite, chlorine dioxide, chloramines, ozone, reverse osmosis, and peracetic acid. However, these techniques have drawbacks such as high energy consumption, expensive equipment requirements, and the generation of numerous disinfection byproducts (DBPs), adding further concerns. Hence, there is an urgent need for efficient, sustainable, cost-effective, and time-efficient disinfection approaches (Zhang et al., 2024). Implementing such methods would be instrumental in addressing waterborne pathogen challenges and enhancing public health standards, particularly in developing regions.

2.22 Sensing and Monitoring

The presence of various substances, both living and non-living, including disease-causing microorganisms, municipal and industrial waste, sewage discharge, animal waste, and heavy metals, has led to environmental contamination. Monitoring water quality across different scales presents a significant challenge due to the low concentrations of contaminants, complexity, and variability of wastewater matrices. Addressing these challenges requires fast and efficient techniques.

In recent years, there has been growing interest in the scientific community regarding the development of nanomaterial-based sensors for water quality monitoring. Nanosensors offer exceptional properties, including efficient detection of trace contaminants and rapid analysis. These sensors are defined as devices or materials sensitive to changes in surrounding stimuli, such as heat, chemical and mechanical stress, alterations in volume, concentration, gravitational and magnetic forces, and electrical forces. Nanosensors provide insights into the physical, chemical, or biological information about the behavior and characteristics of nanoparticles from the nanoscale to the macroscopic level.

Nanosensors typically consist of three main components: a recognition component (such as nanometals, nanotubes, nanowires, or nanoparticles), a transducer (utilizing techniques like voltammetric, amperometric, conductometric, or spectrophotometric), and a display for real-time monitoring.

Biological contaminants, such as antibiotic-resistant pathogens and their antibiotic-resistance genes (ARGS), are increasingly prevalent in wastewater, posing significant public health risks. Methicillin-resistant *S. aureus* (MRSA) is a well-known antibiotic-resistant pathogen, with its mecA gene being responsible for antibiotic resistance. In addressing this concern, Riquelme et al. (2016) synthesized gold nanosensors functionalized with oligonucleotides for environmental monitoring of mecA ARG. These nanosensors underwent ARG spike detection tests in wastewater treatment plant effluent, demonstrating high selectivity for ARG with a detection limit of 70 ppm.

2.23 Barriers and Risks Associated with Nanotechnology

Nanotechnology has emerged as a promising avenue in wastewater treatment, offering innovative solutions to address various challenges. However, the realization of its potential faces significant barriers that need to be addressed for effective implementation. Among the most common obstacles are the toxicity of nanomaterials, cost-effectiveness, and social acceptability (Pardhi et al., 2020).

The toxicity associated with nanomaterials poses a critical concern, necessitating comprehensive studies to understand and mitigate potential adverse effects on the environment and human health. Strategies for safe handling, disposal, and risk assessment protocols are essential to ensure the responsible use of nanotechnology in wastewater treatment.

Moreover, achieving cost-effectiveness is paramount to justify the investment in nanotechnology solutions. While nanomaterials offer unique advantages, their production, deployment, and maintenance costs must be balanced against the benefits they provide. Innovations in manufacturing processes and scale-up efforts can help optimize cost-efficiency in nanotechnology-based wastewater treatment systems.

Additionally, social acceptability plays a crucial role in the adoption of nanotechnology solutions. Public awareness campaigns, transparent communication about risks and benefits, and stakeholder engagement are vital to foster trust and acceptance within communities. Addressing ethical concerns and incorporating public feedback in decision-making processes can enhance the societal acceptance of nanotechnology applications in wastewater treatment.

In short, while nanotechnology holds immense promise for improving wastewater treatment processes, overcoming barriers such as toxicity, cost-effectiveness, and social acceptability is imperative. By proactively addressing these challenges through interdisciplinary collaboration, research, and stakeholder engagement, we can harness the full potential of nanotechnology to safeguard water resources and public health.

2.24 Automatic Valuable Filtration (AVF) Technology

Automated Variable Filtration (AVF) technology represents a cutting-edge approach to wastewater treatment, employing a unique process that combines upward influent flow with downward filter media flow. This innovative system eliminates the need for additional filter media cleaning or the use of fresh water during treatment, as the filter media is self-cleaned by the filtered influent.

The AVF process involves two sets of media filters that can be operated either in series or in parallel. In a two-stage series configuration, the technology produces remarkably high-quality filtrate, making it particularly well-suited for refining secondary wastewater for reuse. Compared to conventional treatment systems, AVF technology stands out for its simplicity and effectiveness. It utilizes specially engineered media that does not require rejuvenation, achieving performance levels comparable to expensive micro-filtration membranes that necessitate frequent replacements. Additionally, with lower capital and operating costs, often as low as one-third of membrane technologies, AVF presents a compelling economic advantage. Its straightforward operation is further enhanced by the elimination of additional filter media cleaning processes.

Equipped with actuated valves, sensors, and programmable logic controllers, the AVF process automatically adjusts from serial to parallel mode during wet weather conditions or other predetermined operating conditions. The system offers several key benefits, including improved efficiency, reduced maintenance costs, and enhanced environmental sustainability. AVF technology represents a state-of-the-art solution for wastewater treatment, offering numerous advantages over conventional systems. Its simplicity, cost-effectiveness, and high performance position it as an attractive option for businesses and municipalities seeking to enhance their wastewater treatment processes.

Some notable features of AVF technology include:

- Higher solids capacity
- Continuously cleaned media beds
- Elimination of ancillary equipment
- Even flow distribution
- Cost-effectiveness in installation and low operating and maintenance costs
- Average reject rate of 5–15%
- Extremely low power consumption
- Ease of operation and maintenance.

Some important benefits of AVF are:

- No shutdown for backwash cycles
- Elimination of backwash equipment
- No flow control valves, splitter boxes or backwash controls
- No short cuttings
- Optimum media-washing efficiency
- Superior filter quality

- Reduce operation attention
- Minimizes over all pressure-drop
- Reduce potential for pluggage
- Significantly reduces maintenance
- Can be easily maintained without filter shutdown
- Upto 70% less compressed air versus other self-cleaning filers.

2.24.1 Working Principle of AVF

The AVF units operate by pumping feed water into the system, which then flows downward upon introduction at the top. The influent feed is directed into the bottom of the media bed through a network of feed radials. As the influent moves upward through the downward-moving media bed, both organic and inorganic impurities are captured by the media.

The purified filtrate continues its upward movement and exits the filter through the effluent pipe. To maintain the efficiency of the filtration process, a small volume of compressed air is introduced at the bottom of the airlift. This action draws the media into the airlift pipe, where it undergoes a scouring process that dislodges any solid particles attached to the media grains. This scouring mechanism is significantly more effective than traditional sand filtration backwash methods (Park et al., 2015). The resulting dirty slurry is then propelled to the top of the airlift and directed into the reject compartment.

As the media descends through the concentric stages of the washer, it encounters a small quantity of polished filtrate moving upward. The clean, recycled media is subsequently deposited at the top of the media bed, initiating a new cycle of influent cleaning and eventual migration to the bottom of the filter (Fig. 2.19).

The AVF units employ a highly efficient filtration process that effectively captures both organic and inorganic impurities. The incorporation of compressed air and airlift technology ensures thorough scouring of the media, leading to superior cleaning outcomes compared to conventional sand filtration backwash methods. The recycled media is then reintroduced into the media bed, where it can continue its effective filtration of the influent feed.

2.24.2 Applications of AVF Technology

- Municipal/Township level drinking water treatment
- Sewage water recycle and reuse
- Industrial Waste water recycle and reuse
- Cooling water side stream filtration
- Desalination projects water filtration
- Algae water filtration

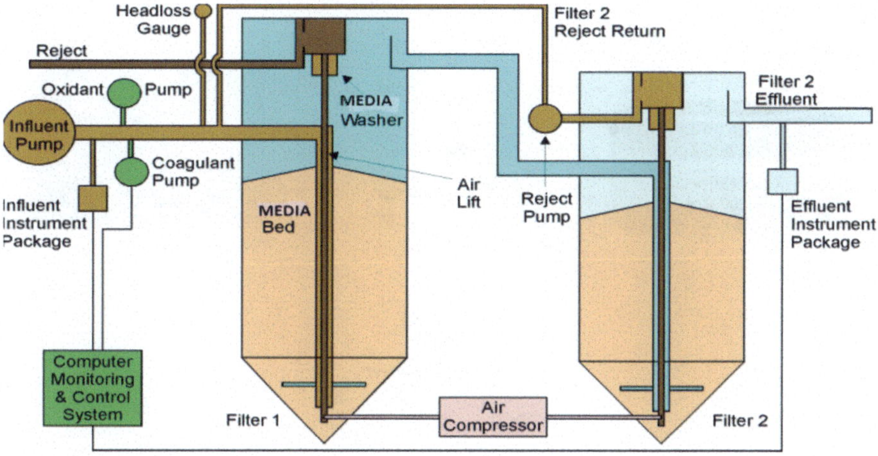

Fig. 2.19 Working of AVF

- Fly ash pond water filtration
- Replacement of slow sand filters
- Industrial Side Stream Filtration.

2.25 Automated Chemostat Treatment (ACT)

Automated Chemostat Treatment (ACT) represents a groundbreaking advancement in wastewater treatment, protected by a patent (ACT patent), and offers a unique solution to the challenges encountered in conventional methods. The approach of ACT involves a meticulous selection and cultivation of bacteria sourced from the client's pre-treated water, leveraging natural bacterial antagonism under site-specific conditions. This process results in the enrichment of bacteria suited to the water type and environmental conditions at the site, while depleting those ill-suited (Todorovych & Syrotina, 2017). Importantly, the bacteria utilized in the Biological Process Control (BPC) system are naturally occurring and are not genetically engineered or modified (Fig. 2.20).

ACT introduces a transformative approach to wastewater treatment, providing a highly efficient and tailored solution to address individual customer needs. This personalized approach, backed by a patent, ensures the uniqueness and protection of this technology.

Automated Chemostat Treatment (ACT) represents a revolutionary method for sludge treatment, offering unparalleled flexibility, efficiency, and integration ease. This cutting-edge technology is fully automated, controllable, and significantly more efficient than conventional practices. At its core, ACT employs an appropriate bacterial cocktail for specific polluted water types within an innovative chemostat system.

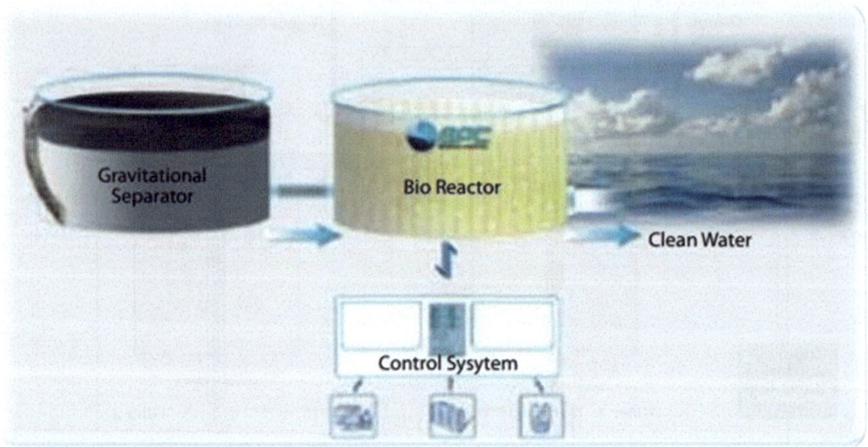

Fig. 2.20 Automated continuous flow

By maintaining a balanced state of bacterial growth and organic compound degradation, ACT prevents aggregate formation, allowing each bacterium to function as a single cell. This enhances surface area availability for the process, leading to substantially improved biodegradation efficiency.

The outcomes of ACT are truly remarkable, yielding virtually sludge-free water outputs suitable for direct return to the environment or further processing. ACT effectively reduces hydrocarbons, total organic carbon (TOC), chemical oxygen demand (COD), and suspended solids from greasy and oily waters, ensuring effluents meet or exceed industry regulatory standards. By analyzing a refinery water sample, ACT identifies the most effective solution tailored to address specific refinery challenges. Furthermore, the hydraulic and bacterial ages are harmonized, necessitating a lower density of single-cell bacteria.

Operating as a continuous flow reactor without activated sludge, ACT offers exceptional flexibility and integration simplicity into existing infrastructure. This capability enables on-site application, resulting in substantial savings in operational and maintenance costs. With its unmatched efficiency, adaptability, and integration ease, ACT stands as a transformative solution in sludge treatment.

2.25.1 Implementation

The modular approach of the Automated Chemostat Treatment (ACT) technology lends itself seamlessly to on-site implementation, making it particularly well-suited for treating stripped sour water streams in refinery settings. This modular design allows for flexible integration with existing infrastructure, offering two distinct implementation options.

In the first option, the stripped sour water undergoes preliminary treatment stages such as API (American Petroleum Institute) and Dissolved Air Flocculation (DAF) treatments before entering the ACT bioreactor for biological treatment. Following biological treatment, the treated water can be reintroduced to the main bioreactor for further biodegradation, filtration, and eventual discharge.

Alternatively, in the second option, the stripped sour water is treated independently through API and DAF processes before undergoing biological treatment in the ACT bioreactor. Subsequently, the treated water undergoes filtration and can be discharged directly into the environment.

The segregation of the stripped sour water from the main stream offers several significant advantages. Firstly, the treated water can be recycled for additional use in various refinery processes, contributing to resource conservation and operational efficiency. Secondly, by reducing the total organic load in the main stream, the treatment capacity and efficiency of the overall system are enhanced.

The fully automated ACT system is equipped with a comprehensive array of online sensors that provide real-time monitoring and feedback on various critical parameters, including Total Petroleum Hydrocarbons (TPH), nitrogen, dissolved oxygen, Total Organic Carbon (TOC), and temperature. This continuous monitoring ensures optimal process management, balancing factors such as flow rate, bacterial growth, additives, and organic compound degradation to achieve efficient and effective treatment outcomes.

2.25.2 ACT System Key Features

- Customized for optimal cost-effective performance
- Increased capacity to prevent shortfalls during peak time processing
- Fully-automated process
- Process stability maintained, even during inlet fluctuation.

2.25.3 ACT Benefits for Refineries

- Eliminates the need to reactivate bio-sludge
- Creates a more streamlined and efficient process by eliminating many complex operations
- Reduces costs by as much as 50%
- Reduces sludge generation and handling significantly as well as corresponding chemical usage.

2.25.4 Advantages

The Automated Chemostat Treatment (ACT) system stands out for its streamlined process, which minimizes the reliance on bio sludge and chemicals, thereby reducing operational complexities and costs. By minimizing the generation of black sludge, ACT contributes to a cleaner and more sustainable treatment process.

One of the key strengths of ACT lies in its modular and flexible design, which enables it to effectively handle varying capacities and contamination levels. Furthermore, the system can be easily modified and scaled up to accommodate increased capacity requirements, providing scalability and adaptability to changing operational needs.

An important advantage of ACT is its ability to produce virtually sludge-free output, surpassing even the most stringent disposal standards. This ensures compliance with environmental regulations and promotes responsible waste management practices.

The environmentally conscious approach of ACT aligns with sustainable principles, making it a "green" solution for wastewater treatment. Its versatility extends to various industrial sites, including oil refineries, storage farms, drilling sites, marine ports, contaminated reservoirs, and storage tanks, offering a comprehensive solution for diverse wastewater treatment needs.

Overall, ACT is dedicated to delivering professional and reliable wastewater treatment solutions that prioritize efficiency, effectiveness, and environmental sustainability. Through its innovative technology and commitment to excellence, ACT aims to make a positive impact on both industry operations and environmental conservation efforts.

2.26 Anaerobic Baffle Reactor

Anaerobic baffled reactors (ABRs) represent an innovative upgrade from traditional septic tanks, offering enhanced removal and digestion of organic matter through a combination of physical and biological treatments. The design of an ABR incorporates a series of baffles within the treatment chamber, creating compartments that facilitate the flow of wastewater and residual sludge. This design promotes increased contact between fresh wastewater and microorganisms responsible for anaerobic digestion (Barber & Stuckey, 1999).

One of the key advantages of ABRS is their ability to separate solids retention time from hydraulic retention time, allowing for efficient anaerobic treatment at short retention times. Despite the rapid treatment rates for solids, ABRS typically produce low amounts of sludge. Moreover, ABRS are known for their simplicity in construction and operation, as well as their resilience to hydraulic and organic shock loading.

Table 2.6 Silent features of ABR

Working principle	Vertical baffles in the tank force the pre-settled wastewater to flow under and over the baffles guaranteeing contact between wastewater and resident sludge and allowing an enhanced anaerobic digestion of suspended and dissolved solids; at least 1 sedimentation chamber and 2–5 up-flow chambers
Capacity/ adequacy	Community (and household) level; for pre-settled domestic or (high-strength) industrial wastewater of narrow COD/BOD ration. Typically integrated in DEWATS systems; not adapted for areas with high ground-water table or prone to flooding
Performance	70–95% BOD; 80–90% TSS; low pathogen reduction Hydraulic retention time (HRT): 1–3 days
Costs	Generally low-cost; depending on availability of materials and economy of scale
Self-help compatibility	Requires expert design, but can be constructed with locally available material
O&M	Should be checked for water tightness, scum and sludge levels regularly; sludge needs to be dug out and discharged properly (e.g. in composting or drying bed); needs to be vented
Reliability	High resistance to shock loading and changing temperature, pH or chemical composition of the influent; requires no energy
Main strengths	Strong resistance; built from local material; biogas can be recovered
Main weakness	Long start-up phase

While ABRS effectively remove and digest organic matter, it is important to note that further treatment of both sludge and effluent is still necessary. Nonetheless, ABRS remain a highly effective and efficient solution for wastewater treatment, particularly suited for influents with high organic loads. They are commonly integrated into decentralized wastewater treatment systems (DEWATS), typically combined with other treatment steps such as anaerobic filters and aerobic treatment units like constructed wetlands and maturation ponds.

The benefits of using ABRS in wastewater treatment are significant. They can achieve up to 90% reduction in BOD, outperforming conventional septic tanks. Additionally, their versatility allows them to be deployed in various settings, making them an ideal solution for diverse wastewater treatment applications. Overall, ABRS offer an effective and adaptable approach to addressing organic pollution in wastewater (Table 2.6).

2.26.1 Design Consideration

Anaerobic baffled reactors (ABRs) combine principles from various wastewater treatment technologies, including septic tanks, moving bed reactors (MBRs), and up-flow anaerobic sludge blanket (UASB) reactors. Unlike MBRs and UASBS, ABRs

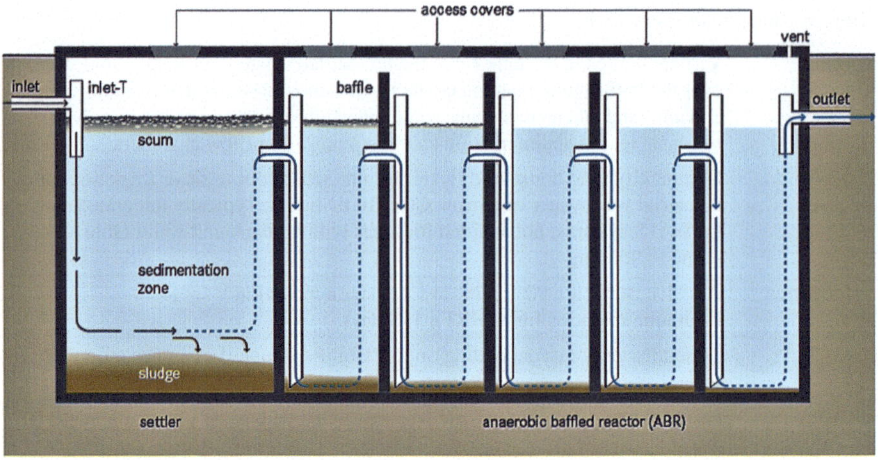

Fig. 2.21 Schematic of the anaerobic baffled reactor. *Source* Tilley et al. (2014)

do not rely on a floating sludge blanket, and effluent retention is unnecessary as active sludge is trapped in the subsequent chamber. Sedimentation chambers are employed to remove settleable solids before entering the ABR, either as an integrated compartment in small-scale units or as a separate settler in larger systems (Fig. 2.21).

Typical inflows for ABRS range from 2 to 200 m^3 per day, with critical design parameters such as hydraulic retention time (HRT), upflow velocity, and the number of upflow chambers. The connection between chambers can utilize vertical pipes or baffles, and accessibility for maintenance through access ports is essential. While biogas produced during anaerobic digestion is typically not collected due to insufficient amounts, venting the tank is necessary to control the release of odorous and potentially harmful gases.

The reactor system effectively treats wastewater by first removing larger solids and impurities through a settling chamber, followed by a series of up-flow chambers. The wastewater enters at the bottom and passes through the sludge to move up to the next compartment, allowing particles to settle against the up-stream. This process facilitates intensive contact between the wastewater and active biomass, ensuring effective treatment.

To ensure equal distribution of entering liquid, chambers should be designed as relatively short compartments with specific dimensions recommended by experts. It is crucial to maintain an appropriate upflow velocity, ensuring effective treatment without compromising reactor performance. Organic load should be controlled, with higher loading rates possible under certain conditions.

During anaerobic digestion, biogas is produced and can be recovered for various purposes, such as kitchen use or driving equipment. The methane concentration increases from the first to the last compartment, with methane production activity influenced by substrate availability and environmental conditions. Proper management of biogas is essential for decentralized systems, either by recovery or venting to prevent gas release.

Overall, ABRs offer a cost-effective and efficient solution for wastewater treatment, particularly suitable for applications with high organic loads and decentralized systems. Proper design and operation ensure effective treatment while minimizing environmental impact.

2.26.2 *Treatment Performance*

Anaerobic Baffled Reactors (ABRs) demonstrate impressive treatment performance, achieving COD removal rates ranging from 65 to 90%, corresponding to approximately 70–95% of Biological Oxygen Demand (BOD) removal (Barber & Stuckey, 1999). This represents a significant improvement over conventional septic tanks, which typically achieve only 30–50% COD removal. ABRs are specifically designed to remove settleable solids in the sedimentation chamber at the outset of the process, typically accounting for 50% of the total volume of Total Suspended Solids (TSS). Furthermore, the unique design of ABRs facilitates enhanced treatment of non-settleable solids, with up to 90% TSS removal achievable.

The series arrangement of tanks in ABRs aids in the digestion of substances that are difficult to degrade, particularly in the rear part of the process after easily degradable matters have been digested in the front part. However, recycling effluent may slightly diminish treatment quality. ABRS can be customized for daily inflows ranging from a few cubic meters per day to several hundred cubic meters per day. The Hydraulic Retention Time (HRT) in ABRS is relatively short, varying from only a few hours up to two or three days.

Overall, ABRs offer an efficient and flexible solution for wastewater treatment, effectively removing organic pollutants and suspended solids while accommodating a wide range of inflow rates. Their unique design and operational characteristics make them suitable for various applications, from small-scale decentralized systems to larger treatment plants.

2.26.3 *Operation and Maintenance*

The Anaerobic Baffled Reactor (ABR) is indeed a highly effective wastewater treatment system, but its successful operation requires a start-up period of several months to establish the slow-growing anaerobic biomass in the reactor. To expedite this process, inoculating the ABR with anaerobic bacteria, typically by adding fresh cow dung or septic tank sludge, can help jumpstart the establishment of active biomass.

During the start-up phase, it is beneficial to begin with a quarter of the daily flow and gradually increase loading rates over three months. This gradual approach allows sufficient time for bacterial multiplication before suspended solids are washed out. However, it is important to note that ABR technology may not be suitable for immediate treatment needs due to its extended start-up time.

Moreover, the delicate ecology of the ABR necessitates careful handling to avoid disrupting its microbial balance. Harsh chemicals should not be discharged into the ABR, and regular monitoring of scum and sludge levels is essential to ensure proper functioning. While process operation is generally straightforward, maintenance involves removing accumulated sludge and scum every 1–3 years. To minimize health risks, specialized equipment such as Motorized Emptying and Transport technology or Human-powered Emptying and Transport Technology should be used for desludging.

The desludging frequency depends on various factors, including pre-treatment steps and ABR design. When emptying the tanks, it is crucial to leave some active sludge in each compartment to maintain a stable treatment process.

2.26.4 Applicability

The Anaerobic Baffled Reactor (ABR) technology is indeed versatile and can be implemented across various scales, from household-level to larger catchment areas. It is particularly suitable for areas with a consistent generation of blackwater and greywater, making it ideal for households, small neighborhoods, or even industrial wastewater treatment.

In decentralized wastewater treatment systems (DEWATS), ABRs can effectively treat industrial wastewaters, handling capacities of several hundred cubic meters per day. However, community organization is crucial to ensure proper usage and maintenance of the ABR systems.

ABRs are also suitable for (semi-)centralized installations, especially when there is existing conveyance infrastructure like simplified sewers in place. They require relatively small land areas, as the tanks are typically installed underground. However, access for vacuum trucks is necessary for regular sludge removal, particularly from the settling compartment.

Pre-fabricated ABRs, such as those developed by BORDA (Bremen Overseas Research and Development Association) using fiberglass construction, offer rapid deployment options, especially for emergency sanitation situations. While the start-up phase of ABRs typically takes several months, pre-fabricated models can be quickly constructed and provide long-term solutions once operational.

Fiberglass construction materials are widely available and affordable, making them accessible for deployment in various regions. However, it is essential to note that regardless of the construction method, the start-up phase of ABRs requires sufficient time for bacterial biomass establishment, typically around three months.

ABR technology offers flexible and effective wastewater treatment solutions across different settings, with options for both permanent installations and rapid deployment in emergency situations. Proper community engagement, maintenance, and adherence to environmental considerations are essential for the successful implementation of ABR systems.

2.26.5 *Advantages*

- Resistant to organic and hydraulic shock loads
- No electrical energy is required
- Low operating costs
- Long service life
- High reduction of BOD
- Low sludge production
- Moderate area requirement (can be built underground)
- Simple to operate.

2.26.6 *Disadvantages*

- Long start-up phase
- Requires expert design and construction
- Low reduction of pathogens and nutrients
- Effluent and sludge require further treatment and/or appropriate discharge
- Needs strategy for faecal sludge management (effluent quality rapidly deteriorates if sludge is not removed regularly)
- Needs water to flush
- Clear design guidelines are not available yet.

2.27 Soil Biotechnology (SBT)

Soil Biotechnology (SBT), also known as Constructed Soil Filter (CSF), is a highly effective terrestrial system for wastewater treatment that leverages the principle of trickling filtration. This innovative approach combines physical processes such as sedimentation and infiltration with biochemical processes to remove suspended solids, organic and inorganic contents from wastewater (Kamble et al., 2017).

The components of SBT include a suitable mineral constitution, a culture containing native micro-flora, and bio-indicator plants. These elements work together to facilitate the treatment process within the system.

The SBT system is typically constructed using materials like reinforced concrete (RCC), stone masonry, and soil bunds. It consists of several key components, including a raw water tank, bioreactor containment, treated water tank, piping, and pumps. This setup allows for efficient wastewater processing while minimizing environmental impact.

One of the notable features of SBT is its eco-friendliness, as it utilizes natural processes and materials for waste processing. It employs bacteria, earthworms, and mineral additives within a garden-like setup to treat both solid organic wastes and wastewater.

By reinforcing the carbon and nitrogen cycles in nature, SBT improves the quality of treated water and reduces Chemical Oxygen Demand (COD), contributing to environmental sustainability. Moreover, it offers a sustainable and cost-effective solution for wastewater treatment, making it increasingly popular in various settings.

Overall, Soil Biotechnology represents an innovative and environmentally friendly approach to wastewater treatment, providing numerous benefits for both the environment and communities in need of effective wastewater management solutions.

2.27.1 The Purpose and Need of SBT

Implementing a sewage management solution is critical to prevent pollution of groundwater resources, lakes, and rivers from sewage contamination, especially in areas lacking a civil drainage system. Adopting a natural method for sewage treatment, such as Soil Biotechnology (SBT), is essential to avoid the adverse effects associated with chemical treatments.

SBT offers an eco-friendly approach to sewage treatment, leveraging natural processes to effectively treat wastewater without harming the environment. By utilizing native micro-flora, bio-indicator plants, and physical processes like sedimentation and infiltration, SBT can remove suspended solids and organic/inorganic contaminants from sewage while minimizing environmental impact.

One of the key advantages of SBT is its flexibility and scalability. Whether implementing sewage management solutions for small communities or large cities, SBT can be tailored to suit the specific needs and requirements of the area. Its adaptable nature allows for customization to address varying levels of sewage generation and treatment capacity.

Overall, adopting a natural and sustainable solution like SBT for sewage management is crucial for safeguarding water resources and protecting the environment from pollution. Its eco-friendly approach, flexibility, and scalability make it an ideal choice for addressing sewage management challenges in diverse settings.

2.27.2 Working Principle of SBT

The Soil Biotechnology (SBT) system is an ingeniously designed ecosystem that harnesses the synergistic interactions between soil and plant systems within a packed bed reactor. This innovative approach incorporates various grades of media, including stone, gravel, and formulated soil, enriched with microflora and geophagus earthworm culture.

The SBT system operates through a combination of physical, biological, and ecological processes to facilitate both aerobic and anaerobic respiration, as well as complex reactions such as nitrification, denitrification, and acidogenesis, depending on the composition of the waste load being treated.

Organic waste is effectively removed through adsorption and filtration mechanisms, followed by biological conversion to CO_2 by indigenous soil microflora. Suspended solids are settled in the primary settling unit, while dissolved solids undergo adsorption, biodegradation, and uptake by green plants.

The media and additives within the SBT system provide ample sites for biological transformation, while the presence of earthworm culture helps regulate microbial ecology, enhancing treatment efficiency. Moreover, the health and vitality of green plants serve as reliable bio-indicators of the treatment process.

The SBT system's unique combination of soil and plant systems, coupled with carefully selected media and additives, creates a robust ecosystem that effectively promotes biological transformation and removes waste through a diverse array of processes. It stands as a testament to the potential of innovative engineering and ecological design in wastewater treatment.

2.27.3 Procedure

Soil Biotechnology represents a pioneering approach that capitalizes on the inherent bioenergy within organic matter by integrating key elements of a productive soil ecosystem, including selected soil bacteria, earthworms, specific plant species, and essential mineral nutrients. However, the presence of litter pests can serve as an indicator of an overloaded system, necessitating corrective actions.

By harnessing the biocarbon energy embedded in waste, Soil Biotechnology effectively releases plant nutrients from minerals, fixes atmospheric nitrogen, and generates metabolites that can be readily assimilated by plants. This process prevents the wasteful dissipation of carbon energy and promotes sustainable nutrient cycling within the ecosystem. Additionally, the presence of select aerobic bacteria and mineral additives contributes to odor control, enhancing the overall efficiency and effectiveness of the treatment process.

In essence, Soil Biotechnology offers a sustainable and efficient approach to harnessing the potential of organic matter. By capitalizing on the natural processes inherent in soil ecosystems, this innovative technology delivers a range of benefits for both environmental conservation and agricultural productivity.

2.27.4 Input Parameter

SBT, or Submerged Biological Treatment, stands as a highly effective method for the comprehensive removal of both organic and inorganic substances from wastewater. Originally designed for sewage treatment, this technology serves a dual purpose by disposing of domestic wastewater while simultaneously rendering it suitable for reuse.

Beyond its primary application in sewage treatment, SBT boasts versatility that extends to various other domains. Its capabilities encompass arsenic and iron removal, processing hospital waste, and treating industrial wastewater and air pollution. This adaptability positions SBT as an indispensable tool in addressing a diverse array of environmental challenges, making it a cornerstone of sustainable development initiatives worldwide.

Whether optimizing sewage treatment processes or pioneering innovative solutions for industrial waste management and pollution control, SBT stands out as a proven and dependable technology capable of facilitating progress toward environmental sustainability goals.

2.27.5 Major Elements of SBT

(1) The media that supports the micro and macro organisms of soil ecology for aerobic respiration, including nitrogen fixation and active oxygen production, is crucial for the success of the bioreactor.
(2) The inclusion of geophagus worms in the culture serves as the top carnivore to regulate soil population, ensuring a healthy and balanced ecosystem.
(3) The addition of a natural mineral mixture acts as a pH regulator, promoting optimal conditions for the microorganisms to thrive.
(4) The presence of green plants serves as a bioindicator of the health of the process, providing valuable insights into the overall well-being of the bioreactor. The ecology of soil and engineered natural oxygen supply create a highly aerobic environment within the bioreactor.

The success of the bioreactor depends on a carefully curated combination of media, culture, additives, and green plants. By maintaining a healthy and balanced ecosystem, we can ensure optimal conditions for the microorganisms to thrive and produce the desired results.

Water Reuse Possibilities after Treated Wastewater from Soil Biotechnology

- Gardening
- Toilet Flushing
- Construction activities and Industrial floor washing
- Car Washing
- For HVAC (Heating, ventilation and air conditioning) Cooling Tower

- Ground Water Recharge after getting desired norms/standards
- Release to Water Bodies.

2.27.6 Industrial Application of SBT

- Effective COD handling
- Polishing unit for existing CETPs (Common Effluent Treatment Plants)
- MEE (Multiple Effect Evaporation) condensate treatment for Pharma and Chemical Industries
- Pharmaceutical Effluent (Formulation and API) Chemical Research Organizations (CRO's), Biotech companies
- Ammonia rich wastewater: Fertilizer Industries
- Petrochemicals, Pesticides, Sugar Effluents, Distillery Condensate recycling
- Arsenic removal upto an extent of 98.6%
- Iron removal upto 97.5%
- Phosphorous removal upto 93%.

In Industry, technology is being extended for zero liquid discharge for following applications

- Pharma (API and Formulations)
- Petrochemicals
- Pesticides
- Textile
- Dye Intermediates
- Wineries, Sugar Industries and Distilleries
- Pulp and Paper Mills.

2.27.7 Advantages

The soil biotechnology system (SBT) represents a highly efficient and sustainable approach to wastewater treatment, offering a plethora of advantages over conventional methods including:

1. **Operational Flexibility**: SBT can operate in both batch and continuous modes, providing adaptability to varying wastewater treatment needs.
2. **Low Maintenance**: SBT requires no mechanical aeration and produces no sludge, resulting in minimal maintenance requirements and reducing operational costs.
3. **Energy Efficiency**: With low energy consumption and mechanization, SBT ensures zero downtime, contributing to overall energy savings.

4. **Long Lifespan**: The specially constituted filtering media bed of SBT can last for tens of years, due to natural living agents that adapt to the chemical and biological load in the system.
5. **Zero Discharge**: SBT has the potential for zero discharge, as there are no disposables or process residues, aligning with sustainable waste management practices.
6. **Environmentally Friendly**: SBT utilizes natural ingredients to combat foul odors, eliminating air and sound pollution and promoting a healthy environment.
7. **Scalability**: SBT can be implemented in operations of any size, making it suitable for decentralized wastewater treatment solutions.
8. **High Water Recovery Rate**: Implementing a SBT-based wastewater treatment system ensures a minimum water recovery rate of 90%, contributing to water conservation efforts.
9. **Biomass and Bio-fertilizer Production**: The system produces valuable by-products in the form of biomass and bio-fertilizer, which can be utilized for various purposes, including horticulture.
10. **Ecological Protection**: By preventing unsafe sewage disposal, SBT helps preserve the local ecology and minimize environmental damage.
11. **Low Carbon Footprint**: With its low-energy operation and absence of toxic waste or foul odors, SBT significantly reduces carbon emissions and promotes environmental sustainability.

2.28 Reed Bed Technology (RBT)

Reed beds represent a highly effective technique for dewatering solids within a confined area, employing common reed plants, specifically *Phragmites communis*, which is closely related to the typical marsh plant (Rathoure et al., 2019). The configuration of these beds can be tailored to accommodate existing land conditions and areas. The construction of these beds involves creating specially designed ponds equipped with underdrains covered by a mixture of sand and gravel. These ponds are then populated with reed plants. Additionally, modified sludge drying beds are a viable alternative and can be adapted for this purpose. These beds come pre-equipped with side walls, sand and gravel layers, an underdrain system, and an impervious membrane liner.

The dewatering process is facilitated through a combination of evaporation, plant transpiration, and decantation. Solids are pumped into the reed beds, where the decanted water percolates through the bottom of the bed and through the sand and gravel layers into the underdrains. This water is then returned to the wastewater treatment plant for secondary treatment. During the dewatering process, the solids undergo a transformation from a liquid state to a solid cake. Specifically, six inches of solids and water compress into a half-inch thick solid cake left in the bed, and the cycle is repeated.

2.28.1 Working of RBT

The reeds in a bed are typically planted one foot apart from each other. To ensure successful growth, aerobically stabilized sludge or effluent is uniformly applied through a grid-perforated tile (Al-Ajalin et al., 2020). *P. communis* is a highly adaptable plant and is one of the most widespread flowering plants globally. It thrives in polluted waters and can subsist on sludge. This reed exhibits a remarkable ability to consume water and is tolerant of low oxygen levels and waterlogged conditions. The reeds anchor themselves in the soil through roots and rhizomes, forming an intricate network of underground stems. These stems give rise to new plants, which, in turn, provide air passages through the sludge. This process creates a hospitable environment for various biological communities to develop and continue to mineralize the sludge (Al-Ajalin et al., 2020).

Reed beds perform three basic functions:

1. Dewater the sludge
2. Transform it into mineral and humus-like components, and
3. Store sludge for several years.

Dewatering is a process that involves the removal of water from sludge. This can be achieved through various methods, including evaporation, transpiration, and filtration. In sludge drying beds, evaporation is the primary method used to remove water. Transpiration, on the other hand, involves the movement of water through the plant's root, stem, and leaf structures. Filtration occurs through the sand and gravel layers of the bed, as well as the plant's root system. To prevent drainage backup, the plants need to be harvested annually, ensuring that the system remains efficient and effective. The harvested vegetation can be composted or burned, depending on the specific needs of the treatment plant. The leachate, which is the liquid that drains from the dewatering bed, must be channeled back to the treatment plant through the underdrain. This prevents any potential environmental contamination and ensures that the treatment process remains safe and effective.

Overall, dewatering is a critical process in the treatment of sludge. By utilizing a combination of evaporation, transpiration, and filtration, treatment plants can effectively remove water from sludge, ensuring that the treatment process remains efficient and effective.

2.28.1.1 Microbial Activity

Microorganisms adhere to the surface of gravel or soil particles, as well as to plants and their roots. These organisms metabolize pollutants, breaking them down and mineralizing them. Reed beds are frequently employed for sewage degradation, and with extended retention times, even persistent compounds such as polycyclic aromatic hydrocarbons (PAH), polychlorinated biphenyls (PCB), dyestuffs, amines, and glycols can be effectively treated.

2.28.1.2 The Role of Plants

The carefully selected plant species, such as *Phragmites karka* and *Typha angustus*, within the reed bed serve three primary functions.

(i) They facilitate secondary restructuring of the soil system. The extensive root and rhizome system create channels for water to pass through and maintain soil openness through continuous growth.
(ii) The roots of the aquatic plants introduce atmospheric oxygen into the soil, promoting the proliferation of diverse microorganisms around the plant roots. This is crucial for the effective breakdown of various organic and inorganic compounds.
(iii) The plants have the capacity to absorb a certain amount of nutrients from the wastewater itself.

2.28.1.3 Choosing Media

The selection of bed media, whether gravel, soil, or sand, depends on specific application requirements. Gravel, though less active microbiologically, facilitates a faster throughput of water. Consequently, it is widely used in secondary and tertiary sewage treatment applications, as well as in mine wastewater treatment, where plants assist in oxygenating the water, promoting the deposition of insoluble metal ion precipitates. On the other hand, soil has traditionally been utilized in the primary and secondary treatment of industrial effluents. Certain soil minerals promote the deposition of metal ions, phosphate, and sulfate, enabling the engineering of soils to treat specific effluent streams. Moreover, the ability of clay particles and humic materials to trap polluting chemicals ensures that the soil system can handle shock loads effectively (Masi et al., 2017).

2.28.2 Features of Reed Bed Technology

The reed bed system stands out as a highly efficient and sustainable wastewater treatment solution, offering numerous benefits. Firstly, it is operation and maintenance-free, devoid of mechanical or moving parts, thus eliminating the energy wastage typically associated with mechanical aeration. Additionally, the system operates cleanly and efficiently, as all chemical, physical, and biological transformations occur underground, resulting in minimal odors or sludge formation. Furthermore, the reed bed system is self-sustaining and durable, capable of sustaining itself for over 40 years post-installation and operation. It exhibits high adaptability, able to handle concentrations ranging from a few mg/L to 20,000 mg/L of Chemical Oxygen Demand (COD) and 400 mg/L of nitrogen, with a capacity spanning from about 1 m^3/day to more than 10,000 m^3/day. Moreover, the system can effectively manage challenging effluents due to the diversity of microbes and plants it encompasses. It can adapt to various types and fluctuating shock loads of effluents, including those containing organic compounds like chlorinated hydrocarbons, dyes, and sulfur-containing aromatics, nitrogen compounds such as ammonia and nitrates, sulfur compounds like sulfides and sulfates, as well as heavy metals and pathogens.

2.28.3 *Design Consideration*

When constructing a reed bed, adherence to specific guidelines is essential to ensure its effectiveness and longevity. First, an impermeable liner or base must be installed to protect groundwater. Then, the outer walls should be constructed using concrete with footers, ideally 6–8 ft. high. To facilitate the loading of liquid biosolids, manifold and risers should be installed along the long sides of the bed, typically three per side. The beds themselves should not exceed 100 ft. in length and 50 ft. in width (Masi et al., 2017). Providing adequate access for bed clean-out is crucial, and the beds should be sized based on allowable loading rates and biosolids production during the winter and early spring dormant phase. Furthermore, it is advisable to design the beds with enough redundancy so that, after six years, one bed at a time can be taken out of service to dry out, remove solids, and re-establish the plants.

By adhering to these guidelines, reed beds can be constructed to effectively treat wastewater while minimizing the risk of environmental contamination.

2.28.4 *Reed Bed Based on Type of Construction*

2.28.4.1 Surface Horizontal Flow (SHF)

The innovative design of the system facilitates the seamless flow of water across the bed's surface, traversing through the firmly rooted stems of the reed plants within the soil. Typically, the clarity of the water is observable up to a depth of approximately 150 mm (Jehawi et al., 2020). This design demonstrates remarkable efficiency in solid separation before undergoing subsequent treatment or regulating the flow into successive reed bed stages (Fig. 2.22).

2.28.4.2 Sub-surface Horizontal Flow (SSHF)

The innovative design of this system facilitates subsurface water flow through a layer of gravel media beneath the reed bed. Strategically planted within the gravel, the reed plants create a visually appealing and odor-free environment. During winter, natural dieback of the reed plants forms a warm composted layer, providing protection to the underlying biofilm.

Fig. 2.22 Horizontal flow wetland

This design has demonstrated high effectiveness in reducing Suspended Solids (SS), Biological Oxygen Demand (BOD), Chemical Oxygen Demand (COD), and partially removing ammonia (Ribadiya & Mehta, 2014). Furthermore, the system has shown success in removing hydrocarbons, certain heavy metals, and nitrates. Overall, this innovative design offers a sustainable and efficient solution for wastewater treatment, while also providing an aesthetically pleasing and safe environment for the public.

2.28.4.3 Down Flow or Vertical Flow (VF)

The design of this system requires dosing the bed's surface via a network of pipes, utilizing either a pumping or siphon system. The objective is to flood the surface of the reed bed several times a day. As water flows down through the bed, it draws in air, creating the ideal bacterial environment. Vertical flow (VF) reed beds are highly effective in removing Biological Oxygen Demand (BOD), ammonia, and some heavy metals, and they require less space for similar treatment compared to horizontal flow systems (Ribadiya & Mehta, 2014).

The efficiency of both SSFH and VF reed beds can be enhanced by adding specific chemicals to the water during treatment. This dosing technique can be employed for Chemical Oxygen Demand (COD) or phosphorus removal in industrial process water, for example. Water can be treated progressively through multiple reed bed stages, and some or all of the aforementioned systems can be integrated into a comprehensive treatment system (Fig. 2.23).

2.28.5 Advantages

The utilization of Reed Bed Treatment (RBT) systems offers a multitude of advantages over conventional biological treatment systems.

Fig. 2.23 Vertical flow wetland

(1) **Reduced Operational Costs**: Reed beds require minimal operational costs compared to traditional biological treatment systems.
(2) **No Sludge Production**: Unlike conventional biological systems, reed beds do not produce sludges, which can be a significant problem in the operation of such systems.
(3) **Low Technical Requirements**: The low-tech nature of reed beds means that highly trained operators are not required to operate the system.
(4) **Odor-Free**: The degradation of the organic content of the effluent occurs within a solid matrix, resulting in an odor-free process.
(5) **High Adaptability**: Reed bed systems exhibit a high diversity of microorganisms, allowing them to adapt to diverse types and varying shock loads of effluents. This adaptability extends to challenging wastewaters containing organic compounds such as chlorinated hydrocarbons, dyes, sulfur-containing aromatics, heavy metals, and pathogens.

2.28.6 Disadvantages

(1) The capital costs of reed beds may vary, but they offer a significant reduction in staff hours and costs required for dewatering compared to conventional sand drying beds or mechanical systems like belt presses or centrifuges.
(2) Dewatering equipment requires a separate building and additional expenses for hauling sludge away and landfilling.
(3) Running even a small belt press can be costly, making reed beds a viable option if you have available land.

(4) The potential for odor during the spring when the ice melts. However, this issue can be mitigated by using a concrete bottom instead of a PVC liner. Operators must also exercise caution during bed clean-out to avoid damaging the liner.
(5) It is crucial to protect the perforated pipes and drains from damage during maintenance or clean-out. Even a small front-end loader can crush the underdrains, causing significant damage.
(6) One common misconception about reed beds is that the reeds can spread and become invasive. However, this is not a problem if the soils are well-drained.

2.29 Microbial Fuel Cells (MFC)

Microbial fuel cells (MFCs) represent a groundbreaking technology enabling the direct extraction of electrical energy from organic matter within the waste stream. This process involves electron transfer capturing the energy produced by microorganisms, cultivated as a biofilm on an electrode. A proton exchange membrane separates the electron donor from the electron acceptor, establishing an electrical current. While still in the developmental stage, significant advancements in process efficiency and economics are necessary before widespread adoption for direct electrical energy production from wastewater organic matter. However, MFCs offer notable advantages such as energy savings from wastewater aeration and sludge treatment, as they produce less sludge compared to aerobic activated sludge (AS) treatment systems and anaerobic digesters. Moreover, they exhibit reduced sensitivity to temperature changes, with limited electrical installations required for sludge treatment plants.

The potential of MFCs is immense, with the capacity to revolutionize energy production and consumption. Further research and development are crucial to fully unlock the technology's capabilities. The global energy crisis has accelerated the development of green energy technologies like MFCs, applicable to wastewater treatment and bioenergy generation. Electricigens in wastewater serve as catalysts for breaking down organic pollutants and generating bioelectricity through bacterial metabolism.

Despite the potential benefits, challenges persist in real-world application, particularly in balancing harvesting efficiency and system upscaling.

2.29.1 MFC Installation in a Biological Treatment System

When considering the functional chain of a wastewater treatment system, it is feasible to replace or modify elements of the secondary waste and sludge treatment using a Microbial Fuel Cell (MFC) (Fig. 2.24). However, replacing components responsible for nutrient and heavy metal removal may present challenges. Several alternatives have been proposed by experts in the field (AlSayed et al., 2020; Leicester et al., 2023).

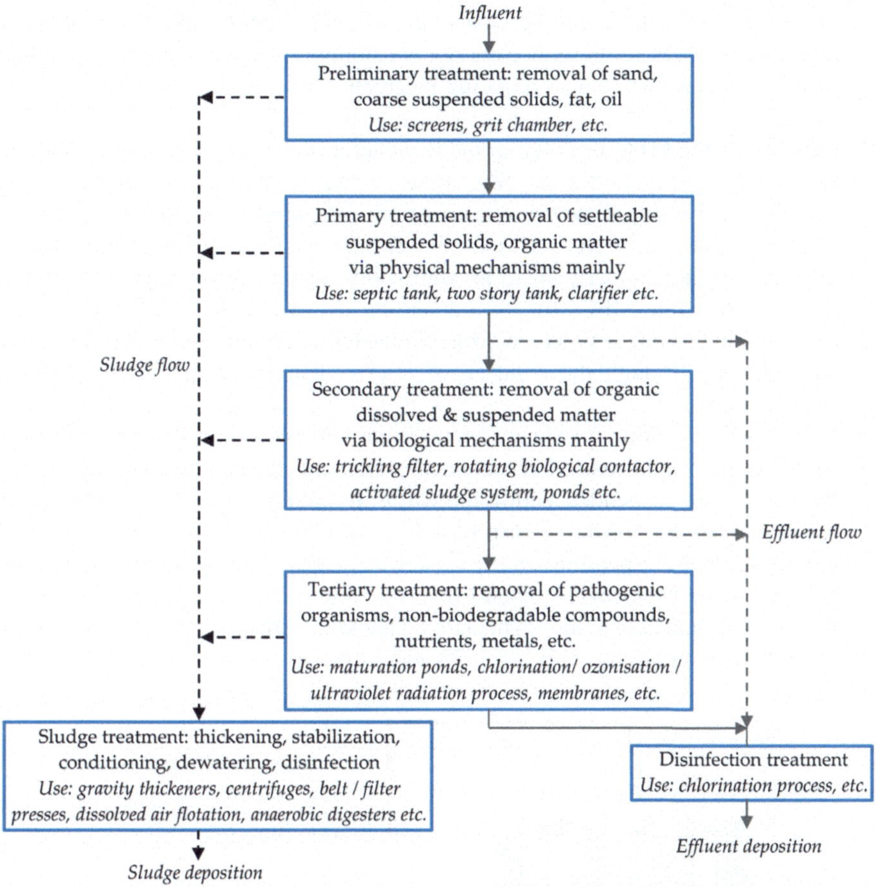

Fig. 2.24 Flow chart of a biological treatment installation: the dashed "black" line represents the sludge flow; the continuous "gray" line represents the treated waste flow; and the dashed "gray" line in grey represents the alternative flow of treated waste

(1) One potential application of the MFC is as a replacement for the Activated Sludge (AS) system or biological filter. In this scenario, the MFC functions similarly to a biological filter, forming a biofilm effective in removing organic loads diluted within waste, but less effective in removing organic solids. A procedure for solids retention must be included for proper treatment of domestic sewage. Unlike a biological filter, the MFC operates anaerobically, eliminating the need for treated waste recirculation at the MFC inflow. Design of the solid contact tank and settler depends on the MFC solid effluence, similar to the solid effluence of the microbial biomass in a biological filter.
(2) Another potential application is combining the MFC with a membrane bioreactor. The reactor, comprising membranes for waste channeling, solids filtration, and biomass retention, serves as both treatment and settling element.

However, this method can be costly due to energy consumption for aeration and filtering. MFCs can reduce energy consumption as power producers rather than consumers, while also reducing production of organic solids and oxygen demand.

(3) Developing the MFC as a Membrane Bioreactor involves constructing cathodes that function as cathodes and filter tubes. Benefits include a smaller reaction chamber and reduced power consumption compared to standard membrane reactors. However, there is a risk of overemphasis, and standard bioreactors operate with atmospheric air to prevent solids accumulation caused by bacterial growth, while MFCs function anaerobically. Ventilated MFC bioreactors are being explored, as most electrogenic bacteria can function aerobically and anaerobically, though the solution of intermediate aeration requires further examination.

(4) Developing the MFC for Heavy Metal Removal involves utilizing specialized MFCs similar to those used in secondary treatment stages, provided the respective metals are suspended within the waste. If solid particles form and require removal in sludge form, the corresponding MFC must be placed in the sludge treatment stage. Depending on the wastewater metal type, series installation of MFCs may be necessary, as each MFC may remove specific metals due to the use of additional chemicals or specialized bacteria with different retention times (Nancharaiah et al., 2019) (Figs. 2.25 and 2.26).

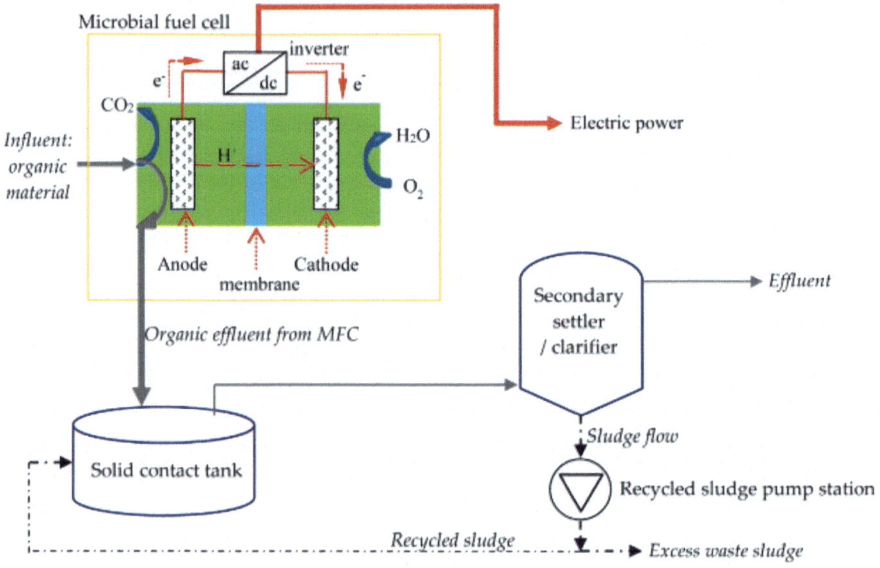

Fig. 2.25 Flow chart of a biological treatment installation of a microbial fuel cell at the secondary treatment stage, placing downstream a solid contact tank, a secondary settler, tube and recycled sludge pump station

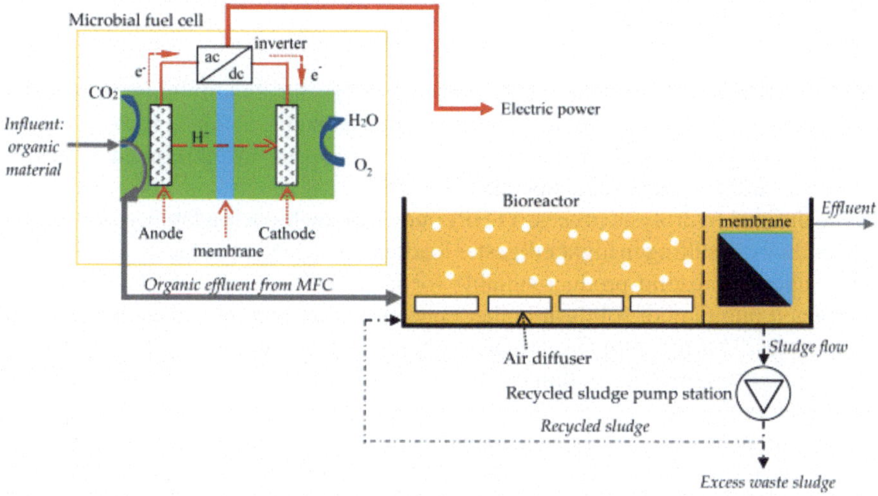

Fig. 2.26 Flow chart of a microbial fuel cell installation at the secondary treatment stage with the use of a membrane bioreactor receiving the required power from the cell (AlSayed et al., 2020)

2.29.2 Reasons for a Microbial Fuel Cell Installation in a Biological Treatment System

The advantages of Microbial Fuel Cells (MFCs) over conventional bioreactors have been extensively documented in the literature (AlSayed et al., 2020; Leicester et al., 2023), summarized as follows:

1. **Power Generation**: MFCs are capable of producing electric current, dependent on wastewater quantity, organic load, and Coulombic efficiency (CE). Alternatively, MFCs can produce hydrogen.
2. **Reduction of Waste Aeration**: In an MFC with a single chamber and air cathode, necessary oxygen transfer is achieved passively. This significantly reduces air supply compared to secondary AS treatment systems, typically consuming around 50% of the electric power in waste treatment plants.
3. **Reduction of Solids Production**: Operating anaerobically, MFCs limit bacterial biomass production compared to aerobic AS systems or biological filters, reducing the cost of sewage sludge treatment installations.
4. **Potential for Odor Control**: MFCs' anaerobic operation mitigates the need for extensive open-air areas in biological filters and large air flows in AS system aeration basins, significantly reducing disturbing odor release to the surroundings.

2.30 Thermal Hydrolysis

Thermal hydrolysis represents an advanced process for conditioning wastewater solids, offering both economic and environmental advantages. This innovative technology, already prevalent in the UK, is gaining traction in the United States as wastewater utilities explore its potential.

The thermal hydrolysis process (THP) improves the biodegradability of wastewater residuals, resulting in higher digester loading rates, increased biogas production, and a superior biosolid product suitable for land fertilization. THP facilitates the mixing and pumping of solids at higher concentrations, thereby enhancing digester loading rates. This capability is particularly beneficial for facilities needing to process greater quantities of solids within existing systems or reduce the size and number of new digesters required.

Thermal hydrolysis is a transformative technology that offers numerous benefits for wastewater treatment facilities. Its ability to enhance biodegradability, improve digester loading rates, and produce valuable biosolids makes it an appealing option for those aiming to optimize their operations. As this technology gains momentum in the US, it is poised to become an increasingly popular choice for wastewater utilities aiming to enhance efficiency and sustainability.

2.30.1 Working of Thermal Hydrolysis

The management of wastewater has evolved significantly since the ancient Babylonians introduced the cesspit around 4000 BCE (Ngo et al., 2021). From the development of latrines by the Romans in 100 BCE to the present day, our approach to wastewater handling has undergone notable changes. In modern, developed communities, wastewater is conveyed from residential, commercial, and industrial sources to wastewater treatment plants (WWTPs). At these facilities, a series of biological, chemical, and physical processes are employed to separate suspended and soluble matter from the liquid wastewater. The resulting organic solids are then directed to anaerobic digestion systems, where they undergo conversion into biogas. The residual byproduct, known as biosolids, finds application as a soil amendment (Ngo et al., 2021). However, the efficiency of the anaerobic digestion process can be further enhanced through the implementation of thermal hydrolysis processing (THP).

THP employs heat and pressure to more efficiently convert anaerobic digestion feedstocks into biogas. Although various process variations exist, the standard procedure involves introducing steam into a reactor containing highly concentrated organics to hydrolyze and solubilize long-chained carbon molecules. THP offers numerous advantages, including increased biogas production and enhanced biosolids quality. Through the utilization of modern technology and innovative techniques like THP, we can continue to enhance the efficiency and sustainability of wastewater management practices.

2.30.2 Key Benefits of THP Include

- Accelerated digestion process compared to conventional treatment methods.
- Intensified anaerobic digestion process resulting in reduced volume requirements.
- Enhanced conversion of organics into biogas.
- Production of a finer residual product devoid of pathogens and with significantly fewer pathogen indicators.
- Generation of a high-quality biosolids product with improved dewatering capabilities, reaching 28% dry solids with 90% less odor.
- Reduction in the quantity of reusable product, leading to increased concentration of valuable nutrients and lower operating costs.

References

Al-Ajalin, F. A. H., Idris, M., Abdullah, S. R. S., Kurniawan, S. B., & Imron, M. F. (2020). Evaluation of short-term pilot reed bed performance for real domestic wastewater treatment. *Environmental Technology & Innovation, 20*, 101110. https://doi.org/10.1016/j.eti.2020.101110

AlSayed, A., Soliman, M., & Eldyasti, A. (2020). Microbial fuel cells for municipal wastewater treatment: From technology fundamentals to full-scale development. *Renewable and Sustainable Energy Reviews, 134*, 110367. https://doi.org/10.1016/j.rser.2020.110367

Ardern, E., & Lockett, W. T. (1914). Experiments on the oxidation of sewage without the aid of filters. *Journal of the Society of Chemical Industry, 33*(10), 523–539. https://doi.org/10.1002/jctb.5000331011

Azizi, A., Dargahi, A., & Almasi, A. (2021). Biological removal of diazinon in a moving bed biofilm reactor—Process optimization with central composite design. *Toxin Reviews, 40*(4), 1242–1252. https://doi.org/10.1080/15569543.2021.1900153

Bairagi, S., & Ali, S. W. (2020). A hybrid piezoelectric nanogenerator comprising of KNN/ZnO nanorods incorporated PVDF electrospun nanocomposite webs. *International Journal of Energy Research, 44*(7), 5545–5563. https://doi.org/10.1002/er.5075

Barber, W. P., & Stuckey, D. C. (1999). The use of the anaerobic baffled reactor (ABR) for wastewater treatment: A review. *Water Research, 33*(7), 1559–1578. https://doi.org/10.1016/S0043-1354(98)00445-0

Cao, L., Su, C., Lu, Y., Wu, J., Wei, L., Liao, J., … Gao, S. (2023). Long-term performance study applying a tandem AnSBR-ASBR to treat wastewater containing N,N-dimethylformamide: Sludge physicochemical properties, microecology, and functional genes. *Journal of Environmental Chemical Engineering, 11*(2), 109447. https://doi.org/10.1016/j.jece.2022.109447

Chadwick, E. (1842). *Report on the sanitary condition of the labouring population of Great Britain*. Her Majesty's Stationery Office.

Chen, B., Chen, S., Zhao, H., Liu, Y., Long, F., & Pan, X. (2019). A versatile β-cyclodextrin and polyethyleneimine bi-functionalized magnetic nanoadsorbent for simultaneous capture of methyl orange and Pb(II) from complex wastewater. *Chemosphere, 216*, 605–616.

Cheremisinoff, P. N. (2019). *Handbook of water and wastewater treatment technology*. Routledge.

Collins, G., Foy, C., McHugh, S., Mahony, T., & O'Flaherty, V. (2005). Anaerobic biological treatment of phenolic wastewater at 15–18 °C. *Water Research, 39*(8), 1614–1620. https://doi.org/10.1016/j.watres.2005.01.019

Droste, R. L., & Gehr, R. L. (2018). *Theory and practice of water and wastewater treatment*. Wiley.

Fazli, N., Mutamim, N. S. A., & Ibrahim, S. A. (2020). Bioelectrochemical cell (BeCC) integrated with granular activated carbon (GAC) in treating spent caustic wastewater: Effects of mixed liquor suspended solid (MLSS). *IOP Conference Series: Materials Science and Engineering, 736*(7), 072010. https://doi.org/10.1088/1757-899X/736/7/072010

Goh, P. S., Ahmad, N. A., Lim, J. W., Liang, Y. Y., Kang, H. S., Ismail, A. F., & Arthanareeswaran, G. (2022). Microalgae-enabled wastewater remediation and nutrient recovery through membrane photobioreactors: Recent achievements and future perspective. *Membranes, 12*(11), 1094. https://doi.org/10.3390/membranes12111094

He, C., Fang, Z., Li, Y., Jiang, C., Zhao, S., Xu, C., … Shi, Q. (2021). Ionization selectivity of electrospray and atmospheric pressure photoionization FT-ICR MS for petroleum refinery wastewater dissolved organic matter. *Environmental Science: Processes & Impacts, 23*(10), 1466–1475. https://doi.org/10.1039/D1EM00145K

Hu, X., Liu, Y., Dong, S., Li, G., Sun, Z., & Liu, Y. (2022). Comparison study of conventional and advanced exergy analysis on cascade high temperature heat pump system based on experiment. *Case Studies in Thermal Engineering, 40*, 102552. https://doi.org/10.1016/j.csite.2022.102552

Hube, S., Eskafi, M., Hrafnkelsdóttir, K. F., Bjarnadóttir, B., Bjarnadóttir, M. Á., Axelsdóttir, S., & Wu, B. (2020). Direct membrane filtration for wastewater treatment and resource recovery: A review. *Science of the Total Environment, 710*, 136375.

Jafarinejad, S. (2017). Cost estimation and economical evaluation of three configurations of activated sludge process for a wastewater treatment plant (WWTP) using simulation. *Applied Water Science, 7*(5), 2513–2521. https://doi.org/10.1007/s13201-017-0588-2

Jauffur, S., Isazadeh, S., & Frigon, D. (2014). Should activated sludge models consider influent seeding of nitrifiers? Field characterization of nitrifying bacteria. *Water Science and Technology, 70*(9), 1526–1532. https://doi.org/10.2166/wst.2014.408

Jehawi, O. H., Abdullah, S. R. S., Kurniawan, S. B., Ismail, N. I., Idris, M., Al Sbani, N. H., … Hasan, H. A. (2020). Performance of pilot hybrid reed bed constructed wetland with aeration system on nutrient removal for domestic wastewater treatment. *Environmental Technology & Innovation, 19*, 100891. https://doi.org/10.1016/j.eti.2020.100891

Kamble, S. J., Chakravarthy, Y., Singh, A., Chubilleau, C., Starkl, M., & Bawa, I. (2017). A soil biotechnology system for wastewater treatment: Technical, hygiene, environmental LCA and economic aspects. *Environmental Science and Pollution Research, 24*, 13315–13334. https://doi.org/10.1007/s11356-017-9084-4

Khan, A. A. P., Khan, A., & Asiri, A. M. (2019). Nanocarbon and its composites for water purification. In *Nanocarbon and its composites* (pp. 711–731). Woodhead Publishing. https://doi.org/10.1016/B978-0-08-102086-1.00025-5

Khan, A. H., Rudayni, H. A., Chaudhary, A. A., Imran, M., & Vambol, S. (2022). Modern use of modified sequencing batch reactor in wastewater treatment. *Ecological Questions, 33*(4), 99–110. https://doi.org/10.12775/EQ.2022.026

Konvensional, B. T. (2017). A review of oilfield wastewater treatment using membrane filtration over conventional technology. *Malaysian Journal of Analytical Sciences, 21*, 643–658.

Leicester, D. D., Settle, S., McCann, C. M., & Heidrich, E. S. (2023). Investigating variability in microbial fuel cells. *Applied and Environmental Microbiology, 89*(3), e02181–e02222. https://doi.org/10.1128/aem.02181-22

Masi, F., Bresciani, R., Martinuzzi, N., Cigarini, G., & Rizzo, A. (2017). Large scale application of French reed beds: Municipal wastewater treatment for a 20,000 inhabitant's town in Moldova. *Water Science and Technology, 76*(1), 134–146. https://doi.org/10.2166/wst.2017.207

Metcalf & Eddy. (2014). *Wastewater engineering: Treatment and resource recovery* (5th ed.). McGraw-Hill Education.

Moharir, S. R., Patni, P. A., Datar, A. V., & Suryawanshi, T. S. (2020). Study of treatment of laundry wastewater. *Journal of Emerging Technologies and Innovative Research, 7*(9), 574.

Moharrami, P., & Motamedi, E. (2020). Application of cellulose nanocrystals prepared from agricultural wastes for synthesis of starch-based hydrogel nanocomposites: Efficient and selective

nanoadsorbent for removal of cationic dyes from water. *Bioresource Technology, 313*, 123661. https://doi.org/10.1016/j.biortech.2020.123661
Muga, H. E., & Mihelcic, J. R. (2008). Sustainability of wastewater treatment technologies. *Journal of Environmental Management, 88*(3), 437–447. https://doi.org/10.1016/j.jenvman.2007.03.004
Nancharaiah, Y. V., Mohan, S. V., & Lens, P. N. L. (2019). Removal and recovery of metals and nutrients from wastewater using bioelectrochemical systems. In *Microbial electrochemical technology* (pp. 693–720). Elsevier. https://doi.org/10.1016/B978-0-12-813314-5.00022-6
Nasir, S., Islam, S., Gul, T., Shah, Z., Khan, M. A., Khan, W., ... Khan, S. (2018). Three-dimensional rotating flow of MHD single wall carbon nanotubes over a stretching sheet in presence of thermal radiation. *Applied Nanoscience, 8*, 1361–1378. https://doi.org/10.1007/s13204-018-0742-1
Neoh, C. H., Noor, Z. Z., Mutamim, N. S. A., & Lim, C. K. (2016). Green technology in wastewater treatment technologies: Integration of membrane bioreactor with various wastewater treatment systems. *Chemical Engineering Journal, 283*, 582–594. https://doi.org/10.1016/j.cej.2015.07.036
Ngo, P. L., Udugama, I. A., Gernaey, K. V., Young, B. R., & Baroutian, S. (2021). Mechanisms, status, and challenges of thermal hydrolysis and advanced thermal hydrolysis processes in sewage sludge treatment. *Chemosphere, 281*, 130890. https://doi.org/10.1016/j.chemosphere.2021.130890
Osakabe, Y., Yamaguchi-Shinozaki, K., Shinozaki, K., & Tran, L. S. P. (2014). ABA control of plant macroelement membrane transport systems in response to water deficit and high salinity. *New Phytologist, 202*(1), 35–49. https://doi.org/10.1111/nph.12630
Pardhi, D. M., Karaman, D. Ş., Timonen, J., Wu, W., Zhang, Q., Satija, S., ... Rosenholm, J. M. (2020). Anti-bacterial activity of inorganic nanomaterials and their antimicrobial peptide conjugates against resistant and non-resistant pathogens. *International Journal of Pharmaceutics, 586*, 119531. https://doi.org/10.1016/j.ijpharm.2020.119531
Park, N. S., Kang, M. S., Jeong, W., & Kim, J. O. (2015). Experimental determination of the characteristics of physico-chemical particles in air-scouring-membrane (microfiltration) backwash water produced during drinking water treatment. *Chemical Engineering Research and Design, 94*, 714–720. https://doi.org/10.1016/j.cherd.2014.12.012
Passoni, C., Marini, A., Belleri, A., & Menna, C. (2021). Redefining the concept of sustainable renovation of buildings: State of the art and an LCT-based design framework. *Sustainable Cities and Society, 64*, 102519. https://doi.org/10.1016/j.scs.2020.102519
Pasveer, A. (1955). Research on activated sludge: VI. Oxygenation of water with air bubbles. *Sewage and Industrial Wastes*, 1130–1146.
Peralta, M. E., Mártire, D. O., Moreno, M. S., Parolo, M. E., & Carlos, L. (2021). Versatile nanoadsorbents based on magnetic mesostructured silica nanoparticles with tailored surface properties for organic pollutants removal. *Journal of Environmental Chemical Engineering, 9*, 104841. https://doi.org/10.1016/j.jece.2020.104841
Piotrowski, R., Paul, A., & Lewandowski, M. (2019). Improving SBR performance alongside with cost reduction through optimizing biological processes and dissolved oxygen concentration trajectory. *Applied Sciences, 9*(11), 2268. https://doi.org/10.3390/app9112268
Rathoure, A. K., Bagrecha, D., & Bhardwaj, P. (2019). Advances in wastewater treatment technologies. *Octa Journal of Environmental Research, 13.*
Ray, S. K., Cho, J., & Hur, J. (2021). A critical review on strategies for improving efficiency of BaTiO3-based photocatalysts for wastewater treatment. *Journal of Environmental Management, 290*, 112679.
Ribadiya, B. M., & Mehta, M. J. (2014). Treatment of municipal and industrial wastewater by reed bed technology: A low cost treatment approach. *International Journal of Engineering Research and Applications, 12*, 15–18.
Rikta, S. Y. (2019). Application of nanoparticles for disinfection and microbial control of water and wastewater. *Nanotechnology in Water and Wastewater Treatment*, 159–176.

Riquelme, M. V., Leng, W., Carzolio, M., Pruden, A., & Vikesland, P. (2016). Gold nanosensor enabled detection of antibiotic resistance genes for environmental monitoring. In *Improved monitoring of emerging environmental biocontaminants through (nano) biosensors and molecular analyses* (pp. 70–86).

Rout, P. R., Zhang, T. C., Bhunia, P., & Surampalli, R. Y. (2021). Treatment technologies for emerging contaminants in wastewater treatment plants: A review. *Science of the Total Environment, 753*, 141990. https://doi.org/10.1016/j.scitotenv.2020.141990

Safwat, S. M. (2019). Moving bed biofilm reactors for wastewater treatment: a review of basic concepts. *International Journal of Research, 6*(10), 85–90.

Sajjadi, M., Ahmadpoor, F., Nasrollahzadeh, M., & Ghafuri, H. (2021). Lignin-derived (nano) materials for environmental pollution remediation: Current challenges and future perspectives. *International Journal of Biological Macromolecules, 178*, 394–423. https://doi.org/10.1016/j.ijbiomac.2021.03.065

Sayed, D. M., El-Deab, M. S., & Allam, N. K. (2021). Multi-walled vanadium oxide nanotubes modified 3D microporous bioderived carbon as novel electrodes for hybrid capacitive deionization. *Separation and Purification Technology, 266*, 118597. https://doi.org/10.1016/j.seppur.2021.118597

Shammas, N. K., & Wang, L. K. (2009). Biosolids composting. In *Biological treatment processes* (pp. 669–714).

Singh, S., Naik, T. S. S. K., Thamaraiselvan, C., Behera, S. K., Nath, B., Dwivedi, P., ... Ramamurthy, P. C. (2023). Applicability of new sustainable and efficient green metal-based nanoparticles for removal of Cr(VI): Adsorption anti-microbial, and DFT studies. *Environmental Pollution, 320*, 121105. https://doi.org/10.1016/j.envpol.2023.121105

Sun, L. P., Chen, J. F., Guo, W. Z., Fu, X. P., Tan, J. X., & Wang, T. J. (2015). Study of the sludge reduction in an oxic–settling–anaerobic activated sludge process based on UNITANK. *Water Science and Technology, 71*(1), 111–116. https://doi.org/10.2166/wst.2014.442

Sutton, P. M., Melcer, H., Schraa, O. J., & Togna, A. P. (2011). Treating municipal wastewater with the goal of resource recovery. *Water Science and Technology, 63*(1), 25–31. https://doi.org/10.2166/wst.2011.007

Tchobanoglus, G., Burton, F., & Stensel, H. D. (2003). Wastewater engineering: Treatment and reuse. *American Water Works Association Journal, 95*(5), 201.

Tilley, E., Ulrich, L., Lüthi, C., Reymond, P., & Zurbrügg, C. (2014). *Compendium of sanitation systems and technologies*. Swiss Federal Institute of Aquatic Science and Technology.

Todorovych, O. S., & Syrotina, I. O. (2017). Innovation technologies for drinking water treatment. *Pure Water: Fundamental, Applied and Industrial Aspects, 54*.

Tong, T., & Elimelech, M. (2016). The global rise of zero liquid discharge for wastewater management: Drivers, technologies, and future directions. *Environmental Science & Technology, 50*(13), 6846–6855. https://doi.org/10.1021/acs.est.6b01600

Van der Hoek, J. P., de Fooij, H., & Struker, A. (2016). Wastewater as a resource: Strategies to recover resources from Amsterdam's wastewater. *Resources, Conservation and Recycling, 113*, 53–64. https://doi.org/10.1016/j.resconrec.2016.05.011

Verma, P., Singh, P. K., Sinha, R. R., et al. (2020). Assessment of groundwater quality status by using water quality index (WQI) and geographic information system (GIS) approaches: A case study of the Bokaro district, India. *Applied Water Science, 10*(1), 27. https://doi.org/10.1007/s13201-019-09792-6

Wacławek, S., Lutze, H. V., Grübel, K., Padil, V. V., Černík, M., & Dionysiou, D. D. (2017). Chemistry of persulfates in water and wastewater treatment: A review. *Chemical Engineering Journal, 330*, 44–62. https://doi.org/10.1016/j.cej.2017.07.020

Wang, J., Lu, H., Chen, G. H., Lau, G. N., Tsang, W. L., & Van Loosdrecht, M. C. (2009). A novel sulfate reduction, autotrophic denitrification, nitrification integrated (SANI) process for saline wastewater treatment. *Water Research, 43*(9), 2363–2372.

Wu, L. N., Liang, D. W., Xu, Y. Y., Liu, T., Peng, Y. Z., & Zhang, J. (2016). A robust and cost-effective integrated process for nitrogen and bio-refractory organics removal from landfill leachate via short-cut nitrification, anaerobic ammonium oxidation in tandem with electrochemical oxidation. *Bioresource Technology, 212*, 296–301. https://doi.org/10.1016/j.biortech.2016.03.082

Zhang, Q. H., Yang, W. N., Ngo, H. H., Guo, W. S., Jin, P. K., Dzakpasu, M., ... & Ao, D. (2016). Current status of urban wastewater treatment plants in China. *Environment International, 92*, 11–22.

Zhang, J., Eggen, M., Peruzzi, S., Klokkerengen, R., Sundfør, E., Odei, D. K., … Mota, V. C. (2024). Effects of prolonged application of peracetic acid-based disinfectant on recirculating aquaculture systems stocked with Atlantic salmon parr. *Science of The Total Environment,* 173762. https://doi.org/10.1016/j.scitotenv.2024.173762

Zhou, C., Yan, C., Zhou, Q., Wang, H., & Luo, W. (2016). Producing a synthetic zeolite from secondary coal fly ash. *Environmental Technology, 37*(22), 2916–2923. https://doi.org/10.1080/09593330.2016.1154747

Chapter 3
Drinking Water Treatment Process

3.1 Introduction

Water, as a vital natural resource, exhibits uneven distribution both temporally and spatially. It serves as a cornerstone for sustaining human life and economic activities, fulfilling crucial roles such as providing drinking water, supporting irrigation, and facilitating industrial operations. Despite its fundamental importance, the availability of water is increasingly jeopardized by escalating human populations and heightened demands for high-quality water for various domestic and economic endeavors. The extraction of water for agricultural, industrial, and other purposes, coupled with inadequate treatment practices, can detrimentally impact water quality and quantity, thereby risking both aquatic ecosystems and the provision of safe drinking water.

The rapid expansion of industrialization and urbanization has led to the proliferation of large urban centers, industrial hubs, and intensive agricultural practices, exacerbating the demand for water and pollution levels in rivers due to untreated discharges. Consequently, the self-purification capacity of water bodies is compromised, necessitating urgent measures to ensure access to clean water.

Presently, the quest for clean water stands as one of the most pressing environmental challenges worldwide. Alarmingly, over 1.2 billion individuals lack access to safe drinking water, with half of the global populace lacking adequate sanitation services. Contaminated water serves as a vector for many diseases, with diarrheal illnesses alone claiming approximately hundreds of lives daily.

Addressing this critical issue and guaranteeing universal access to clean water and sanitation services is imperative to avert dire consequences for human well-being and economic stability. The quality of both surface and groundwater is influenced by a confluence of natural phenomena and human interventions. Unchecked human activities, such as industrialization, agriculture, and urbanization, disrupt the natural equilibrium of dissolved substances, thereby impeding water quality. Hence, adopting sustainable practices that mitigate the adverse impacts of human endeavors

S. A. Bandh and B. Mushtaq, *Wastewater Treatment Technology*, Springer Water,
https://doi.org/10.1007/978-3-031-86684-5_3

on water quality is paramount. This necessitates the implementation of robust policies, regulations, and best management practices to foster responsible water usage and conservation.

Aquatic environments are not mere reservoirs but intricate ecosystems requiring judicious management to ensure sustainable functioning. Effective management mandates an understanding of the intricate interplay between ecosystem dynamics and human activities, which can perturb physical, chemical, and biological processes driving ecosystem functioning.

The Millennium Development Goals underscore the importance of providing safe water to global populations and promoting sustainable water resource management. Recognizing the inextricable links between ecosystems and human well-being, the international community is committed to tracking progress toward minimizing environmental impacts and enhancing access to safe water. This necessitates ongoing monitoring of water quality and quantity at local, national, and international levels.

Water quantity monitoring typically involves remote hydrological monitoring stations that record water level, discharge, and velocity, necessitating minimal human intervention. Conversely, water quality assessment typically relies on periodic sampling by technical teams, with the associated costs often resulting in less frequent monitoring. However, such data is indispensable for discerning spatial and temporal trends in water resources.

Sustainable Development Goal 6 (SDG 6) aims to ensure universal access to water and sanitation services by 2030, necessitating prioritized monitoring of water quality and quantity. Enhanced understanding of water resources enables informed decision-making to secure their sustainability. Investing in simultaneous monitoring of pertinent parameters is indispensable for realizing SDG 6 and ensuring equitable access to clean water.

The World Health Organization projects that by 2030, water scarcity will displace 700 million individuals, while desertification will imperil the livelihoods of one billion people in 100 countries by 2050. Addressing these challenges requires collaborative efforts and innovative solutions to ensure equitable access to clean water for all. Modern water treatment technologies serve as indispensable barriers against waterborne diseases, safeguarding public health. However, comprehensive approaches encompassing source protection, treatment processes, and distribution system management are imperative to ensure the safety and quality of drinking water. By adopting holistic strategies, stakeholders can uphold the integrity of the water supply chain and guarantee access to safe and high-quality drinking water for all.

3.2 Physical Parameters of Drinking Water

3.2.1 *Turbidity*

Turbidity, characterized by the cloudiness of water, serves as a vital metric for assessing light penetration in aquatic environments. It arises from suspended materials such as clay, silt, organic matter, plankton, and other particulates. Turbidity can also result from sewage or industrial discharges, and high concentrations of microorganisms. Beyond its aesthetic implications, turbidity poses significant risks to both water consumers and aquatic ecosystems.

The impact of turbidity on drinking water is multifaceted and substantial. Elevated turbidity levels can escalate the costs associated with water treatment across various sectors. Particulate matter provides refuge for harmful microorganisms, impeding disinfection processes. Furthermore, suspended materials can impair fish gills, reducing their disease resistance, growth rates, and reproductive success, thereby affecting fisheries productivity. Additionally, suspended particles serve as adsorption media for heavy metals and hazardous organic pollutants, exacerbating water contamination concerns. Elevated turbidity levels also elevate water temperatures and absorb more heat, reducing dissolved oxygen concentrations, which are vital for aquatic life.

Turbidity is quantified using a nephelometric turbidimeter, with readings typically expressed in Nephelometric Turbidity Units (NTU) or Turbidity Units (TU), where one TU equals 1 mg/L of silica in suspension. Regular monitoring of turbidity levels is essential to ensure water safety and prevent adverse impacts on aquatic ecosystems and human health. Turbidity exceeding 5 NTU may be discernible to the average observer, while levels exceeding 100 NTU indicate muddy water conditions. Groundwater typically exhibits low turbidity due to natural filtration processes during infiltration through soil.

3.2.2 *Temperature*

The assessment of water temperature, while often overlooked, holds paramount importance due to its profound implications for various water-related processes and the overall quality of the water resources. While controlling water temperature in large-scale water supply projects may not be feasible, understanding and monitoring temperature variations are crucial for ensuring water safety and quality.

Potable water ideally maintains a temperature of around 10 °C, as temperatures exceeding 25 °C are deemed unfavorable. Elevated water temperatures create favorable conditions for bacterial proliferation, necessitating rapid cooling of water samples collected for bacteriological analysis. Additionally, recording air temperature concurrent with water sampling provides valuable context for interpreting temperature data.

Measurement of water temperature typically employs ordinary thermometers, enabling the determination of key water characteristics such as density, viscosity, vapor pressure, and surface tension. Temperature influences the saturation levels of solids and gases dissolved in water, as well as the rates of chemical, biochemical, and biological activities. Furthermore, temperature impacts water palatability, viscosity, solubility, odors, and chemical reactions. Optimal water temperature, ranging from 10 to 15 °C, is considered most palatable by the majority of individuals.

Beyond its sensory aspects, water temperature profoundly influences various processes critical to water management, including sedimentation, chlorination, biological oxygen demand (BOD), and the biosorption of dissolved heavy metals. By comprehensively understanding and monitoring water temperature, stakeholders can make informed decisions to safeguard water quality and ensure its suitability for various uses.

3.2.3 Color

The presence of organic and inorganic matter in water can impart an unsightly color, posing aesthetic concerns and potential nuisances. While not typically indicative of health risks, discolored water can stain fabrics and disrupt industrial processes. To quantify water color, a tintometer is commonly employed, featuring an eyepiece with two apertures—one for a standard colored water slide and the other for the water under examination.

Color intensity is assessed using an arbitrary scale, often based on the cobalt scale, where one unit represents the color produced by one milligram of platinum cobalt in one liter of distilled water. For public water supply, the cobalt scale reading should ideally not exceed 20 and preferably remain below 10. Color examination typically involves comparing the water sample against standard color solutions or colored glass disks. Each color unit corresponds to the color generated by a 1 mg/L solution of platinum (potassium chloroplatinate, K_2PtCl_6).

Water color assessment can be reported as apparent color, representing the overall color of the water sample, inclusive of both dissolved and suspended components, or true color, determined after filtering the sample to remove all suspended material. Color is graded on a scale ranging from 0 (clear) to 70 color units, with pure water being colorless, equivalent to 0 color units.

While color comparison with standard references suffices for most purposes, it is crucial to acknowledge that personal factors and lighting conditions may influence color perception. By systematically evaluating water color and adhering to standardized measurement procedures, stakeholders can effectively monitor water quality and ensure its suitability for various applications.

3.2.4 Taste and Odor

The taste and odor of water can be influenced by a variety of foreign substances, including organic materials, inorganic compounds, and dissolved gases, sourced from natural, domestic, or agricultural origins. These contaminants can render water unpalatable for drinking purposes, necessitating thorough testing and remediation measures.

To assess water taste and odor, an osmoscope is commonly utilized, featuring two tubes—one inserted into a flask containing diluted water and the other into a flask containing the water under evaluation. Furthermore, the taste and odor of water can be quantified using the threshold number method that involves diluting the water sample with odor-free water until the mixture's odor becomes detectable. The point at which odor becomes perceptible indicates the threshold number. It is important to note that test results may be subject to variation based on the observer's sensitivity.

For public water supply, it is recommended that the threshold number does not exceed 3 units. Adhering to these guidelines ensures that drinking water is safe and devoid of unpleasant tastes and odors. By employing systematic testing protocols and maintaining stringent quality standards, stakeholders can safeguard public health and ensure the potability of drinking water resources.

3.2.5 Solids

Solids present in water or wastewater, whether suspended or dissolved, can significantly impact water quality and effluent characteristics. These solids originate from diverse sources, including household activities, runoff from various surfaces, and natural erosion processes.

Household activities contribute to the solid content in wastewater, with food waste, toilet paper, and human waste being the common constituents. In older combined sewer systems, runoff may introduce additional solids, including soil particles eroded from surfaces.

Elevated levels of dissolved solids can detrimentally affect water taste and may induce adverse physiological reactions in consumers. Regulatory guidelines often stipulate that dissolved solids should not exceed 500 mg/L to maintain safe drinking water standards. Furthermore, highly mineralized waters are unsuitable for many industrial applications due to their adverse effects. Water containing high levels of suspended solids may also be aesthetically unpleasing for recreational activities such as bathing.

To manage water quality effectively and ensure compliance with regulatory standards, it is imperative to analyze the content of solids. This enables control over biological and physical wastewater treatment processes and facilitates adherence to wastewater effluent limitations set by regulatory agencies. By implementing robust monitoring and control measures for solids, stakeholders can safeguard water quality and protect public health and environmental integrity.

3.2.6 Total Solids

Total solids represent the residual material remaining in a vessel following the evaporation and drying of a sample in an oven at a specified temperature. This category encompasses both "total suspended solids," which are captured by a filter, and "total dissolved solids," which pass through the filter. In wastewater analysis, total solids denote the remaining substances subsequent to water evaporation, typically conducted at temperatures ranging from 103 to 105 °C.

Total solids can be further categorized into filtrable and nonfiltrable fractions. The filtrable fraction comprises colloidal particles and dissolved solids that traverse a filter during laboratory procedures. In contrast, the nonfiltrable fraction encompasses settleable and non-settleable components that did not pass through the filter.

Numerous factors influence the separation of suspended and dissolved solids, including the characteristics of the filter holder (e.g., type, pore size, porosity, area, thickness), as well as the physical attributes of the material (e.g., nature, particle size, deposition quantity) deposited on the filter.

A profound comprehension of total solids is indispensable in wastewater treatment and analysis. Accurate measurement of total solids enables a deeper understanding of wastewater composition, facilitating the development of effective treatment strategies to mitigate environmental impact and ensure water quality.

3.2.7 Dissolved Solids

Dissolved solids and suspended solids represent distinct components of waterborne solids, each with unique characteristics and origins. Dissolved solids refer to those particles capable of passing through a filter with a nominal pore size of 2.0 μm or smaller under specific conditions. These solids are typically microscopic and invisible to the naked eye, requiring laboratory testing for detection. They originate from diverse sources including natural minerals, industrial processes, and agricultural activities.

In contrast, suspended solids are particles that are retained on a filter and are often visible to the naked eye. These solids can cause turbidity in water, diminishing its clarity and aesthetic appeal. Suspended solids originate from various sources such as soil erosion, wastewater discharge, and stormwater runoff.

Understanding the distinction between dissolved and suspended solids is paramount in water quality management and treatment. Accurate identification and measurement of these solids enable the implementation of appropriate treatment methods to ensure the provision of safe and clean water for various uses, safeguarding public health and environmental integrity.

3.2.8 *Electrical Conductivity (EC)*

Electrical conductivity (EC) serves as a critical indicator of a solution's ability to conduct electrical current, directly influenced by the concentration of ions present within the solution. This property holds particular significance in assessing water suitability for purposes such as irrigation and firefighting.

EC is measured using different units depending on the region or standard used. In the United States, it is typically expressed in micromhos per centimeter (μS/cm), while in the International System of Units (SI), it is measured in MilliSiemens per meter (mS/m) or DeciSiemens per meter (dS/m). Notably, 1 dS/m equals 1000 μS/cm or 10 μmho/cm.

Understanding the EC of water is essential for ensuring its safety and effectiveness for intended applications. By analyzing this parameter, it becomes possible to assess whether water meets the required standards for different uses, enabling appropriate actions to maintain water its quality and suitability.

3.3 Chemical Properties of Water

3.3.1 *pH*

The pH level is indeed a fundamental parameter in assessing water quality, as it provides insight into the acidity or alkalinity of a solution. Defined as the negative logarithm of the hydrogen ion concentration, pH is a dimensionless number that indicates the strength of an acidic or basic solution.

Water with a pH of 7 is considered neutral, while values below 7 indicate acidity and values above 7 indicate alkalinity. Pure water typically has a pH close to 7 at 25 °C, while normal rainfall is slightly acidic with a pH of approximately 5.6 due to atmospheric carbon dioxide.

For safe drinking water, the pH range is generally recommended to be between 6.5 and 8.5 for domestic use, as organisms thrive within this range. It is worth noting that a change of 1 unit on the pH scale represents a 10-fold change in the hydrogen ion concentration.

Various methods, including electrometric and colorimetric techniques, are available for determining pH levels. Monitoring and maintaining the pH level of water are crucial as it significantly influences chemical and biological processes within water. Ensuring the pH level remains within acceptable ranges is essential for safeguarding both human health and environmental sustainability.

3.3.2 Alkalinity

Alkalinity is a critical property of water, representing its capacity to resist changes in pH when acid is added. It is quantitatively defined as the sum of all titratable bases in the water, measured by the amount of acid required to lower the pH to a certain level, typically around pH 4.5. The significant contributors to alkalinity are carbonates, bicarbonates, and hydroxide ions, which neutralize hydrogen ions.

Derived from the decomposition of minerals in soil or the atmosphere, alkalinity acts as a buffer, stabilizing the pH of water against acidic or basic changes. High alkalinity indicates a greater ability to maintain pH stability when exposed to acid, such as acid rain or spills. However, excessive alkalinity can impart a bitter taste to water and potentially lead to sedimentation issues, causing clogging in pipes and water supply networks.

Alkalinity measurements are vital in analyzing natural waters to assess their buffering capacity, crucial for maintaining stable pH levels. In water and sewage treatment, alkalinity serves as a key process control variable. Understanding alkalinity is therefore essential for ensuring the quality and safety of our water supply systems.

3.3.3 Chloride

Chlorides are a common water pollutant, and while they occur naturally in water sources, elevated levels can pose challenges for drinking water supplies. Desalination technologies have helped mitigate this issue, particularly in regions where chloride concentrations in ocean water are high. Surface water contamination by chlorides can stem from various sources, including salt storage, industrial discharge, and natural geological formations rich in chloride salts. In well water, chlorides may leach from surrounding soil, contributing to pollution. Although chlorides generally have mild effects on organisms, excessive intake can lead to health concerns. The recommended limit of chloride in drinking water is typically set below 250 mg/L to prevent adverse health effects. Titration with silver nitrate is a common method for measuring chloride concentrations in water due to its accuracy and reliability. This method involves adding silver nitrate solution to a water sample until all chloride ions have reacted, forming a white precipitate of silver chloride, which indicates the endpoint of the titration. By monitoring and managing chloride levels in water sources, we can ensure the safety and quality of drinking water for human consumption and environmental sustainability.

3.3.4 Chlorine Residual

Chlorine is not naturally present in water but is commonly added to water and wastewater for disinfection purposes (Kumar & Singh, 2015). Maintaining an optimal residual concentration of chlorine in drinking water, typically around 0.2 mg/L, is crucial for ensuring good sanitary quality throughout the distribution system (Srivastav & Kaur, 2020). However, chlorine can react with organic matter in water to form toxic compounds known as trihalomethanes (THMs), including carcinogens such as chloroform ($CHCl_3$).

The residual chlorine in water is typically measured using a color comparator test kit or a spectrophotometer. There are three forms of residual chlorine in water treatment: free, combined, and total. Free residual chlorine consists of dissolved hypochlorite ions, hypochlorous acid, and chlorine gas, while combined residual chlorine comprises chloramines formed by the reaction of chlorine with ammonia. Total residual chlorine is the sum of free and combined residual chlorine.

Although chlorine is an effective disinfectant for water and wastewater, its use requires careful monitoring to mitigate potential health risks associated with disinfection byproducts. Operators in water treatment and distribution plants must ensure that the proper amount of residual chlorine is present as the water reaches the end of the system to prevent microbial contamination and ensure water safety (Srivastav & Kaur, 2020).

Additionally, while chlorine is effective in disinfection, it can react with certain substances in water distribution systems, such as lead, leading to increased corrosion. Therefore, maintaining proper residual chlorine levels at various points in the distribution system is crucial to prevent corrosion and ensure the overall safety and quality of drinking water.

3.3.5 Water Hardness

Water hardness, a term describing the sum of polyvalent cations in water, primarily comprises calcium and magnesium ions, which are predominant in natural waters (Edzwald, 2011). Excessive hard water can escalate soap and synthetic detergent usage for laundry and contribute to scaling in boilers and industrial equipment.

Expressed in milligrams per liter (mg/L), water hardness is primarily due to calcium and magnesium ions, termed total hardness (TH), further divided into carbonate and noncarbonate hardness. Carbonate hardness, or temporary hardness, can be removed by heating, causing insoluble carbonates to precipitate and form bottom deposits in hot water heaters.

Hard water typically contains over 120 mg/L $CaCO_3$ equivalent, while soft water contains less than 60 mg/L (Edzwald, 2011). Freshwater animals can tolerate both soft and hard water within the potable range, but sudden changes can lead to health issues. Monitoring hardness is critical, especially when introducing or transferring animals.

Hard-water minerals, like calcium carbonate, buffer pH shifts, whereas soft water is less buffered, making it susceptible to acidification. Most supply waters maintain a pH range of 6.5–8.0, within which common aquatic vertebrates can adapt.

Permanent hardness, or noncarbonate hardness, remains even after heating. It is the total hardness exceeding alkalinity, with no noncarbonate hardness if alkalinity equals or exceeds total hardness. Understanding water hardness aids in managing water quality for various purposes.

3.3.6 Dissolved Oxygen

Dissolved oxygen (DO) is crucial for assessing water quality in streams, rivers, and lakes, serving as a key indicator of pollution. Higher DO concentrations generally signify better water quality. However, oxygen solubility in water is sensitive to temperature, with saturation concentrations varying. For instance, at 20 °C, it is around 9 mg/L, while at 0 °C, it rises to 14.6 mg/L (Antanasijević et al., 2020). While DO doesn't directly affect public health, low oxygen levels can make water unpalatable.

DO is essential for aquatic organisms' survival, and levels below 5 mg/L can be lethal for fish and other organisms. Cold water generally holds more DO than warm water, and at sea level and 20 °C, freshwater typically holds 9.1 mg/L of DO. However, DO decreases with higher elevations.

The introduction of organic waste, especially from domestic, industrial, and agricultural sources, can significantly reduce DO levels, as the bacteria decompose organic waste and consuming oxygen in the process. For instance, oxidizing 3 mg/L of carbon requires 9 ppm of soluble oxygen. Monitoring DO is vital for pollution control and water quality management.

3.3.7 Biochemical Oxygen Demand

Biochemical Oxygen Demand (BOD), also known as Biological Oxygen Demand, quantifies the amount of oxygen required by aerobic microorganisms to decompose organic matter in a water sample over a specific time period and temperature. This process depletes dissolved oxygen crucial for aquatic life, impacting the ecosystem.

Dissolved oxygen in water, vital for aquatic life, can decrease when organic matter is present, as microbes consume oxygen to break down the organic substances. BOD serves as an indicator of water quality, with elevated levels often indicating pollution, either from environmental factors or human activities.

BOD values are expressed as milligrams of oxygen consumed per liter of sample during a five-day incubation at 20 °C. Pristine rivers typically have a BOD below

1 mg/L, while moderately polluted ones range from 2 to 8 mg/L. Rivers with BOD exceeding 8 mg/L are considered severely polluted.

Understanding BOD is crucial for effective water management. Monitoring BOD helps identify organic pollution levels and guides mitigation efforts to protect ecosystems and aquatic life.

3.3.8 Chemical Oxygen Demand

Chemical Oxygen Demand (COD) is a crucial measure of the oxidation susceptibility of both organic and inorganic materials in water bodies, municipal, and industrial wastes. It quantifies the oxygen equivalent of the organic matter in a water sample that can be oxidized by a strong chemical oxidant. Unlike the Biological Oxygen Demand (BOD) test, which only oxidizes biologically reactive carbon, COD converts all organic matter to carbon dioxide.

It is important to recognize that the COD test doesn't identify the specific oxidizable material or differentiate between organic and inorganic substances. Additionally, it doesn't indicate the total organic carbon present. Therefore, COD values are generally higher compared to BOD. However, COD is a valuable and rapidly measurable parameter, taking only 3 h compared to the 5 days required for a BOD test.

COD concentrations in surface water resources typically range from 20 mg/L or less in unpolluted waters to over 200 mg/L in waters receiving effluents. Industrial wastewaters may exhibit COD levels ranging from 100 to 60,000 mg/L. These variations emphasize the importance of monitoring COD levels in water bodies and wastewater treatment plants. Despite its limitations, COD remains a valuable tool for assessing the oxidation susceptibility of organic and inorganic materials in water and wastewater, providing valuable insights for water quality management.

3.4 Conventional Water Treatment Process

3.4.1 Coagulation

The presence of extremely fine suspended particles in wastewater presents a significant challenge for their removal via traditional sedimentation methods. These particles have a low settling velocity, requiring an impractical detention period for their removal. Additionally, their continuous movement, known as Brownian motion, hinders settling by gravitational force. Moreover, these particles carry electric charges, further complicating their removal.

To address these challenges, coagulation, also known as forced sedimentation, is employed prior to sedimentation. Coagulation involves the addition of chemical compounds such as ferric chloride, ferric sulfate, and alum to the wastewater. These compounds form a gelatinous precipitate called floc when thoroughly mixed. The fine mud particles and other colloidal matter present in the wastewater are absorbed into these flocculated particles, forming larger agglomerates.

The coagulated sewage is then directed through a sedimentation tank where the flocculated particles settle out and are subsequently removed. Coagulants with a positive charge neutralize the negatively charged particles in the water, causing suspended solids to bind together into larger flocculated aggregates. The larger the size of the particles, the faster the flocculation process and settling.

Coagulation effectively eliminates various pollutants that render water dirty or toxic. It is a critical process for the removal of very fine suspended particles in wastewaters, enhancing their size for easier removal through sedimentation. By employing coagulants, water treatment facilities can efficiently remove pollutants and ensure the provision of clean and safe water.

3.4.2 Properties of Common Coagulants

When selecting a coagulant, it is important to consider several key properties.

(i) The coagulant should react quickly in water to produce spongy, gelatinous flocs.
(ii) It should be an affordable material that is easy to handle and store.
(iii) It should not deteriorate in quality over time.
(iv) The coagulant should be electrolytic in nature, releasing positively charged electric ions that attract negatively charged colloidal impurities.
(v) It should also produce high valence ions to ensure maximum absorptive efficiency.
(vi) It should be able to react effectively across a wide range of pH levels.

3.4.3 Types of Coagulants

3.4.3.1 Organic Coagulants

Organic coagulants are indeed highly effective for solid-liquid separation and are preferred for their ability to reduce sludge generation. Their organic nature offers several advantages, including the ability to operate at lower doses and without significantly impacting the pH of water (Garg & Garg, 1999; Sun et al., 2019). Organic coagulants are commonly formulated based on the following compositions.

Poly AMINEs and PolyDADMACs

Cationic coagulants are extensively employed organic agents that function primarily through charge neutralization. Prominent examples include polyamines and poly-DADMACs, which act by neutralizing the negative charges on colloidal particles in water, leading to the creation of a spongy aggregation known as a "microfloc." However, it is important to note that these coagulants do not facilitate the sweep-floc mechanism, a process discussed in further detail within the scope of inorganic coagulants.

Melamine Formaldehydes and Tannins

Natural coagulants operate similar to inorganic coagulants, with both capable of coagulating colloidal material in water and producing their own precipitated floc (Cui et al., 2020). This sweep-floc precipitate can adsorb organic materials, such as oils and greases, while simultaneously coagulating undesirable particles in the water. These coagulants are particularly beneficial for operations generating hazardous sludge, such as those in oil refineries, as the precipitate effectively dewaters the sludge to a low moisture content.

Utilizing natural coagulants allows industries to efficiently manage wastewater and reduce hazardous sludge volumes. This practice not only benefits the environment but also fosters sustainable industrial operations. Natural coagulants present a cost-effective and efficient solution for managing wastewater and mitigating the environmental impact of industrial activities.

3.4.3.2 Inorganic Coagulants

Inorganic coagulants provide a cost-effective solution for diverse water treatment applications, often being less expensive than organic alternatives. They are particularly effective for treating raw water with low turbidity. When introduced into water, inorganic coagulants form aluminum or iron precipitates that absorb impurities as they settle, a process known as the "sweep-floc" mechanism. However, this process can lead to an increase in sludge volume, necessitating proper treatment and disposal (Hammer, 2000; Saxena et al., 2018). Commonly used inorganic coagulants include alum, and iron salts such as ferric chloride, ferric sulfate, ferrous sulfate, and sodium aluminate. The choice of coagulant should be tailored to the specific treatment requirements and the characteristics of the water being processed.

Use of Alum [$Al_2(SO_4)_3 \cdot 18H_2O$]

Alum is extensively used in water treatment due to its rapid reaction time and ability to form stable flocs. When introduced in the raw water, alum reacts with the natural alkalinity of the water. In cases of insufficient alkalinity, lime may be added to facilitate the formation of a gelatinous aluminum hydroxide precipitate. The use of alum can slightly increase the sulfate hardness and corrosiveness of the water.

A variant known as Black Alum, containing 2–5% activated carbon, is particularly advantageous in water treatment. Black Alum enhances coagulation and assists in the removal of taste and color from the water. Incorporating Black Alum into the treatment process results in a cleaner and more palatable water supply.

$$Al_2(SO_4)_3 \cdot 18H_2O + 3Ca(HCO_3)_2 \rightarrow 2Al(OH)_3 \downarrow + 3CaSO_4 + 6CO_2 \uparrow + 18H_2O$$

$$Al_2(SO_4)_3 \cdot 18H_2O + 3Ca(OH)_2 \rightarrow 2Al(OH)_3 \downarrow + 3CaSO_4 + 18H_2O$$

$$Al_2(SO_4)_3 \cdot 18H_2O + 3Na_2CO_3 \rightarrow 2Al(OH)_3 \downarrow + 3Na_2SO_3 + 3CO_2 + 18H_2O$$

The quantity of alum necessary for coagulation is contingent upon the turbidity and coloration of the raw water. The dosage of alum can range from 5mg/L for relatively clear water to 85 mg/L for highly turbid waters (Metcalf & Eddy, 2004; Sun et al., 2019).

Use of Sodium Aluminate ($Na_2Al_2O_4$) as Coagulant

This compound, classified as an alkaline substance, is sometimes used as a coagulant, particularly in water lacking natural alkalinity. Upon dissolution and mixing with water, it reacts with calcium and magnesium salts present in the raw water, leading to the formation of calcium or magnesium aluminate precipitates (Permitsky & Edzwald, 2006). This reaction effectively reduces both carbonate and noncarbonate hardness during the coagulation process, making it valuable for treating hard, turbid waters.

Despite being approximately 1.5 times more expensive than alum, its use is typically avoided for ordinary public water supplies. However, it is highly advantageous for treating water deficient in natural alkalinity, where pure alum is ineffective (Henze et al., 2006; Kumar et al., 2017; Metcalf & Eddy, 2004).

$$Na_2Al_2O_4 + Ca(HCO_3)_2 \rightarrow CaAl_2O_4 \downarrow + Na_2CO_3 + CO_2 \uparrow + H_2O$$

$$Na_2Al_2O_4 + CaCl_2 \rightarrow CaAl_2O_4 \downarrow + 2NaCl$$

Use of Iron Salts as Coagulants

Iron salts are frequently employed in treating soft water with low pH due to their effectiveness in color removal. Unlike alum, iron salts can react over a broader pH range, except for ferrous sulfate. These salts are cost-effective, react swiftly, and form dense flocs. However, their corrosive and deliquescent nature poses challenges in handling and storage. Additionally, iron salts can encourage the growth of iron bacteria in distribution systems.

Despite these drawbacks, ferric salts are powerful oxidizing agents capable of effectively removing hydrogen sulfide, along with unpleasant tastes and odors from water. While primarily used as coagulants in sewage treatment rather than potable water treatment, iron salts are invaluable in specific applications. There are two main types of iron salts commonly utilized as coagulants: ferric chloride and ferric sulfate.

Use of Copperas as Coagulants

The use of copperas as a coagulant in water treatment is a common practice. Copperas is also known as ferrous sulphate and has a chemical formula of $FeSO_4 \cdot 7H_2O$. Copperas is typically added to raw water in conjunction with lime, although the order of addition may vary, with lime being added either before or after copperas. When lime is added before copperas, the following reaction occurs:

$$FeSO_4 \cdot 7H_2O + Ca(OH)_2 \rightarrow CaSO_4 + Fe(OH)_2 \downarrow +7H_2O$$

This ferrous hydroxide [$Fe(OH)_2$] further gets oxidized to ferric hydroxide.

$$4Fe(OH)_2 + O_2 + 2H_2O \rightarrow 4Fe(OH)_3 \downarrow$$

The $Fe(OH)_3$ forms the flocs, which aids in sedimentation. The quantity of copperas required is typically similar to that of alum. By utilizing copperas as a coagulant, water treatment facilities can effectively remove impurities and improve the overall quality of the water supply.

Use of Chlorinated Copperas as Coagulant

Chlorinated copperas, along with lime, is a potent coagulant. The constituents of chlorinated copperas, ferric sulphate, and ferric chloride, react with calcium hydroxide to form calcium sulfate and ferric hydroxide, which then forms the flocs and aids in sedimentation. Ferric sulphate is most effective in the pH range of 4–7 and above, while ferric chloride is most effective in the pH range of 3.5–6.5 and above 8.5.

Chlorinated copperas is a valuable coagulant that can effectively remove colors from water. Its combination with lime and the resulting formation of ferric hydroxide flocs aids in sedimentation, making it a highly effective treatment option for low pH waters.

$$6FeSO_4 \cdot 7H_2O + 3Cl_2 \rightarrow 2Fe_2(SO_4)_3 + 2FeCl_3 + 42H_2O$$

The resultant combination of ferric sulphate and ferric chloride is known as chlorinated copperas and is a valuable coagulant for removing colours, especially where the raw water has a low pH value.

Both constituents of chlorinated copperas along with lime are effective coagulants.

$$Fe_2(SO_4)_3 + 3Ca(OH)_2 \rightarrow 3CaSO_4 + 2Fe(OH)_3 \downarrow$$

$$2FeCl_3 + 3Ca(OH)_2 \rightarrow 3CaCl_4 + 2Fe(OH)_3 \downarrow$$

Auxiliary Chemicals

While not primarily utilized for coagulation, lime and sodium carbonate play pivotal roles in facilitating other treatment processes, particularly in regulating water pH. Interestingly, their application has been shown to enhance floc formation while reducing the consumption of coagulants. This finding holds significant implications for the water treatment industry, providing a more efficient and cost-effective approach to achieving optimal water quality.

3.4.4 Advantages and Disadvantages of Coagulation

The coagulation process is not commonly employed in modern sludge treatment plants due to several factors.

(i) Contemporary biological secondary treatments in sewage treatment are comprehensive and often do not require coagulation. In fact, coagulation can hinder some of these processes.
(ii) The chemicals used in coagulation react with sewage and may disrupt certain microorganisms essential for sludge digestion, complicating the digestion process.
(iii) The cost of chemicals adds an extra expense to sedimentation without significant benefits, making the treatment more costly.
(iv) The coagulation process and subsequent sedimentation result in larger quantities of sludge compared to plain sedimentation, exacerbating sludge disposal challenges.
(v) The coagulation process necessitates skilled supervision and chemical handling, increasing the complexity of the treatment process.

3.4.5 Sedimentation

Sedimentation, an important process in water treatment, involves the settling of suspended particles under the influence of gravity, forming sediment or sludge (Carlsson, 1998; Hirom & Devi, 2022). Consolidation occurs when a thick layer of sediment settles continuously, while thickening involves mechanically assisting sediment or sludge consolidation.

In water treatment, sedimentation is commonly employed to reduce suspended particle concentrations before coagulation, thus minimizing the required coagulating chemicals. Alternatively, sedimentation may follow coagulation and flocculation to decrease suspended solids concentrations, facilitating more efficient filtration.

Other options for solids-liquid separation include dissolved air flotation and various filtration methods, all falling under clarification processes (Jahanshahi & Taghizadeh, 2018).

3.4.5.1 Principle of Sedimentation

The majority of suspended impurities, particularly organic matter, in water have a specific gravity exceeding that of water itself. Consequently, in stagnant sewage, these impurities naturally settle to the tank bottom due to gravity. However, in flowing sewage, turbulence keeps these impurities suspended. By providing storage to sewage and reducing turbulence, impurities settle to the tank bottom. This sedimentation principle is well-established and extensively studied by experts (Al-Mafraji & Al-Mussawy, 2021; Gregory & Edzwald, 2010). Understanding this principle is crucial for effective water treatment and management.

3.4.5.2 Theory of Sedimentation

The settling of particles by gravity in liquid is a multifaceted process influenced by various factors. When particles come to rest, they encounter resistance from opposing factors that must be considered:

1. **Flow Velocity**: Flow velocity is pivotal in carrying particles horizontally. A larger flow area reduces velocity, facilitating easier settling.
2. **Water Viscosity**: Viscosity, inversely proportional to temperature, affects particle travel. Warmer water, being less viscous, offers less resistance to settlement.
3. **Particle Characteristics**: Particle size, shape, and specific gravity significantly influence settling. Greater specific gravity promotes quicker settling, while particle size and shape affect settling rates.
4. **Sedimentation Tanks**: These tanks are crucial for sewage clarification via sedimentation, providing conditions conducive to particle settling.

Understanding these factors is vital in wastewater treatment and environmental engineering. In wastewater treatment, particle settling tendencies are controlled by three forces, with only two, flow velocity and particle characteristics, manipulable in settling tanks. The third force, viscosity or temperature, remains uncontrolled due to practical limitations (Al-Mafraji & Al-Mussawy, 2021).

To reduce flow velocity, the length of travel can be increased, allowing particles to settle over a longer period in sedimentation basins. Additionally, chemical additives can alter particle size and shape, primarily in chemical precipitation processes. Ultimately, sedimentation basins are designed to facilitate particle settlement by reducing flow velocity or detaining sewage within them.

Effective control of these forces is crucial for wastewater treatment efficiency. By understanding these mechanisms and implementing appropriate measures, efficient and sustainable wastewater treatment processes can be ensured.

3.4.5.3 Types of Sedimentation Tanks

Horizontal Flow Tank

Sedimentation is a vital process in township water treatment. While the simplest sedimentation involves allowing particles to settle in a jar or tank, practical applications require continuous operation (Al-Mafraji & Al-Mussawy, 2021; Hirom & Devi, 2022). To achieve effective sedimentation, rectangular tanks with horizontal flow are commonly employed.

In these tanks, water with suspended particles enters at one end and flows to the other, allowing particles to settle on the tank floor before water exits. Tanks are designed with gently sloping floors leading to hoppers at the inlet end (Hirom & Devi, 2022).

For sediment removal, tanks are equipped with mechanisms scraping sediment from the outlet end back to the inlet, where it is discharged hydraulically. Tank design ensures uniform water flow from end to end, often requiring multi-layer tanks with two or three decks due to their large footprint (Almeida et al., 2020).

Radial Flow Tank

Radial flow tanks, circular in shape with a central inlet and peripheral outlet, play a crucial role in water treatment. Careful design of the inlet is essential to ensure uniform flow distribution throughout the tank, facilitating efficient sediment settling. Sediment is directed towards a central hopper for effective discharge.

In addition to standard features, some circular tanks may incorporate specialized components like flocculation designs for optimal mixing or premix-recirculation designs for settled particle recirculation. These additional features significantly enhance the efficiency and effectiveness of radial flow tank systems.

Inclined Settling

The size of an unhindered settling tank is determined by the time required for particles to settle through the water depth. Efficiency of particle removal relies on available settlement area. To maximize area with minimal footprint, inclined plates or tubes are commonly employed (Gregory & Edzwald, 2010). These lightweight structures can be easily positioned in concrete or steel tanks.

Flow between inclined plates can be co-current, counter-current, or cross-flow. Co-current involves downward water flow between plates, aligning with particle settlement. In counter-current, water flows upward against particle settlement direction. Cross-flow entails horizontal water flow across plates, perpendicular to settlement direction. Optimizing water flow distribution within and between plates is key for efficient particle removal.

Inclined tubes, used in co-current or counter-current modes, typically favor the latter. Tube modules offer flexibility in construction, allowing for various tube cross-sectional shapes (Juraev et al., 2022).

Ballasted Sedimentation

The small density difference between water and particles generated during coagulation, known as flocs, results in slow settling rates. To overcome this, sedimentation methods like horizontal, radial, and inclined sedimentation are preceded by flocculation—a slow mixing process.

Flocculation aims to promote collision and adhesion of coagulated particles, forming larger particles that settle faster. Polyelectrolytes, high molecular weight polymers, aid in this process. However, while flocculation doesn't increase particle density, the density of floc particles decreases as their size increases.

To boost floc particle density, ballasting agents like Bentonite or fine sand are employed. Fine sand, typically 80–200 μm, can be recycled by passing sludge collected from the sedimentation tank through hydrocyclones. Floc particles ballasted with fine sand settle three or more times faster than non-ballasted ones. These particles settle in a tank with a counter-current inclined plate system over a hopper, where settled sludge is collected.

Utilizing these methods, water treatment facilities can enhance process efficiency and ensure production of high-quality water.

Flock Blanket Sedimentation

Floc blanket is preferred over "sludge blanket" since the former acknowledges the role of floc in removing particles from incoming water. In sewage treatment, sediment accumulation on the tank floor is commonly referred to as a sludge blanket. The incoming suspension is directed downwards into the apex of the pyramidal hopper through a single pipe, with the blanket occupying most of the hopper. The success of floc blanket settling leads to the development of premix-recirculation tanks that include floc blanket zones. However, more cost-effective designs and more efficient developments lead to the use of flat-bottomed rectangular tanks that are fed by an arrangement of multiple inlet pipes. These pipes are either candelabra-like or horizontal, with multiple orifices located close to the tank floor.

3.4.6 Filtration

Filtration, a vital process in sewage treatment, involves the use of filter units typically consisting of open beds of coarse aggregates. These beds are intermittently sprinkled with sewage, providing a necessary contact surface for the growth of aerobic bacteria. The aggregates in the bed also facilitate natural aeration. Filtration in sewage treatment is essential for ensuring the safe and effective disposal of wastewater. By utilizing appropriate filter units, sewage treatment facilities can effectively remove harmful contaminants, thus protecting public health and the environment. There are several types of filters commonly used in sewage treatment, either singly or in combination, to provide secondary treatment. The primary goal of these filters is to oxidize unstable organic matter into stable forms, primarily through aerobic decomposition. The three main types of filters used in sewage treatment are contact beds, intermittent sand filters, and trickling filters, each with unique characteristics and advantages, making them suitable for different applications.

3.4.6.1 Contact Beds

Contact beds, also known as contact filters, are a type of sewage filtration system that involves retaining the sewage in contact with the filtering media for a specific period. As the sewage percolates through the filtering media, fine colloidal organic matter becomes trapped in the voids between the particles, resulting in the formation of an organic film around the particles. This film contains a large number of bacteria that carry out the aerobic decomposition of the organic matter present in the film. To ensure sufficient oxygen for the aerobic bacteria, the filter is left empty for a certain period. During this time, the filter absorbs oxygen from the atmosphere, completing the oxidation process. The stabilized organic matter produced during this cycle is then washed away by fresh sewage in the next cycle of operation (Shih & Chang, 2015).

While contact beds produce satisfactory effluent and consume little head, they are generally not adopted due to their low rate of loading, which requires large land areas for installation. Additionally, their intermittent operation necessitates continuous attendance, making them unsuitable for modern use (Shih & Chang, 2015). Consequently, while contact beds were once a popular sewage filtration system, their limitations have led to the development of alternative filtration systems to meet the demands of contemporary wastewater treatment needs.

3.4.6.2 Intermittent Sand Filters

The earlier forms of biological units for sewage treatment, while once commonly used, are now rarely employed due to their large area requirements. However, in certain circumstances, such as treating sewage from hospitals, they may still be utilized to produce highly nitrified and polished effluent (Wu et al., 2022). These units are similar to contact beds but utilize finer contact media and lack a concrete lining around the filter media. The effluent produced by these filters is of exceptional quality, with suspended solids less than 10 ppm and BOD less than 5 ppm (Huidobro-López et al., 2023; Wu et al., 2022). In fact, this effluent is superior to that obtained from any other ordinary type of sewage treatment and can be disposed of without dilution. However, due to their low rate of loading, these filters require large land areas and significant quantities of sand for installation, making them economically unfeasible for larger plants.

3.4.6.3 Trickling Filters

Trickling filters, also known as sprinkling or percolating filters, have become a widely adopted method for providing secondary treatment to sewage. Conventional trickling filters and their improved forms, known as high-rate trickling filters, are now almost universally used. A trickling filter is essentially a bed of crushed stone, gravel, or slag of relatively large size, to which settled sewage is applied by sprinkling on the surface. The applied sewage trickles in a thin film over the surface of the filtering media, which becomes coated with a zoogloeal film containing bacteria, fungi, protozoa, and algae. Both suspended and dissolved solids come in contact with this film. Due to the aerobic conditions, a large number of aerobic bacteria inhabit the film, reacting with the organic solids in suspended, colloidal, or dissolved form, leading to the reduction of biological oxygen demand (BOD), ammonia, organic nitrogen, and other contaminants.

To ensure the large-scale growth of aerobic bacteria, sufficient quantities of oxygen are supplied by providing suitable ventilation facilities in the body of the filter, and to some extent by the intermittent functioning of the filters. The contact between filter media and wastewater is allowed only for a short time, after which the bed is drained and allowed to rest before the next cycle. A typical cycle requires 12 h, with 6 h for operation and 6 h for resting. Temperature can affect the efficiency of trickling filters, with warmer climates resulting in higher efficiency of BOD removal. The effluent obtained from the filter must be directed to a secondary sedimentation tank to settle out the organic matter that has been oxidized while passing down the filter. It is important to note that the sewage influent entering the filter must undergo pretreatment, including screening and primary sedimentation, to ensure the trickling filter operates at maximum efficiency.

3.4.7 *Flocculation*

Flocculation is a process used to enhance the aggregation of fine particles into larger clusters, called flocs, which can be more easily separated from liquid phases. It is widely applied in water treatment, wastewater management, and various industrial processes to improve sedimentation or filtration of suspended solids.

In the flocculation process, destabilized particles (often pre-conditioned by coagulation) are brought together by gentle mixing, allowing collisions and the formation of flocs. Chemical agents called flocculants—such as aluminum salts, iron salts, or synthetic polymers—are often introduced to promote the aggregation. These flocculants can be charged or neutral, depending on the charge of the particles involved. The particles, which are typically colloidal in nature, are bound together through mechanisms such as charge neutralization, polymer bridging, and Van der Waals forces.

Flocculation is a key step in water treatment plants to remove turbidity, color, and organic contaminants, improving the quality of potable water. In wastewater treatment, flocculation helps in the removal of solids, pathogens, and metals, enhancing the efficiency of downstream processes like sedimentation and filtration.

The effectiveness of flocculation depends on factors such as the type and dosage of flocculants, mixing intensity, water chemistry (pH, ionic strength), and particle characteristics. Understanding these factors is critical for optimizing flocculation performance and achieving efficient solid-liquid separation in industrial and environmental applications.

3.4.7.1 Types of Flocculation

Natural Flocculation

Natural flocculation is a highly sustainable and eco-friendly process widely employed in wastewater treatment. This method utilizes natural substances such as plant seeds like moringa or plantago, which have demonstrated high efficiency in flocculation (Bratby, 2016). Additionally, natural products like starch can serve as effective natural flocculants. One of the significant advantages of employing natural flocculants in wastewater treatment is their cost-effectiveness. Furthermore, these natural flocculants are biodegradable and non-toxic, rendering them safe and preferable for purification (Suresh et al., 2018). The utilization of non-toxic flocculants ensures not only the effectiveness of the purification process but also its environmental friendliness. It is important to note that natural flocculants have a relatively short lifespan, yet this does not diminish their effectiveness in wastewater treatment. Indeed, the use of natural flocculants represents a highly sustainable and eco-friendly approach gaining popularity in the industry.

Chemical Flocculation

Chemical flocculants play a critical role in the coagulation process, ensuring highly effective treatment. These flocculants are chemically synthesized from various monomers, including electrolytes, cationic, anionic, and nonionic polymers. However, it is crucial to acknowledge that these chemical flocculants can be highly toxic, as they produce sludge with high quantities of metal hydroxides that can be harmful to living beings if released into water bodies. Techniques such as copolymerization can occur primarily due to the use of chemical flocculants, as highlighted by Lee et al. (2014) and Hou et al. (2019). Therefore, exercising caution when using chemical flocculants is imperative to avoid adverse effects on the environment and living organisms, prompting exploration of alternative methods that are less harmful and more sustainable in the long run.

3.4.7.2 Applications of Flocculation

Flocculation is a versatile process with applications across various fields:

(i) In emulsions, flocculation aids in cluster formation by bringing together individual droplets. This is utilized in mineral dressing and pharmaceutical product design to manipulate their physical properties.
(ii) Flocculation processes contribute to eutrophication by adsorbing substances from underwater, thereby maintaining freshwater quality under the soil. Colloid formation is instrumental in these processes.
(iii) The cheese industry extensively utilizes flocculation for the hydrolysis of molecules and micro peptides, determining the time required for curd formation during cheese manufacture.
(iv) In construction, flocculation aids in the coagulation of clay particles and polymers through mechanical agitation, leading to the spontaneous formation of structures. This is crucial for civil engineering processes.
(v) Biological and biotechnological processes benefit from flocculation, where microfiltration and flocculation work together, and synthetic flocculants increase in size in the bioreactor, aiding in purification.
(vi) Coagulation, flocculation, and sedimentation processes are employed in treating stormwater, sewage, industrial wastewater, and drinking water.
(vii) In the brewing industry, flocculation facilitates the fermentation of beer by causing yeast to sediment and float down to the base or top, where it can be separated and reused for fermentation.

3.4.8 Advanced Water Treatment Processes

The American Institute of Chemical Engineers defines advanced wastewater treatment as any process capable of reducing impurities in wastewater below levels achievable through conventional secondary or biological treatment. As the identification of contaminants continues to challenge traditional water and wastewater treatment processes, alongside population growth, industrial activities, and declining water resources, the need for advanced wastewater treatment becomes increasingly evident.

Despite conventional secondary treatment, effluent still contains 20–40 mg/L BOD, which may be objectionable in streams. Suspended solids, besides contributing to BOD, may settle on the stream bed and inhibit aquatic life. BOD discharge into low-flow streams can harm aquatic life by reducing dissolved oxygen content. Additionally, secondary effluent contains significant plant nutrients, dissolved solids, and traces of organic chemicals and heavy metals from industrial sources. To address these concerns, various methods are employed in advanced waste treatment, targeting specific goals such as suspended solids, BOD, plant nutrients, dissolved solids, and toxic substances removal. Advanced water treatment technologies offer alternatives for enhanced public health and environmental protection, integrating into any stage of the total treatment process or utilized for comprehensive pollutant removal after secondary treatment (Hiller et al., 2019).

With global interest in preserving essential resources amidst population rise, urbanization, industrialization, and dwindling potable water, advanced wastewater treatment has become essential. New techniques surpass the limitations of conventional treatment systems, improving wastewater quality. This pursuit of resource recovery or conservation is facilitated by advanced treatment technologies, which can also augment domestic water supply availability. A sustainability assessment by Høibye et al. (2008) identified several promising technologies, including sand filtration, ozone treatment, UV disinfection, advanced oxidation processes (AOP), UV combined with advanced oxidation, and nanotechnology. These advancements offer solutions to conventional wastewater treatment challenges, ensuring essential resource availability for future generations.

3.4.8.1 Ozonation

Ozonation is an advanced oxidation process that generates reactive oxygen species (ROS), which effectively degrade organic and inorganic pollutants and disinfect water. Ozone, an unstable gas, breaks down to form ROS capable of attacking a wide range of pathogens (Hu, 2019; Hussain et al., 2022; Wu et al., 2019). Its advantages include simple operation, no sludge production, and the ability to combat disinfection-resistant microorganisms like Giardia and Cryptosporidium, while avoiding harmful chlorination byproducts (Chen et al., 2020; Wolf et al., 2019). Ozone, often combined with hydrogen peroxide or UV radiation, has been widely used for disinfection and oxidation of contaminants in water treatment. Recent innovations, such as microflocculation and ozone-biofiltration, enhance water treatment by improving biodegradability and reducing chlorine demand, making ozone a crucial tool for ensuring water quality and safety (Chen et al., 2020).

3.4.8.2 Types of Ozonation

Ozonation can be classified into several types based on different criteria, including application, configuration, and operational conditions.

Based on Application

- **Drinking Water Ozonation**: Used for disinfection and removal of contaminants in potable water. Ozone effectively inactivates pathogens and reduces taste and odor issues.
- **Wastewater Ozonation**: Applied in the treatment of industrial and municipal wastewater to oxidize organic pollutants, remove color, and reduce toxicity before discharge or reuse.
- **Groundwater Treatment**: Ozone is used to oxidize iron, manganese, and sulfide, improving water quality for agricultural or potable use.

Based on Configuration

- **Batch Ozonation**: Involves treating a specific volume of water in a single batch process, allowing for controlled reaction times and ozone doses.
- **Continuous Ozonation**: Common in large-scale water treatment plants, water continuously flows through the ozonation system, providing ongoing treatment.

Based on Operational Conditions

- **Ozone Contactors**: These systems include various designs such as diffusers, contact tanks, and reactors that ensure efficient mixing of ozone with water.
- **Ozone-Biofiltration**: Combines ozonation with biological filtration, where ozonated water is passed through a biofilter (e.g., granular activated carbon) to enhance biodegradation of organic byproducts.
- **Microflocculation**: Involves using low doses of ozone to aid coagulation and flocculation processes, improving the removal of suspended solids and colloids.

Based on the Combination with Other Technologies

- **Ozone-Hydrogen Peroxide Process**: Combines ozone with hydrogen peroxide to enhance oxidation, generating hydroxyl radicals for more effective contaminant degradation.
- **Ozone-UV Process**: Integrates UV light with ozone to increase the oxidative potential, particularly for the degradation of complex organic compounds.

Based on the Ozone Generation Method

- **Electrical Discharge Ozone Generation**: Uses high-voltage electrical discharge to convert oxygen into ozone. This is the most common method for generating ozone in water treatment.
- **UV Ozone Generation**: Utilizes UV light to produce ozone from oxygen, although this method is less common than electrical discharge systems.

3.4.8.3 Organic Oxidation Products

Ozonation is a widely used method for removing organic compounds from water, producing more biodegradable and less toxic byproducts due to their higher oxygen content. Compared to other oxidizing agents like chlorine and chlorine dioxide, ozone excels in generating oxidized organics such as acids, aldehydes, and alcohols, which can be more easily mineralized during biofiltration. Although ozonation produces organic byproducts, these compounds have simpler molecular structures than their original forms, enhancing their biodegradability. This process also reduces chlorine demand in subsequent treatments, making ozonation a valuable and efficient step in advanced water treatment technologies.

3.4.8.4 Bromate Anion

The presence of bromide ions in water destined for ozonation can lead to the formation of hypobromite ion (OBr^-) and hypobromous acid (HOBr), similar to reactions involving chlorine. HOBr acts as a brominating agent, capable of producing various brominated organics. Chlorination of water containing bromide ions produces mixed bromo-chloro-organics, including two of the trihalomethanes and several haloacetic acids, as well as bromo-organics such as bromoform and mono-, di-, and tribromoacetic acids. During ozonation, water containing bromide ions can form HOBr, which can lead to the production of brominated organics such as bromoform and brominated acetic acids. However, only traces of bromoform have been identified in some ozonated water samples containing high bromide levels.

In the presence of ozone, OBr^- can be further oxidized to bromate ion (BrO^{3-}), a compound identified as carcinogenic to certain laboratory animals. Consequently, the US Environmental Protection Agency (EPA) classifies BrO^{3-} as a probable human carcinogen, establishing a maximum contaminant level (MCL) of 10 μg/L in the Surface Water Treatment Rule (SWTR). Many source waters for water treatment plants contain bromide ions, and the higher the bromide ion concentration, the greater the potential for bromate ion formation during ozonation, particularly a pH 6.5 and above.

Understanding the chemistry of bromate formation and the various chemical techniques for minimizing its formation is crucial for water treatment specialists. Bromide ions are oxidized by ozone or chlorine, resulting in a mixture of HOBr and OBr^-.

Ozonation can slowly oxidize OBr^- (but not HOBr) to bromate ions if the pH is above 6.5. At pH 6.5 or below, OBr^- cannot exist, significantly reducing or eliminating bromate ion formation.

To address the potential issue of bromate ion formation in potable water supplies during ozonation, several techniques can be employed:

1. **Conducting Ozonation at Low pH**: By performing ozonation at pH 6.5 or less and subsequently adjusting the pH later in the treatment process, bromate ion formation can be minimized or eliminated.
2. **Adjusting Ozonation Conditions**: Minimizing the levels of residual ozone allows other water contaminants to out-compete OBr^- for the available ozone.
3. **Adding Ammonia**: Introducing a trace of ammonia to the water before ozonation can minimize bromate ion production by inhibiting the oxidation of OBr^-.

Implementing these techniques is essential to ensure the safety and quality of potable water supplies. By minimizing or eliminating bromate ion formation, water treatment facilities can provide a safer and healthier water supply for consumers.

3.4.8.5 Ultraviolet (UV) Disinfection

UV therapy is a highly effective physical process that causes a molecular rearrangement of the genetic material of microorganisms, primarily their DNA. This rearrangement renders the microorganisms dormant and unable to infect humans. UV disinfection systems are utilized for a wide range of purposes, from purifying drinking water in individual homes to sterilizing entire township water supplies (Chen et al., 2023; Das, 2001).

The treatment of water with UV radiation is a safe, environment friendly, cost effective and efficient method for disinfection (Bilenko et al., 2019; Koutahzadeh et al., 2016; Modak et al., 2020) and is applied in both domestic and industrial wastewater. In this method, UV radiation penetrates the cell wall of microorganisms and damages their genetic material (Guo et al., 2019), preventing the reproduction of cells. Proper quantities of UV radiation can destroy nearly 99.9% of microorganisms present within a water stream. In UV water disinfection technology, ultraviolet light (wavelength 253.7 nm) is used for the disinfection of bacteria, viruses, molds, algae, and other microorganisms (Chen et al., 2023). UV disinfection technology destroys the DNA of microorganisms, leaving them dead and unable to grow further. It has diverse applications, including drinking water disinfection, process water disinfection, wastewater disinfection, and surface disinfection. In addition to disinfection applications, this technology can also be used for Total Organic Carbon (TOC) removal and ozone destruction. UV disinfection offers an efficient and environmentally friendly solution to microbial contamination, ensuring the safety and quality of water supplies for various uses.

3.4.8.6 Ultraviolet Process Description

A UV disinfection system is a powerful tool that utilizes electromagnetic energy from a mercury arc lamp to target an organism's genetic material, including DNA and RNA. By penetrating the cell wall of an organism, UV radiation effectively destroys its ability to reproduce. The efficacy of a UV disinfection system depends on several factors, including the characteristics of the wastewater, the intensity of UV radiation, the duration of exposure, and the reactor configuration.

For any given treatment plant, the success of disinfection is directly related to the concentration of colloidal and particulate constituents in the wastewater. These constituents can shield microorganisms from UV radiation, reducing the effectiveness of the disinfection process. The primary components of a UV disinfection system include mercury arc lamps, a reactor, and ballasts. The source of UV radiation is typically a low-pressure or medium-pressure mercury arc lamp with low or high intensities. The optimal wavelength for effectively inactivating microorganisms falls within the range of 250–270 nm (Chen et al., 2023; Yang et al., 2022).

The intensity of the radiation emitted by the lamp decreases as the distance from the lamp increases. Low-pressure lamps emit essentially monochromatic light at a wavelength of 253.7 nm, with standard lengths of 0.75 and 1.5 m and diameters of 1.5–2.0 cm. The ideal lamp wall temperature falls between 95 and 122 °F (Masschelein & Rice, 2016). Medium-pressure lamps are typically used for larger facilities, as they offer approximately 15–20 times the germicidal UV intensity of low-pressure lamps. The medium-pressure lamp disinfects faster and has greater penetration capability due to its higher intensity. However, these lamps operate at higher temperatures and consume more energy.

Two types of UV disinfection reactor configurations include contact type and non-contact type. In both configurations, wastewater can flow either perpendicular or parallel to the lamps. In contact systems, the UV lamps are submerged directly in the water, allowing for direct exposure to the UV light. Non-contact systems, on the other hand, place the UV lamps in a protective sleeve, and the water flows around the sleeve. The configuration choice depends on various factors, including the specific application, water quality, and maintenance requirements.

The effectiveness of UV disinfection systems in removing pathogens and other microorganisms from water is well-documented. This technology is crucial for ensuring safe drinking water, treating wastewater, and supporting various industrial processes where microbial contamination must be controlled or eliminated. By understanding and optimizing the factors influencing UV disinfection, treatment plants can enhance their operational efficiency and ensure higher water quality standards (Bilenko et al., 2019; Koutahzadeh et al., 2016; Modak et al., 2020).

3.4.8.7 Applicability

When selecting a UV disinfection system, it is crucial to consider three key areas to ensure optimal performance and efficiency. These areas include hydraulic properties of the reactor, intensity of the UV radiations, and wastewater characteristics. Each of these factors plays a significant role in determining the effectiveness of the UV disinfection process.

Hydraulic Properties of the Reactor

The hydraulic properties of the reactor are crucial in determining the efficiency of a UV disinfection system. Ideally, the reactor should ensure a uniform flow with sufficient axial motion to maximize exposure to UV radiation. The path an organism takes through the reactor affects the amount of UV radiation it receives before inactivation. Proper reactor design should eliminate short-circuiting and dead zones, which can lead to inefficient power use and reduced contact time with UV radiation. Ensuring that the flow is evenly distributed and that all microorganisms receive adequate exposure to UV light is essential for achieving effective disinfection (Chen et al., 2023).

Intensity of the UV Radiations

The intensity of the UV radiation is another critical factor to consider. Several factors can affect the intensity, including the age of the lamps, lamp fouling, and the configuration and placement of the lamps within the reactor. Over time, UV lamps can lose intensity, and fouling of the lamp surface by substances in the water can further reduce UV output. Therefore, regular maintenance and proper lamp placement are vital to maintain the desired intensity of UV radiation. High-intensity UV lamps ensure that microorganisms receive sufficient radiation to be effectively inactivated (Yang et al., 2022).

Wastewater Characteristics

The characteristics of the wastewater being treated also significantly impact the effectiveness of a UV disinfection system. These characteristics include flow rate, suspended and colloidal solids, initial bacterial density, and other physical and chemical parameters. High concentrations of total suspended solids (TSS) and particle-associated microorganisms can shield bacteria and viruses from UV radiation, reducing the disinfection efficiency. Consequently, it is essential to account for these factors when designing and operating a UV disinfection system. Pretreatment steps to reduce TSS and other interfering substances can enhance the overall effectiveness of the UV system (Bilenko et al., 2019; Modak et al., 2020).

3.4.8.8 Advantages of UV Disinfection

UV disinfection offers numerous potential advantages for water treatment, making it a highly reliable and efficient technology. These advantages include:

1. **Exceptional Effectiveness Against Pathogens**

UV disinfection is particularly effective in inactivating resilient pathogens such as Cryptosporidium and Giardia. These microorganisms are resistant to traditional disinfection methods like chlorination but are highly susceptible to UV radiation, making UV disinfection a crucial technology for ensuring water safety (Yang et al., 2022).

2. **No Significant Alteration of Water Quality**

UV disinfection does not significantly alter the chemical composition of water. It does not affect total organic carbon (TOC), pH, corrosivity, disinfection byproduct (DBP) formation potential, or turbidity. This ensures that the treated water maintains its quality and safety for consumption and other uses (Chen et al., 2023).

3. **Cost-Effective Solution**

Compared to other disinfection options, especially for protozoa, UV disinfection has relatively low capital and operating costs. The simplicity of the technology and the lack of need for chemical additives contribute to its cost-effectiveness (Koutahzadeh et al., 2016).

4. **Ease of Operation**

UV disinfection systems are straightforward to operate. The equipment can be easily adjusted based on variations in water flow, quality, and other operational parameters, ensuring consistent performance without requiring extensive manual intervention (Modak et al., 2020).

5. **Small Footprint**

UV disinfection systems require minimal space, making them suitable for retrofitting into existing water treatment facilities. This small footprint is particularly advantageous for facilities with limited space or for upgrading existing treatment processes (Bilenko et al., 2019).

6. **Chemical-Free Process**

UV disinfection does not require the addition of chemicals, making it an environmentally friendly option. This eliminates the risks associated with chemical handling, storage, and disposal, and prevents the formation of harmful chemical byproducts in the treated water (Das, 2001).

7. **Rapid Disinfection**

The disinfection process with UV radiation is extremely fast, with contact times typically in the range of a few seconds. This rapid action allows for high-throughput treatment and makes UV disinfection suitable for various applications, including emergency responses and high-demand situations (Chen et al., 2023).

3.4.8.9 Disadvantages

While UV disinfection is effective in eliminating harmful microorganisms from water, it has several drawbacks that need to be considered:

1. **Lack of Residual Disinfection Capacity**

UV disinfection does not provide a residual disinfectant in the water distribution system. This means that once water has been treated with UV, there is no continuing disinfection effect. As a result, utilities often need to add chlorine or chloramines post-UV treatment to maintain a disinfectant residual in the distribution system to prevent microbial regrowth and contamination during distribution (Koutahzadeh et al., 2016).

2. **Inability to Continuously Monitor UV Dose**

One significant challenge with UV disinfection is the inability to directly monitor the UV dose in real-time. Operators must rely on secondary measurements such as sensor readings, UV transmittance, and water flow rates to infer the delivered UV dose. This indirect monitoring can sometimes lead to uncertainties in ensuring that the correct dose of UV radiation is consistently applied to inactivate pathogens (Chen et al., 2023).

3. **Presence of Mercury Lamps**

Most UV reactors use mercury lamps, which can pose a hazard if a lamp breaks. Mercury is a toxic substance, and breakage can lead to contamination and require special handling and cleanup procedures. This risk necessitates careful maintenance and monitoring of the UV equipment to prevent accidental breakage and potential environmental contamination (Das, 2001).

4. **Sensitivity to Electrical Power Supply**

UV disinfection systems are dependent on a consistent electrical power supply. Interruptions in power can cause UV lamps to extinguish, typically for 1–5 min, until they can restart. This interruption can lead to periods of under-disinfection, which may allow microorganisms to pass through untreated. Therefore, reliable power supply and backup systems are critical for the continuous operation of UV disinfection units (Yang et al., 2022).

3.5 Advanced Oxidation Processes (AOP's)

Advanced Oxidation Processes (AOPs) represent a cutting-edge technology for water treatment, particularly effective for removing bio-recalcitrant organic pollutants and inactivating pathogenic microorganisms resistant to conventional methods. First introduced for drinking water treatment in the 1980s, AOPs have since been extensively studied and applied to various wastewater treatment scenarios (Bethi et al., 2016; Glaze et al., 1987). AOPs are oxidation processes that generate reactive oxygen species (ROS), such as hydroxyl radicals (HO·), in sufficient quantities to reclaim effluents. These HO· radicals possess high redox potential and non-selectively attack organic compounds through various pathways, including hydrogen abstraction, radical combination, and electron transfer (Deng & Zhao, 2015). This results in the degradation and mineralization of organic pollutants present in water. AOPs are typically applied as tertiary treatment, where HO· radicals are generated in situ through different methods. These include the use of oxidizing agent mixtures (e.g., ozone and hydrogen peroxide), ultrasound (US) or irradiation (UV), and catalysts (Cho et al., 2005; Lai et al., 2013). Titanium dioxide (TiO_2) is a commonly used catalyst, with its particles excited by UV light to generate HO· radicals upon contact with water. However, the recovery and recycling of suspended TiO_2 particles can be challenging and costly, leading to the development of immobilized catalyst systems (Darby et al., 1993; Hassaan & Nemr, 2017). AOPs have found applications in various wastewater treatment processes, targeting both organic pollutants and pathogenic microorganisms. Their non-selective nature makes them effective against a wide range of contaminants, offering a promising solution for achieving water quality standards and environmental sustainability. Despite their efficacy, challenges remain in optimizing AOPs for practical applications, including catalyst recovery and system scalability. Future research focuses on improving catalyst immobilization techniques, exploring novel catalyst materials, and optimizing process parameters to enhance efficiency and reduce costs.

3.5.1 Selection of Appropriate AOPs

In recent years, there has been a noticeable increase in small concentrations of inorganic, organic, and mineral compounds in the aquatic environment, attributed to human activities such as excessive and rapid industrialization, urban encroachment, and improved agricultural operations (Huang et al., 1993; Sivagami et al., 2018). To eliminate organic pollutants from wastewater, one feasible option is the application of Advanced Oxidation Processes (AOPs) or their combinations with other treatments. These methods have been widely recognized for their ability to remove recalcitrant contaminants or to transform contaminants into shorter-chain compounds that can be treated by traditional biological processes (Bethi et al., 2016). However, it is crucial to select the appropriate AOP or combination of AOPs for the remediation of a specific industrial or urban wastewater. Factors such as wastewater characteristics, technical applicability, regulatory requirements, economical aspects, and long-term environmental impacts must be considered (Huang et al., 1993; Sivagami et al., 2018). For instance, AOPs such as photocatalysis and photo-Fenton have been proposed as tertiary treatments for urban effluents due to their ability to detoxify wastewater streams containing persistent contaminants (Bethi et al., 2016).

3.5.2 Industrial Wastewater Treatment Through AOPs

The treatment of industrial wastewater (IWW) effluents is a complex challenge due to the broad array of contaminants present. While activated sludge is more efficient and less expensive for removing high concentrations of organic compounds, there are some circumstances where AOPs may offer some advantages. AOPs typically have a small footprint and can be easily integrated with other treatment processes. They could be used to remove non-biodegradable substances that persist after biological treatment. In fact, some IWWs are toxic or bio-recalcitrant to activated sludge treatment due to the high dissolved organic carbon (DOC) concentration. Recent studies have demonstrated the efficacy of advanced oxidation processes (AOPs) in partially degrading toxic compounds, resulting in effluents with higher biodegradability prior to biological treatment (Waldemer et al., 2007). As a result, there has been a significant increase in research on the combined use of AOPs and biological technologies for treating complex industrial wastewater (IWW) in recent years. This has led to the possibility of reusing IWW as a safe water resource under appropriate sanitary conditions. To this end, novel treatments based on AOPs and their integration with conventional methods are being evaluated for a wide range of IWWs generated by various processing industries (Sivagami et al., 2018). By leveraging the benefits of AOPs, it is possible to partially degrade toxic compounds in IWW, thereby increasing its biodegradability and making it more amenable to biological treatment. This approach has the potential to significantly reduce the environmental impact of industrial activities while also providing a valuable source of water for various applications. As such, continued research in this area is crucial for developing sustainable solutions to the challenges posed by industrial wastewater. At present, Advanced Oxidation Processes (AOPs) are considered as alternative technologies for water treatment due to their ability to eliminate a wide range of contaminants and microorganisms from wastewater. However, these treatments are being investigated to reduce operational costs and ensure their feasibility. The primary AOPS studied for water purification include Tiog, photocatalysis, UV/O_3, UV/H_2O_2, Fenton, and photo-Fenton (Giannakis et al., 2016).

3.5.3 Types of Advanced Oxidation Processes

3.5.3.1 Hydroxyl Radical Based AOP's

Hydroxyl Radical-Based Advanced Oxidation Processes (AOPs) are highly effective in water treatment due to the exceptional reactivity of hydroxyl radicals. With an oxidation potential ranging from 2.8 V (pH 0) to 1.95 V (pH 14) versus SCE (saturated calomel electrode), hydroxyl radicals are the most reactive oxidizing agents in water treatment. These radicals exhibit non-selective behavior and react rapidly with a wide range of species, with rate constants ranging from 10^8 to 10^{10} M^{-1} S^{-1}. Hydroxyl radicals attack organic pollutants through four basic pathways: radical addition, hydrogen abstraction, electron transfer, and radical combination (Scaria & Nidheesh, 2022). These reactions produce carbon-centered radicals (R· or R·-OH), which can be transformed into organic peroxyl radicals (ROO·) when combined with O_2. The resulting radicals further react, leading to the formation of more reactive species such as H_2O_2 and superoxide, ultimately resulting in the chemical degradation and mineralization of organic compounds. Due to their short lifetime, hydroxyl radicals are only produced in situ during application through various methods, including a combination of oxidizing agents (such as H_2O_2 and O_3), irradiation (such as ultraviolet light or ultrasound), and catalysts (such as Fe^{2+}) (Huang et al., 1993; Scaria & Nidheesh, 2022). Hydroxyl Radical-Based AOPs are a highly effective method for water treatment due to the exceptional reactivity of hydroxyl radicals. These radicals exhibit non-selective behavior and rapidly react with a wide range of species, ultimately leading to the chemical degradation.

3.5.3.2 Ozone Based AOP's

Ozone-based Advanced Oxidation Processes (AOPs) are a subset of AOPs that utilize ozone (O_3) as the primary oxidant for the degradation of organic pollutants in water and wastewater treatment. These processes are often employed to treat contaminants that are not easily removed by conventional treatment methods due to their recalcitrant nature.

Ozone is a powerful oxidizing agent that can react with a wide range of organic compounds through direct ozone attack or the generation of highly reactive hydroxyl radicals (·OH) through ozone decomposition in the presence of water. These hydroxyl radicals are particularly effective in breaking down organic molecules into smaller, less harmful byproducts through processes such as oxidation, hydroxylation, and fragmentation.

$$3O_3 + H_2O \rightarrow 2OH \cdot +4O_2 \tag{3.1}$$

One of the main advantages of ozone-based AOPs is their ability to efficiently degrade a variety of organic pollutants, including pharmaceuticals, pesticides, industrial chemicals, and emerging contaminants, without the formation of harmful byproducts such as chlorinated compounds found in some other oxidation processes. Additionally, ozone can be generated on-site using ozone generators, providing a cost-effective and environmentally friendly treatment option.

Several ozone-based AOPs have been developed and studied for water and wastewater treatment applications, including ozonation alone, ozonation combined with hydrogen peroxide (known as peroxone), ozonation combined with ultraviolet (UV) radiation (known as ozonation-UV), and ozonation combined with heterogeneous catalysts.

$$H_2O_2 \rightarrow HO^2 + H + O_2 \tag{3.2}$$

$$HO_2^- + O_3 \rightarrow OH + O_2^- + O_2 \quad (3.3)$$

$$O_3 + H_2O + h\upsilon \rightarrow H_2O_2 + O_2 \quad (3.4)$$

3.5.3.3 UV Based AOP's

Hydroxyl radicals, which are highly reactive and powerful oxidants, can be generated through the interaction of photons with catalysts or oxidants. The most commonly used catalyst for this purpose is titanium dioxide (TiO_2), a semiconductor of the RO type. When TiO_2 particles are excited, they produce positive holes in the valence band (hv + vb) with an oxidative capacity, and negative electrons at the conduction band (e − cb) with a reductive capacity. This process is represented by the following equation:

$$TiO_2 + h\upsilon \rightarrow e - cb + H_2O_2 + hv + vb \quad (3.5)$$

The holes and electrons generated by TiO_2 can further react with OH^-, H_2O, and $O_2{}^-$ at the surface of TiO_2 to form hydroxyl radicals. The reactions are as follows:

$$H\upsilon + vb + OH^- \rightarrow OH^- \quad (3.6)$$

$$H\upsilon + vb + H_2O \rightarrow OH^- + H^+ \quad (3.7)$$

$$e^- + vb + O_2 \rightarrow O_2 \quad (3.8)$$

These hydroxyl radicals are highly reactive and can be used for various applications, such as water treatment and air purification.

In the presence of oxidants such as H_2O_2 or O_3, additional OH· may be generated under UV irradiation. For instance, a H_2O_2 molecule can be cleaved by UV irradiation to produce two OH·, as shown in Eq. (3.10):

$$H_2O_2 + hv \rightarrow 2OH\cdot \quad (3.9)$$

Moreover, OH· can also be produced through photolysis of H_2O at a wavelength less than 242 nm, as represented in Eq. (3.11):

$$H_2O + hv \rightarrow OH \cdot + H^+ \quad (3.10)$$

3.5.3.4 Fenton Related AOP's

Iron has the ability to activate hydrogen peroxide (H_2O_2) and produce hydroxyl radicals in water, a property that has made it the most commonly used metal in the Fenton process. In this process, H_2O_2 reacts with Fe^{2+} to generate highly reactive species, traditionally recognized as hydroxyl radicals (Mohapatra et al., 2021).

The classical Fenton radical mechanisms primarily involve a series of reactions. These reactions are of great interest to researchers due to their potential applications in various fields. However, it is important to note that the Fenton process can also produce harmful byproducts, such as hydrogen peroxide and iron ions, which can have negative effects on the environment. While the process has potential benefits, it is important to consider the potential negative effects and explore alternative methods to minimize environmental impacts.

$$\mathrm{Fe_2 + H_2O_2 \rightarrow Fe_3 + OH + H^-} \tag{3.11}$$

$$\mathrm{Fe_2 + H_2O_2 \rightarrow Fe_2 + HO_2 + H^+} \tag{3.12}$$

$$\mathrm{OH + H_2O_2 \rightarrow HO_2 + H_2O} \tag{3.13}$$

$$\mathrm{OH + Fe_2 \rightarrow Fe_3 + OH^-} \tag{3.14}$$

$$\mathrm{Fe_3 + HO_2 \rightarrow Fe_2 + O_2H^+} \tag{3.15}$$

$$\mathrm{Fe_2 + HO_2 + H \rightarrow Fe_3 + H_2O_2} \tag{3.16}$$

$$\mathrm{2HO_2 \rightarrow H_2O_2 + O_2} \tag{3.17}$$

Hydroxyl radicals (OH·) are generated through electron transfer in accordance with Eq. (3.12). However, the scavenging of OH· can occur through either of the Fenton reagents as demonstrated in Eqs. (3.13) and (3.14). Therefore it is crucial to experimentally determine the optimal molar ratio of iron ion to hydrogen peroxide to minimize unwanted scavenging. While Eq. (3.13) suggests that the produced Fe^{3+} from Eq. (3.12) can be reduced to Fe^{2+}, the iron cannot act as a catalyst in the Fenton system due to the rate constant in Eq. (3.13) being several orders of magnitude less than that in Eq. (3.12). As a result, Fe^{3+} forms iron sludge under typical water and wastewater treatment conditions, which must be separately disposed of increasing treatment complexity and operational costs.

It is important to note that the generation of hydroxyl radicals during the Fenton reaction is most effective under acidic pH conditions, which limits the practical application of the Fenton reaction for wastewater treatment. To address this limitation, three modified Fenton processes have been proposed: the Fenton-like system, photo-Fenton system, and electro-Fenton system. In the Fenton-like reaction, ferric ion (Fe^{3+}) replaces Fe^{2+}, and the series of reactions in the Fenton system are initiated from Eq. (3.13), rather than Eq. (3.12) in traditional Fenton treatment. The photo-Fenton reaction involves the application of overline UV irradiation with the traditional Fenton system to enhance the UV-induced reduction of dissolved Fe^{3+} to Fe^{2+}. In the electro-Fenton reaction, either or both of the Fenton reagents may be generated through electrochemical methods.

The Fenton reaction is a powerful tool for wastewater treatment, but its limitations must be considered. The modified Fenton processes offer promising solutions to overcome these limitations and improve the efficiency and effectiveness of wastewater treatment.

3.6 Membrane Filtration

Membrane filtration is a widely used physicochemical process in water treatment, alongside rapid granular filtration. The primary objective of membrane filtration is to remove microorganisms and other particles from water, but the processes differ significantly in their physicochemical mechanisms. Unlike rapid granular filtration, which employs a thick bed of granular material, membrane filtration utilizes a thin synthetic material with tiny pores through which water can pass. During filtration, water passes through the pores, while particles are physically retained at the surface of the material due to their size. The difference in filtration mechanism between straining and depth filtration is a crucial distinction between membrane and rapid granular filtration. Despite being a relatively new concept, membrane filtration has rapidly become a mainstream technology that competes effectively with rapid granular filtration in new water treatment plant designs. However, the rapid evolution of membrane filtration design concepts presents unique challenges for design engineers. Relying solely on previous projects or tried-and-true rules of thumb may fail to capitalize on recent technological advancements. Conversely, undue reliance on manufacturers' claims about unproven technologies may lead to failure. Therefore, it is critical for design engineers to stay up-to-date with technical advancements and provide appropriate guidance to facility owners. To evaluate new technologies effectively, design engineers must have a fundamental understanding of membrane materials, modules, fouling, and performance. This knowledge enables them to capitalize on valuable technological advancements while avoiding unproven alternatives that have an unreasonable chance of failure. Membrane filtration is a highly effective water treatment process that has gained popularity in recent years. It involves the use of a semi-permeable membrane to separate impurities from water. This process is particularly useful in removing contaminants such as bacteria, viruses, and other harmful particles that may be present in water. When compared to rapid granular filtration, membrane filtration offers several advantages. For instance, it is more efficient in removing smaller particles and requires less space. Additionally, it is more flexible and can be used in a wider range of applications. The equipment used in membrane filtration is relatively simple and consists of a membrane module, a feed pump, and a control system. The membrane module is the heart of the system and is responsible for separating impurities from water. Particle capture is a critical aspect of membrane filtration, and it involves the removal of particles from the membrane surface. This process is essential in maintaining the efficiency of the system and preventing membrane fouling. Flow hydraulics is another crucial consideration in membrane filtration, involving the optimization of flow rates to ensure that the system operates at maximum efficiency. Membrane fouling is a common problem in membrane filtration and can significantly reduce the efficiency of the system. It occurs when particles accumulate on the membrane surface, leading to reduced water flow and increased energy consumption. Sizing of membrane systems is an essential consideration in ensuring that the system is optimized for its intended application.

3.6.1 Classification of Membrane Processes

Membranes are used in two distinct physicochemical processes in water treatment:

(1) Membrane filtration and
(2) Reverse osmosis.

Membrane filtration is a process that utilizes pressure or vacuum to separate particles from a suspension, which is a two-phase system consisting of particles in a fluid. This separation is achieved by straining the fluid as it passes through a porous material. Reverse osmosis, on the other hand, is a pressure-driven membrane separation process that separates dissolved constituents from a solution. This solution is a single-phase system consisting of solutes in a solvent. Preferential diffusion occurs as the solvent and solute molecules pass through a permeable material, which may or may not be porous. The constituents targeted for removal in reverse osmosis are truly dissolved solutes, such as ions and molecules like sodium, chloride, calcium, magnesium, dissolved NOM, and synthetic organic chemicals.

Reverse osmosis membranes are widely used to produce potable water from ocean or brackish water, to soften hard waters by removing calcium and magnesium ions, to reduce the concentration of NOM to control disinfection by-product (DBP) formation, and to remove specific dissolved contaminants like pesticides, pharmaceuticals, arsenic, nitrate, and radionuclides. Membrane filtration systems are marketed as containing either microfiltration (MF) or ultrafiltration (UF) membranes, with UF membranes having a smaller pore size than MF membranes. Reverse osmosis membranes are marketed under various names, including seawater RO, brackish water RO, low-pressure RO, and so on. A particular class of low-pressure RO products is sold as nanofiltration (NF) membranes. Although the distinction between membrane products is somewhat arbitrary, they are loosely identified by the types of materials rejected, operating pressures, and nominal pore dimensions. A "loose" NF membrane marketed by one manufacturer might be substantially similar to a "tight" UF membrane marketed by another manufacturer.

The disparities between membrane filtration and reverse osmosis are significant. Membrane filtration primarily relies on straining or size exclusion as the primary removal mechanism, which theoretically allows for perfect exclusion of particles, regardless of operational parameters such as influent concentration and pressure. On the other hand, reverse osmosis involves a diffusive mechanism for mass transfer, making separation efficiency dependent on influent solute concentration, pressure, and water flux rate. The differences between these two processes are evident in the materials used for the membranes, the configuration of the membrane elements, the equipment used, the flow regimes, and the operating modes and procedures. It is important to note that the membranes are utilized for various purposes in a wide range of fields and industries, and the distinction between membrane products as used in water treatment may not be applicable in other industries.

Understanding the differences between membrane filtration and reverse osmosis is crucial for selecting the appropriate treatment method for a given application. While both processes have their unique advantages and disadvantages, it is essential to consider the specific requirements of the application to determine the most suitable option.

3.6.2 Process Description of Membrane Filtration

Membrane filters operate over a cycle consisting of two stages, just like granular filters:

(i) A filtration stage, during which particles accumulate, and
(ii) A backwash stage, during which the accumulated material is flushed from the system.

As solid particles accumulate against the filter medium, the trans-membrane pressure required to maintain a constant flux increases. Once a preset time interval or maximum pressure is reached, the system undergoes a backwash process. However, fouling can occur due to slow adsorption or clogging of material that cannot be removed during backwash, resulting in a loss of performance. This issue can significantly impact the cost-effectiveness of membrane filtration. To minimize fouling, various operational strategies can be employed, including pretreatment, chemically enhanced backwash (CEB), chemical wash (CW) operations, and clean-in-place (CIP) operations. Additionally, the filtration process includes an integrity testing procedure to validate the reliability of the filtration barrier (Gul et al., 2021). The flow regime of the feed water near the membrane surface can significantly impact permeate flux and fouling. It is crucial to understand the impact of fouling on membrane filtration and to implement effective operational strategies to minimize it. By doing so, we can ensure the reliability and cost-effectiveness of the filtration process. Two filtration strategies have been developed to influence this flow regime: cross-flow filtration and dead-end filtration.

3.6.2.1 Cross Flow Filtration

Cross-flow filtration is a highly effective mode of filtration that involves the continuous flow of feed water through the lumen of inside-out membrane fibers or channels of tubular membranes, parallel to the membrane surface (Peters, 2023). This process is accompanied by a retentate stream that is recycled back into the feed water. The cross-flow velocity, which typically ranges from 0.5 to 1 m/s, is four to five orders of magnitude greater than the superficial velocity of water towards the membrane surface. This high velocity creates a shear force that reduces the development of a surface cake, which is a common issue in other filtration methods (Bera et al., 2022).

One of the key advantages of cross-flow filtration is that many solids are carried away with the retentate instead of accumulating on the membrane surface. This allows the system to be operated at a higher flux or with longer intervals between backwashes (Gul et al., 2021). However, it is important to note that cross-flow filtration requires a substantial recirculation of retentate, as the permeate flow is typically only 15–20% of the feed flow (Peters, 2023).

3.6.2.2 Dead End Filtration

Dead-end filtration is a mode of filtration in which all feed water passes through the membrane without any recirculated retentate stream (Bera et al., 2022). During the filtration cycle, the membrane operates without a defined continuous cross-flow velocity, resulting in the accumulation of all solids on the membrane (Peters, 2023). This accumulation of solids during the filter run may lead to lower average flux values than those achieved with cross-flow filtration (Wang et al., 2022).

In water treatment, the dead-end flow regime is the most common mode of membrane filtration because many industrial feed streams have high solids concentrations, making cross-flow operation critical for achieving reasonable flux and filter run length (Gul et al., 2021). However, surface waters are typically fairly dilute, with many membrane plants operating with feed water turbidity of 100 NTU or less, which corresponds to a solids concentration of about 0.01% (Peters, 2023). As a result, the advantages of cross-flow filtration are less significant in water treatment.

Furthermore, the piping and pumping costs of recirculating a large fraction of the feed water become prohibitive as the facility size gets larger, and water treatment facilities are built with considerably higher capacity than most industrial applications (Thakur et al., 2020). Some cross-flow systems are designed to operate in a dead-end mode by closing a valve in the retentate line when raw-water quality conditions permit (turbidity is low) and switch to a cross-flow mode only when necessary to maintain flux. It is a common mode of membrane filtration in water treatment due to the dilute nature of surface waters and the prohibitive costs of recirculating a large fraction of the feed water. While cross-flow filtration is critical for achieving reasonable flux and filter run length in many industrial applications, it is less significant in water treatment.

3.7 Reverse Osmosis as a Membrane Treatment Processes

Reverse osmosis (RO) is a highly effective membrane treatment process that is utilized to separate dissolved solutes from water (Singh & Mishra, 2020). The membrane itself is a semipermeable material that allows certain components in the feed stream to pass through while blocking others. The feed water is pressurized and some water, known as permeate, passes through the membrane. As the water passes through the membrane, solutes are rejected, and the feed stream becomes more concentrated (Vaseghi et al., 2016). The permeate is relatively free of targeted dissolved solutes and exits at nearly atmospheric pressure, while the remaining water, called concentrate, exits at the far end of the pressure vessel at nearly the feed pressure.

It is important to note that reverse osmosis is a continuous separation process, meaning that there is no periodic backwash cycle (Gul et al., 2021). Other membrane processes used in municipal water treatment include microfiltration (MF), ultrafiltration (UF), and nanofiltration (NF) (Peters, 2023).

Reverse osmosis is a versatile process that can be used for a variety of applications, including desalinating seawater and brackish groundwater, softening water, removing natural organic matter for disinfection by-product control, advanced treatment for potable water reuse, and removing specific contaminants (Thakur et al., 2020). Its effectiveness and efficiency make it a popular choice for water treatment facilities around the world.

3.7.1 Osmotic Pressure and Reverse Osmosis

Understanding the concept of osmotic pressure is crucial to grasp the concept of reverse osmosis (Singh & Mishra, 2020). Osmosis refers to the movement of solvent through a semipermeable membrane, from a less concentrated solution to a more concentrated one (Gottschalk et al., 2009). This process reduces the flux through a reverse osmosis membrane by creating a driving force for flow in the opposite direction (Ikehata & Li, 2018). The physicochemical basis for osmosis is deeply rooted in the thermodynamics of diffusion (Vaseghi et al., 2016). Therefore, a thorough understanding of osmotic pressure is fundamental to comprehend the principles of reverse osmosis.

3.7.2 Diffusion and Osmosis

Consider a vessel that contains two solutions, separated by a removable partition, as depicted in Fig. 3.1a (Jiao et al., 2021). The left side of the vessel is filled with a concentrated salt solution, while the right side contains pure water. The partition is gently removed without disturbing the solutions, resulting in a non-equilibrium

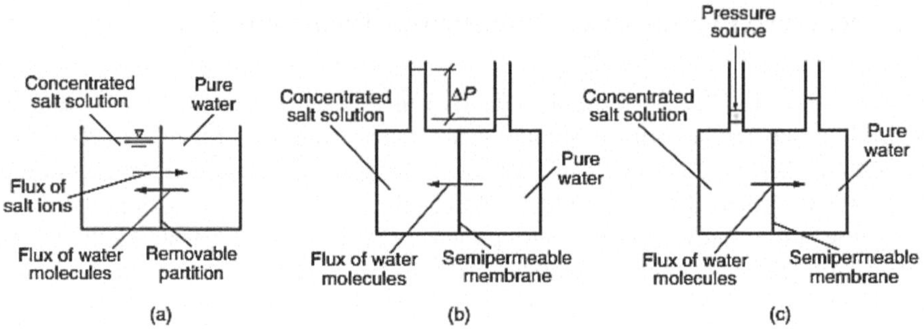

Fig. 3.1 Diffusion sketch for reverse osmosis: **a** diffusion, **b** osmosis, and **c** reverse osmosis

state. Over time, the salt ions will diffuse through the water until the concentration is uniform throughout the vessel. This diffusion process requires a flux of water molecules in the opposite direction to maintain mass conservation. Without this flux, mass would accumulate on the right side of the container, which is not feasible with a continuous water surface. Therefore, equilibrium requires mass transfer in both directions (Fig. 3.1).

In Fig. 3.1b, the top of the vessel has been closed and fitted with manometer tubes, and the removable partition has been replaced with a semipermeable membrane. The semipermeable membrane allows the flow of water but prevents the flow of salt. Once again, the chambers are filled with salt solution and pure water, creating a thermodynamically unstable system that must be equilibrated by diffusion. However, the membrane prevents the flux of salt, causing mass to accumulate in the left chamber. As a result, the water level in the left manometer rises, while the right manometer's water level drops. This flow of water from the pure side to the salt solution is known as osmosis. Despite the difference in hydrostatic pressure that develops due to the difference in manometer levels, water flux occurs.

3.7.3 *Osmotic Pressure*

The driving force behind diffusion is commonly attributed to a concentration gradient, although a more rigorous thermodynamic explanation involves a gradient in Gibbs energy (G) (Crittenden et al., 2012). Thermodynamically speaking, equilibrium is achieved when $G = 0$. When a vessel reaches thermodynamic equilibrium, water ceases to flow from right to left, even though both pressure and concentration are unequal between the chambers. At equilibrium, Gibbs energy remains constant throughout the second vessel but includes components that account for both pressure and concentration differences. This ensures that the concentration gradient in one direction is precisely balanced by the pressure gradient in the opposite direction. The pressure required to balance the difference in concentration of a solute is known as osmotic pressure and is represented by the symbol π. When the vessel in the second experiment reaches equilibrium, the difference in hydrostatic pressure between the manometers is equal and opposite of the difference in osmotic pressure between the two chambers.

An equation for osmotic pressure can be derived thermodynamically using assumptions of incompressible and ideal solution behavior (Salmani et al., 2022). In dilute solution, osmotic pressure can be approximated by the van't Hoff equation. This equation provides a useful tool for predicting osmotic pressure in a variety of applications.

Understanding the thermodynamics of diffusion and osmotic pressure is crucial for a wide range of scientific and engineering applications. By delving deeper into the underlying principles of these phenomena, we can gain a more comprehensive understanding of the world around us.

$$\pi = \frac{\mathrm{nS}}{\mathrm{V}} RT \quad or \quad \pi = CRT$$

where

- π osmotic pressure (bar)
- nS total amount of all solutes in solution (mol)
- V volume of solution (L)
- R universal gas constant, 0.083145 L bar/mol K
- T absolute temperature (K, 273 + °C)
- C concentration of all solutes (mol/L)

3.7.4 Reverse Osmosis

The rate of water flow through a semipermeable membrane is influenced by two opposing forces: the concentration gradient and the pressure gradient (Arens et al., 2017). These forces are harnessed in reverse osmosis (RO) technology. To illustrate this concept, let us consider a new experiment using the apparatus depicted in Fig. 3.1, which has been modified to allow for the application of an external force on the left side, as shown in Fig. 3.1c. When an external force equivalent to the osmotic pressure is applied, the system reaches thermodynamic equilibrium, and no water flows. However, if a force in excess of the osmotic pressure is applied, the system becomes non-equilibrium, with a pressure gradient that exceeds the concentration gradient. This results in water flowing from the concentrated solution to the dilute solution.

This process of causing water to flow from a concentrated solution to a dilute solution across a semipermeable membrane by the application of an external pressure in excess of the osmotic pressure is known as reverse osmosis. It is a highly effective method for purifying water and removing impurities. The opposing forces of the concentration gradient and the pressure gradient play a crucial role in the process of reverse osmosis.

3.8 Ion Exchange Process

In 1850, Thomas and Way conducted pioneering scientific research that revealed the existence of an ion exchange process involving the passage of a solution of ammonium sulfate through soil, resulting in the collection of a filtrate composed of calcium sulfate instead of ammonium sulfate (Bochenek et al., 2011). Although the significance of this discovery in ion exchange terms was not fully comprehended at the time, it was later found that this reaction was reversible (Skipton et al., 2008).

Ion exchange processes can remove various charged atoms or molecules (ions) such as nitrates, fluoride, sulphates, perchlorate, iron and manganese ions, as well as toxic metals (radium, uranium, chromium, etc.) from water. The most common application of ion exchange is the preparation of high purity water for industrial applications, water softening, recovery or removal of metals in the chemical industry.

Ion exchange is a widely used water treatment process that is commonly employed for water softening or demineralization. However, it is also used to remove other substances from water in processes such as dealkalization, deionization, denitrification, and disinfection. With many other overlapping technologies available, it is crucial to determine whether ion exchange is the best choice in a given scenario (Applebaum, 2013).

Ion exchange is a highly effective chemical process utilized in the treatment of water and wastewater to eliminate unwanted dissolved ions. This process involves the exchange of ions with similar charges to remove the undesired ions (cations and anions).

During the water treatment process, positively charged ions are exchanged with other positively charged ions, typically sodium, on the resin surface. This exchange occurs when the cations come into contact with the ion exchange resin. On the other hand, during anion exchange, negatively charged ions are exchanged with other negative ions, usually chloride ions, on the resin surface. This process is crucial in removing contaminants such as nitrate, arsenic, sulfate, and fluoride from the water. It involves the exchange of ions with similar charges to eliminate unwanted dissolved ions. This process is highly effective in removing contaminants and ensuring that water is safe for consumption.

The process involves the exchange of ions between a solid material, such as resin or zeolite, and the liquid water. During this exchange, less desirable compounds are replaced with more desirable ones, which are then loaded onto the resin material. These resins can be used alone or in combination to effectively remove ionic contaminants from water.

Compared to other water treatment technologies, such as continuous electrodeionization, chromatography, ultrafiltration, and biological treatments, ion exchange is particularly effective in removing low concentration pollutants, such as boron, from well water. Overall, it is a reliable and efficient method for removing ionic contaminants from water, and its ability to target specific pollutants makes it a valuable tool in water treatment processes.

3.8.1 Principles of Ion Exchange Process

Ion exchange is a highly effective process that involves the exchange of mobile ions from an external solution with ions that are electrostatically bound to the functional groups within a solid matrix. This process is particularly useful when the functional groups are negatively charged, as it allows for the exchange of cations, or when they are positively charged, which facilitates the exchange of anions (Neumann & Fatula, 2009).

Overall, ion exchange is a powerful tool that can be used to separate and purify a wide range of ionic species. Its versatility and effectiveness make it an essential technique in many fields, including chemistry, biology, and environmental science.

Depending on the type of functional group, ion exchangers can be divided into several types:

- Strong Acid Cation (SAC) Resins
- Strong Base Anion (SBA) Resins
- Weak Acid Cation (WAC) Resins
- Weak Base Anion (WBA) Resins.

SAC resins are highly effective in neutralizing strong bases and converting salts into corresponding acids. These resins are capable of removing all cations and replacing them with hydrogen ions, making them a versatile solution for a wide range of pH levels. On the other hand, SBA resins are designed to neutralize strong acids and convert salts into corresponding bases. These resins are commonly used in water softening and demineralization processes, making them an essential component in many industrial applications.

WAC and WBA resins are both capable of neutralizing strong bases and acids, making them ideal for partial and complete demineralization processes and dealkalization systems. While WAC resins are effective in removing cations to produce carbonic acid, they may not be able to remove all cations in most water supplies. However, their high regeneration efficiency compared to SAC resins makes them a popular choice. WBA resins, on the other hand, are highly effective in removing acids such as sulfuric, nitric, and hydrochloric acid. These resins play a crucial role in many industrial processes, ensuring that the water used is free from harmful contaminants.

The use of resins in water treatment processes is essential for ensuring the purity and safety of water. By understanding the unique properties and capabilities of different types of resins, industries can make informed decisions about which resins to use for their specific needs.

3.9 Disinfection

Disinfection is a crucial component of the comprehensive approach to ensuring that drinking water is safe for consumption. The provision of water that is free from pathogenic organisms is achieved through a combination of strategies that work in tandem. These strategies include selecting a water source that is free from microbiological contamination, such as groundwater, protecting surface water sources to minimize microbiological contamination, treating water to remove microorganisms or eliminating their pathogenicity, and preventing recontamination of water as it is delivered to customers through the distribution system. It is imperative to note that the selection of a water source that is free from microbiological contamination is the first line of defense in ensuring that the water is safe for consumption. This is because the presence of pathogenic organisms in the water source can lead to the spread of waterborne diseases. Protecting surface water sources from contamination is also crucial in minimizing the risk of microbiological contamination. This can be achieved through the implementation of measures such as buffer zones, fencing, and the use of appropriate agricultural practices.

Treating water to remove microorganisms or eliminate their pathogenicity is another critical strategy in ensuring that the water is safe for consumption. This can be achieved through the use of various treatment methods such as chlorination, ozonation, and ultraviolet radiation. However, it is essential to note that the effectiveness of these treatment methods is dependent on the quality of the water source and the level of contamination.

Finally, preventing recontamination of water as it is delivered to customers through the distribution system is crucial in ensuring that the water remains safe for consumption. This can be achieved through the implementation of measures such as regular cleaning and maintenance of the distribution system, the use of appropriate storage facilities, and the implementation of appropriate hygiene practices.

Disinfection is a crucial component of the final stages of water treatment. The treatment process involves the removal of microorganisms, primarily through filtration, and their subsequent inactivation. Inactivation is the process by which microorganisms are transformed to render them incapable of causing disease. This can include eliminating their ability to reproduce in a host organism, rendering them unable to cause harm, even though they may not be dead. This process, also known as primary disinfection, typically occurs at a treatment facility.

Residual maintenance, or secondary disinfection, is another critical aspect of the disinfection process. This occurs in the distribution system and ensures that the water remains free from harmful microorganisms as it travels to its final destination.

3.9.1 Disinfection Agents and Systems

Five disinfection agents commonly used in drinking water treatment are:

1. Free chlorine
2. Combined Chlorine (chlorine combined with ammonia, also called chloramines)
3. Chlorine dioxide
4. Ozone
5. Ultraviolet (UV) light.

The first four are chemical oxidants, whereas UV light involves the use of electromagnetic radiation.

In 1978, the majority of utilities, approximately 91%, utilized chlorine gas to apply free chlorine to the water, while only 7% used sodium hypochlorite, commonly known as bleach. However, by 2007, the use of chlorine gas decreased to 63%, and nearly 40% of utilities transitioned to either bulk liquid or onsite generation of sodium hypochlorite, a shift attributed to safety and security concerns in the use of chlorine gas (AWWA Disinfection Systems Committee, 2008).

The use of chloramines, which is a combination of free chlorine and ammonia for disinfection, increased to 30% by 2007. However, its use is typically limited to residual maintenance, and a different disinfectant is often used for primary disinfection when chloramine is utilized.

Ozone, the strongest of the four oxidants, has seen an increase in usage from less than 1% of utilities in 1989 to 9% in 2007, an increase attributable to its stronger disinfecting properties and its ability to control taste and odor compounds, specifically geosmin and methyl isoborneol (AWWA Disinfection Systems Committee, 2008).

Although ultraviolet light is not frequently used for disinfecting in drinking water applications, only 2% of utilities reported using it in 2007. Its usage is expected to increase in the future due to its lack of by-product generation and its effectiveness against protozoa.

3.9.2 Disinfectant System Design

When designing a disinfection system, three primary activities must be undertaken.

(i) Selection of a suitable disinfectant and determination of appropriate dose: The selection of a suitable disinfectant and dose is a critical step in the design process. Factors such as the type of microorganisms present in the water, the water quality, and the desired level of disinfection must be considered. Once a disinfectant has been chosen, the appropriate dose must be determined to ensure effective disinfection without causing harm to the environment or human health.
(ii) Effective injection or introduction of the disinfectant into the water: The system must ensure that the disinfectant is evenly distributed throughout the water and

that the dose is accurately measured and controlled. Factors such as the flow rate of the water, the type of disinfectant, and the injection method must be carefully considered during the design process.

(iii) Designing of contactors to provide sufficient time for the disinfection reaction: This involves designing a system that allows for adequate contact time between the disinfectant and the microorganisms in the water. The design of the contactors will depend on factors such as the type of disinfectant, the water quality, and the desired level of disinfection.

By selecting a suitable disinfectant and dose, designing an effective injection system, and designing contactors that provide sufficient contact time, it is possible to create a system that effectively disinfects water while minimizing harm to the environment and human health.

3.9.3 Disinfectant Dose

The appropriate dosage of disinfectant depends on its intended use, whether for inactivation, residual maintenance, or both. When chemical disinfectants are added to water, a portion of the chemical is consumed during the rapid oxidation of reduced compounds present in the water, known as the initial demand. Once the initial demand is met, any additional chemical added contributes to the residual concentration in the water.

For residual maintenance, the necessary dosage is determined by the requirement to maintain a specific residual concentration in the far-reaching areas of the distribution system. It is essential to consider that disinfectant concentrations naturally decay over time. Therefore, the dosage required for residual maintenance depends on various conditions within the distribution system, such as temperature, residence time, presence of biofilms, and other water quality parameters, which can impact the rate of decay.

It is crucial to ensure that the appropriate dosage of disinfectant is used to effectively maintain water quality and safety. By understanding the factors that influence disinfectant dosage, water treatment professionals can make informed decisions to optimize disinfection processes and safeguard public health.

3.10 Iron and Manganese Removal

Iron and manganese are naturally occurring metallic elements commonly found in groundwater and some surface waters. They are mainly present in their soluble reduced divalent forms as ferrous (Fe^{2+}) and manganous (Mn^{2+}) ions. These elements can dissolve from minerals in the Earth's crust as water percolates through soil and rocks.

In deep lakes and reservoirs, dissolved iron and manganese may be present due to stratification, which creates anaerobic conditions in the bottom water zone, leading to the dissolution of iron and manganese from floor deposits. During the annual overturn, these dissolved species are dispersed into the general water body.

While iron and manganese are categorized as indicator parameters with no critical health significance, their presence in water can cause staining and impart offensive tastes and appearances. When exposed to air or oxygen, waters containing iron and manganese become cloudy and turbid due to oxidation to the Fe^{3+} and Mn^{4+} states, forming colloidal precipitates. Although the rates of oxidation are not rapid, reduced forms can persist for some time in aerated waters.

The presence of iron and manganese in water presents a challenge for water treatment engineers who must develop effective removal methods. Understanding the behavior and characteristics of these elements in water is crucial for devising reliable treatment strategies.

3.11 Phosphate Treatment

Phosphate serves as a valuable tool in managing elevated iron concentrations in water distribution systems. When the water contains less than 0.3 ppm of iron or 0.1 ppm of manganese, adding phosphate at the source can effectively delay the precipitation of oxidized manganese and iron. This delay significantly reduces the formation of scales on pipes, a process known as sequestration.

Sequestration occurs when the iron or manganese ion is encapsulated by a chain of phosphate molecules, preventing its precipitation in the water. Pyrophosphate, tripolyphosphate, and metaphosphate are all effective sequestering agents for iron and manganese. However, sodium phosphate in low concentrations has proven to be the most effective.

It is essential to determine the proper dose and type of phosphate through bench-scale testing conducted by a qualified technician or consultant. This ensures that the correct amount of phosphate is added to the water to achieve optimal results without any adverse effects.

By utilizing phosphate as a sequestering agent, water distribution systems can maintain their efficiency and longevity while delivering safe and clean water to consumers.

3.12 Removal by Iron and Manganese Filtration

When iron and manganese concentrations exceed certain levels in drinking water, it is advisable to remove them rather than sequestering them, especially if the concentrations surpass 0.3 ppm of iron or 0.05 ppm of manganese (Liu et al., 2016). Softening with lime can effectively eliminate these elements. However, the most common method for removing iron and manganese involves filtration after oxidation, achieved through processes such as air oxidation, potassium permanganate oxidation, or chlorine oxidation. This filtration typically utilizes gravity and pressure filters, with pressure filtration being the more prevalent option.

3.13 Oxidation with Aeration

Aeration plays a crucial role in converting soluble forms of iron and manganese into insoluble forms through oxidation. According to Khadse et al. (2015), specific dissolved oxygen concentrations are required to oxidize iron and manganese effectively. It takes approximately 0.14 ppm of dissolved oxygen to oxidize 1 ppm of iron and 0.27 ppm of dissolved oxygen to oxidize 1 ppm of manganese. However, maintaining optimal conditions during aeration is essential for successful oxidation.

Control of water flow is critical during the aeration process. If the flow rate is too high, insufficient air may be applied to oxidize the iron and manganese effectively. Conversely, if the flow rate is too low, the water may become oversaturated with dissolved oxygen, leading to corrosive effects on the distribution system. Corrosive water can result in elevated levels of lead, posing serious health risks.

During aeration, slime growths may develop on the equipment, potentially causing taste and odor issues in the water if left unchecked. To prevent this, it is vital to control slime growth by adding chlorine at the treatment plant's head. Regular inspections of the process are also necessary to detect and address any issues promptly. Maintaining proper aeration conditions and controlling slime growth are essential for ensuring the effectiveness and safety of the water treatment process.

3.14 Oxidation with Chlorination

The oxidation of iron and manganese using chlorine is indeed a common practice in water treatment. When chlorine is introduced into water containing iron and manganese, it reacts with these metals to form their respective oxidized forms, ferric hydroxide ($Fe(OH)_3$) and manganese dioxide (MnO_2). These oxidized compounds are insoluble and can be effectively removed through filtration processes.

The rate of this oxidation reaction is influenced by several factors, including the concentration of chlorine and the concentrations of iron and manganese in the water. As the treatment plants often aim for a specific ratio of chlorine to iron to optimize the oxidation process, this ratio ensures that there is enough chlorine present to react with the iron and manganese effectively without causing excess residual chlorine in the treated water.

By controlling the chlorine dosage and maintaining the appropriate chlorine-to-iron ratio, water treatment plants can achieve efficient oxidation of iron and manganese, leading to their precipitation and subsequent removal through filtration. This process plays a crucial role in improving water quality and ensuring that treated water meets regulatory standards for safe consumption.

3.15 Oxidation with Permanganate

The utilization of potassium permanganate for the oxidation of iron or manganese is a widely adopted practice in water treatment (Li et al., 2021). This chemical agent effectively transforms iron and manganese into their insoluble states. However, it is crucial to administer the appropriate dosage to ensure complete oxidation of manganese without causing a pinkish hue in the water distribution system. Regular monitoring of the water treatment process is necessary to determine if adjustments to the chemical feeders are necessary. It is noteworthy that potassium permanganate is more efficient in oxidizing manganese compared to aeration or chlorination.

3.16 Defluoridation

Defluoridation is the process of removing excess fluoride from drinking water, which is crucial for preventing dental and skeletal fluorosis, conditions caused by prolonged fluoride exposure. Fluoride, while beneficial at low concentrations for dental health, can become toxic at elevated levels, particularly in areas where natural fluoride concentrations exceed 1.5 mg/L.

Various methods for defluoridation have been developed, including adsorption, ion exchange, and precipitation. Adsorption is one of the most common techniques, utilizing materials like activated alumina, bone char, and modified clay to capture fluoride ions from water (Gupta et al., 2016; Sharma et al., 2017). Ion exchange methods employ resins that selectively remove fluoride ions while releasing less harmful ions, such as hydroxide (Satyavani et al., 2016).

Precipitation methods, including the use of calcium-based compounds, can also effectively reduce fluoride levels by forming insoluble calcium fluoride (Calderon et al., 2020). In addition to these methods, biological approaches involving microbial reduction of fluoride are being explored (Vijayakumar et al., 2021).

The choice of defluoridation method depends on factors such as initial fluoride concentration, water chemistry, cost, and the required final fluoride levels. Successful implementation of defluoridation strategies can significantly improve public health outcomes, particularly in regions suffering from high natural fluoride levels (Bansal et al., 2015; Namasivayam & Sangeetha, 2007).

3.17 Boiling

To ensure safe consumption of water, it is crucial to address potential contaminants, such as harmful parasites and germs, which may be present in water sources and distribution channels. Boiling water is a recommended method of purification, as it effectively eliminates pathogens. Researchers suggest bringing clean water to a rolling boil and maintaining it for 1–3 min. However, individuals in high altitude areas should extend boiling time due to lower boiling temperatures, ensuring thorough pathogen elimination.

After boiling, covering the water and allowing it to cool before consumption is essential. For well water, settling before filtration is advisable. Following these steps ensures water is free from harmful contaminants, reducing the risk of waterborne diseases. Boiling water is not only simple and cost-effective but also readily available, making it a feasible option for households and communities.

3.18 Filtration

Filtration stands out as a highly effective method of water purification, especially when employing multimedia filters. This approach combines chemical and physical processes to eliminate harmful compounds, ensuring water safety for human consumption (Mandoreba et al., 2021).

One of the notable advantages of filtration is its ability to preserve essential mineral salts present in water, rendering filtered water healthier compared to other purification methods. Through a chemical absorption process, filtration effectively removes unwanted compounds, offering a robust solution to water purification. Unlike reverse osmosis, filtration excels at selectively eliminating smaller molecular compounds like chlorine and pesticides. Moreover, filtration is a cost-effective option, requiring less energy than distillation or reverse osmosis, thus minimizing water loss during the purification process.

3.19 Distillation

Distillation emerges as a highly effective water purification method leveraging heat to collect pure water in vapor form. This scientific process capitalizes on water's lower boiling point compared to other contaminants and disease-causing elements. By subjecting water to heat until it reaches its boiling point, it vaporizes and is then directed into a condenser to cool, yielding clean and safe drinking water. Distillation excels in removing various impurities such as bacteria, germs, salts, and heavy metals like lead, mercury, and arsenic. It proves especially beneficial for individuals with access to untreated water.

However, distillation does present certain drawbacks. It is a slow purification process and requires a heat source, which can be costly. While advancements in energy technologies are underway, distillation remains relatively expensive and is thus more suitable for purifying smaller quantities of water. It's less recommended for large-scale commercial or industrial purification due to these limitations.

3.20 Chlorination

Chlorine stands out as a potent chemical with a long history of purifying household water. Its effectiveness lies in its ability to eradicate germs, parasites, and other disease-causing organisms that may lurk in ground or tap water. Chlorine is conveniently available in tablet or liquid form, making it accessible for purchase and application.

As a widely accessible water purification product, chlorine is both cost-effective and efficient. However, caution is warranted when using chlorine liquid or tablets for treating drinking water. Individuals with thyroid problems should seek advice from a medical practitioner before utilizing this product. Additionally, when employing chlorine tablets, it is crucial to dissolve them in heated water, as they dissolve more effectively at temperatures of 21 °C or higher. Chlorine tablets boast the capability to eliminate all bacteria, thereby ensuring that the water is clean and safe for consumption.

While chlorine serves as a powerful tool for water purification, responsible and careful use is paramount. By adhering to appropriate guidelines and precautions, you can effectively purify your water and safeguard your health.

3.21 Desalination

Desalination, a process aimed at removing dissolved mineral salts from water, plays a pivotal role in obtaining fresh water for human consumption and agricultural purposes, especially concerning seawater. This method mirrors the natural desalination process occurring during the water cycle, where seawater evaporation leaves salt behind, contributing to cloud formation and subsequent rain.

Applicable across municipal, industrial, and commercial sectors, desalination processes are increasingly cost-competitive with alternative methods of water production. This process renders water suitable for various uses, including human consumption, irrigation, and industrial applications. However, industrial desalination processes can be costly, making them practical only where obtaining fresh water through other means is economically unfeasible.

Desalination serves as a critical tool for securing fresh water for diverse purposes. As global water demands escalate, it becomes imperative to continuously explore and enhance desalination technologies to meet these needs sustainably.

The global desalination capacity witnessed an annual increase of nearly 12% from 1972 to 1999, with over 8600 desalination plants installed worldwide. Notably, approximately 20% of these plants are located in the United States, representing the highest number in any single country.

Desalination employs two major technologies worldwide, each contributing to addressing water scarcity and meeting the increasing demand for usable water.

3.22 Thermal Technology

Thermal technologies, as the name suggests, entail the heating of saline water and the collection of the condensed vapor (distillate) to generate pure water. This technology includes the following sub-categories

- Membrane Multi-stage Flash Distillation (MSF Technology)
- Multi-effect Distillation (MED)
- Vapor Compression Distillation (VCD).

3.22.1 Multi-stage Flash Distillation (MSF)

Multi-stage Flash Distillation (MSF) is a desalination process that operates through multiple chambers, with each stage functioning at progressively lower pressures. Initially, feed water is heated under high pressure before entering the first flash chamber. Upon entering this chamber, the sudden release of pressure causes rapid boiling or "flashing" of the water. This flashing phenomenon continues in subsequent stages as the pressure decreases progressively.

In each stage, the vapor produced by flashing is condensed into fresh water by heat exchanger tubing, which is cooled by incoming cooler feed water. Only a small percentage of the feed water undergoes vaporization and condensation. MSF distillation plants, in operation since the late 1950s, typically consist of 15–25 stages, with capacities usually not exceeding 15 million gallons per day (mgd).

MSF plants can adopt either a 'once-through' or 'recycled' process. In the once-through design, feed water passes through the heater and flash chambers once before disposal, whereas in the recycled design, feed water for cooling is reused. Both designs can feature either a long tube or cross tube configuration, with tubing arranged parallel or perpendicular to the concentrate flow, respectively.

Despite being a widely used desalination method known for its cost-effectiveness and energy efficiency, MSF plants are susceptible to corrosion, erosion, and impingement attack. Corrosion can be mitigated through the extensive use of stainless steel. MSF plants contribute significantly to global desalination capacity, accounting for approximately 84%, or 3.4 billion gallons per day (gpd).

Primarily constructed in regions with abundant and cost-effective energy resources, such as the Middle East, MSF plants require meticulous design and construction using advanced materials and technologies to address corrosion and erosion risks effectively.

3.22.2 Multi-effect Distillation (MED)

Multi-effect Distillation (MED) is a prominent desalination method that has been utilized since the late 1950s and early 1960s. This process involves a series of vessels, known as effects, which leverage evaporation and condensation principles under reduced ambient pressure conditions. As water boils at lower temperatures under decreased pressure, the vapor generated in one vessel serves as the heating medium for the subsequent vessel, leading to the production of water at progressively lower pressures.

The efficiency of the MED process improves with the inclusion of more vessels or effects. MED units can be categorized based on the arrangement of heat exchanger tubing into horizontal tube, vertical tube, or vertically stacked tube bundles configurations. Notably, several MED plants have been established globally, including in the United States, with successful operations observed in St. Thomas, U.S. Virgin Islands. In St. Thomas, the Virgin Islands Water and Power Authority oversees the operation of three low-temperature MED plants, where desalinated water serves as the primary water supply source. These MED units are integrated with the power plant, with steam directed to evaporators within the desalination units. The condensate from the vapor in each vessel serves as the product water.

The MED process has demonstrated its reliability and efficiency in consistently producing high-quality desalinated water, making it a preferred choice for water treatment facilities worldwide.

3.22.3 Vapor Compression Distillation (VCD)

Vapor Compression Distillation (VCD) is a widely employed process, either in combination with other techniques like Multiple Effect Distillation (MED) or as a standalone method. In contrast to conventional distillation approaches reliant on direct heat exchange from steam generated in a boiler, VCD employs vapor compression to generate the requisite heat for water evaporation. This pioneering method has spurred the creation of diverse VC unit configurations, typically incorporating a mechanical compressor to facilitate the evaporation process.

VC units, often characterized by smaller capacities, find common application in hotels, resorts, and industrial environments. The advantages of VCD encompass heightened energy efficiency, diminished operational expenses, and enhanced water quality. Through its utilization of vapor compression, VCD presents a sustainable and economical solution for water purification and desalination. It stands as an advanced distillation process, offering myriad benefits over traditional methodologies, with its innovative utilization of vapor compression technology paving the way for efficient and effective water purification systems adaptable to a broad array of scenarios.

3.23 Membrane Technologies

The membrane technology of distillation includes the following sub-categories:

- Electrodialysis (ED)
- Electrodialysis reversal (EDR)
- Reverse Osmosis (RO).

3.23.1 Electrodialysis (ED)

Electrodialysis (ED) is a membrane process driven by voltage, employing an electrical potential to transport salts through a membrane, resulting in fresh water as the end product. Although commercially introduced in the 1960s, approximately a decade before reverse osmosis (RO), ED was initially conceptualized for seawater desalination.

ED operates on the fundamental principle that the majority of salts dissolved in water exist as ions, categorized as positively charged (cations) or negatively charged (anions). Due to the repulsion of like charges and attraction of unlike charges, ions migrate toward electrodes possessing opposite electric charges. Selective membranes can be engineered to allow the passage of either anions or cations. In a saline solution, ions such as sodium (+) and chloride (−) migrate to electrodes with opposite charges, traversing specific membranes tailored for either cations or anions. Typically, membranes are arranged alternately, featuring anion-selective membranes followed by cation-selective ones. Throughout this process, the salt content in the water channel is diluted, while concentrated solutions are generated at the electrodes. The spaces between the alternating membranes, bounded by two membranes, are termed cells.

ED units comprise numerous cells interconnected with electrodes, forming a stack. Feed water flows through all cells concurrently, ensuring a continuous supply of desalinated water and a constant discharge of concentrate (brine) from the stack.

3.23.2 Electrodialysis Reversal (EDR)

In the early 1970s, the Electrodialysis Reversal (EDR) process was introduced, representing an advancement in the efficiency of the ED process. The EDR unit operates on the same underlying principle as an ED unit, with the distinction that both the product and concentrate channels are constructed identically. Periodically throughout the day, the polarity of the electrodes is reversed, causing ions to migrate in the opposite direction across the membranes. Subsequent to this reversal, product water is extracted until the lines are flushed and the desired water quality is reinstated. The flushing procedure typically lasts only a few minutes before water production resumes. This reversal mechanism is particularly advantageous for dislodging and expelling scales, slimes, and other deposits in the cells before they accumulate. Moreover, flushing aids in mitigating membrane fouling.

3.23.3 Reverse Osmosis (RO)

Reverse Osmosis (RO), a relatively recent innovation commercialized in the 1970s (Buros, 2000), has become the predominant method for desalination in the United States. This process harnesses pressure as the driving force to propel saline water through a semi-permeable membrane, yielding a product water stream and a concentrated brine stream (Lim et al., 2021). RO technology finds extensive application in desalinating brackish water (TDS > 1500 mg/L) and seawater.

The principle behind RO is rooted in osmosis, a natural phenomenon where water flows from a low salt concentration solution to a higher concentration solution across a semi-permeable membrane. By subjecting the higher salt concentration solution to pressure, water flows in the reverse direction through the semi-permeable membrane, leaving the salt behind. This process constitutes the Reverse Osmosis or RO process.

An RO desalination plant encompasses four primary systems:

(a) **Pretreatment system**
(b) **High-pressure pumps**
(c) **Membrane systems**
(d) **Post-treatment**.

(a) **Pretreatment**: This initial stage in RO is paramount as the cleanliness of the membrane surfaces is imperative for efficient operation (Lim et al., 2021). The objective is to remove all suspended solids and pre-treat the water to prevent salt precipitation or microbial growth on the membranes. Pre-treatment methods may involve conventional techniques such as chemical feed followed by coagulation/flocculation/sedimentation, and sand filtration. Alternatively, pre-treatment may employ membrane processes like microfiltration (MF) and ultrafiltration (UF).
(b) **High-Pressure Pumps**: These pumps are indispensable in the desalination process as they provide the requisite pressure to propel water through the semi-permeable membrane and reject salt. The pressure needed varies depending on the water type, ranging from approximately 150 psi for slightly brackish water to 800–1000 psi for seawater.
(c) **Membrane Systems**: In the RO process, the membrane assembly consists of a pressure vessel housing a semi-permeable membrane that facilitates the passage of feed water. Two primary types of RO membranes are prevalent in desalination:

spiral wound and hollow fiber. Spiral wound elements are crafted from flat sheet membranes, typically composed of materials like cellulose acetate or other composite polymers.

In the spiral wound configuration, the membrane envelope encases a central collecting tube. Pressurized feed water flows in a spiral trajectory within the membrane envelope, with desalinated water collected in the central tube. As water permeates through the membrane, the remaining feed water becomes more concentrated in salts. To prevent salt super-saturation, a portion of the feed water is discharged without passing through the membrane. The percentage of feed water discharged as concentrate varies, ranging from about 20% for brackish water to approximately 50% for seawater.

(d) **Post-treatment**: Following the primary reverse osmosis process, post-treatment is pivotal for ensuring the production of high-quality water. This phase involves additional treatment steps aimed at further refining the water's purity and taste. Various techniques are employed during post-treatment to eliminate any residual impurities, minerals, or contaminants that may have traversed the reverse osmosis membrane. Common post-treatment methods include activated carbon filtration, UV disinfection, and pH adjustment.

References

Al-Mafraji, E. A., & Al-Mussawy, H. A. (2021). Using lower and upper baffle arrangements to enhance sedimentation tank performance. In *4th International Conference on Engineering Sciences*.

Almeida, R. A., de Rezende, R. V. P., Mataczinski, A. K., Khan, A. I., Camilo, R., Ravagnani, M. A. S. S., & Lautenschlager, S. R. (2020). Three-dimensional simulation of a secondary circular settling tank: Flow pattern and sedimentation process. *Brazilian Journal of Chemical Engineering, 37*(2), 333–350. https://doi.org/10.1590/0104-6632.20200372s20200134

Antanasijević, D., Pocajt, V., Perić-Grujić, A., & Ristić, M. (2020). Multilevel split of high-dimensional water quality data using artificial neural networks for the prediction of dissolved oxygen in the Danube River. *Neural Computing and Applications, 32*, 3957–3966. https://doi.org/10.1007/s00500-019-03986-7

Applebaum, S. B. (2013). *Demineralization by ion exchange in water treatment and chemical processing of other liquids*. Elsevier.

Arens, L., Weißenfeld, F., Klein, C. O., Schlag, K., & Wilhelm, M. (2017). Osmotic engine: Translating osmotic pressure into macroscopic mechanical force via poly(acrylic acid) based hydrogels. *Advanced Science, 4*(9), 1700112. https://doi.org/10.1002/advs.201700112

AWWA Disinfection Systems Committee. (2008). Committee report: Disinfection survey, part 2—Alternatives, experiences, and future plans. *Journal of the American Water Works Association, 100*(11), 110–124. https://doi.org/10.1002/j.1551-8833.2008.tb10027.x

Bansal, S., et al. (2015). Defluoridation techniques: A review. *Environmental Science and Pollution Research, 22*(7), 5256–5268. https://doi.org/10.1007/s11356-014-3870-5

Bera, S. P., Godhaniya, M., & Kothari, C. (2022). Emerging and advanced membrane technology for wastewater treatment: A review. *Journal of Basic Microbiology, 62*(3–4), 245–259. https://doi.org/10.1002/jobm.202100137

Bethi, B., Sonawane, S. H., Bhanvase, B. A., & Gumfekar, S. P. (2016). Nanomaterials-based advanced oxidation processes for wastewater treatment: A review. *Chemical Engineering and Processing-Process Intensification, 109*, 178–189. https://doi.org/10.1016/j.cep.2016.07.003

Bilenko, Y., Dobrinsky, A., Smetona, S., Shur, M., Gaska, R., & Bettles, T. J. (2019). U.S. Patent No. 10,442,704. U.S. Patent and Trademark Office, Washington, DC.

Bochenek, R., Sitarz, R., & Antos, D. (2011). Design of continuous ion exchange process for wastewater treatment. *Chemical Engineering Science, 66*(23), 6209–6219. https://doi.org/10.1016/j.ces.2011.08.031

Bratby, J. (2016). *Coagulation and flocculation in water and wastewater treatment*. IWA Publishing.

Buros, O. K. (2000). *The ABCs of desalting* (Vol. 2). International Desalination Association.

Calderon, M., et al. (2020). Fluoride removal from water: A review of methods. *Chemosphere, 248*, 126015. https://doi.org/10.1016/j.chemosphere.2020.126015

Carlsson, B. (1998). *An introduction to sedimentation theory in wastewater treatment*. Systems and Control Group, Uppsala University.

Chen, C. N., Jia, K., Song, X., Zheng, Y., Lan, Y., & Li, Y. (2020). Sulfur-doped copper-yttrium bimetallic oxides: A novel and efficient ozonation catalyst for the degradation of aniline. *Separation and Purification Technology, 236*, 116248. https://doi.org/10.1016/j.seppur.2019.116248

Chen, X., Hao, K., Zong, Y., Chio, M., You, J., He, O., & Zhang, D. (2023). Effects of ultraviolet radiation on microorganism and nitrogen metabolism in sewage under plate background. *Environmental Science and Pollution Research*, 1–18. https://doi.org/10.1007/s11356-023-28757-5

Cho, M., Chung, H., Choi, W., & Yoon, J. (2005). Different inactivation behaviors of MS-2 phage and Escherichia coli in TiO_2 photocatalytic disinfection. *Applied and Environmental Microbiology, 71*(1), 270–275. https://doi.org/10.1128/AEM.71.1.270-275.2005

Crittenden, J. C., Trussell, R. R., Hand, D. W., Howe, K. J., & Tchobanoglous, G. (2012). *MWH's water treatment: Principles and design*. Wiley.

Cui, H., Huang, X., Yu, Z., Chen, P., & Cao, X. (2020). Application progress of enhanced coagulation in water treatment. *RSC Advances, 10*(34), 20231–20244. https://doi.org/10.1039/D0RA01881C

Darby, J. L., Snider, K. E., & Tchobanoglous, G. (1993). Ultraviolet disinfection for wastewater reclamation and reuse subject to restrictive standards. *Water Environment Research, 65*(2), 169–180. https://doi.org/10.2175/WER.65.2.5

Das, T. K. (2001). Ultraviolet disinfection application to a wastewater treatment plant. *Clean Production Processes, 3*, 69–80. https://doi.org/10.1007/s100980100108

Deng, Y., & Zhao, R. (2015). Advanced oxidation processes (AOPs) in wastewater treatment. *Current Pollution Reports, 1*, 167–176. https://doi.org/10.1007/s40726-015-0015-z

Edzwald, J. (2011). *Water quality & treatment: A handbook on drinking water*. McGraw-Hill.

Garg, S. K., & Garg, R. (1999). *Sewage disposal and air pollution engineering*. Khanna Publishers.

Giannakis, S., Voumard, M., Grandjean, D., Magnet, A., De Alencastro, L. F., & Pulgarin, C. (2016). Micropollutant degradation, bacterial inactivation and regrowth risk in wastewater effluents: Influence of the secondary (pre)treatment on the efficiency of advanced oxidation processes. *Water Research, 102*, 505–515. https://doi.org/10.1016/j.watres.2016.06.032

Glaze, W. H., Kang, J. W., & Chapin, D. H. (1987). The chemistry of water treatment processes involving ozone, hydrogen peroxide, and ultraviolet radiation. *Ozone: Science & Engineering, 9*(4), 335–352. https://doi.org/10.1080/01919518708552148

Gottschalk, C., Libra, J. A., & Saupe, A. (2009). *Ozonation of water and wastewater: A practical guide to understanding ozone and its applications*. Wiley.

Gregory, R., & Edzwald, J. (2010). Sedimentation & flotation (Chap. 9). In *Water quality & treatment* (6th ed.). AWWA & McGraw Hill.

Gul, H., Raza, W., Lee, J., Azam, M., Ashraf, M., & Kim, K. H. (2021). Progress in microbial fuel cell technology for wastewater treatment and energy harvesting. *Chemosphere, 281*, 130828.

Guo, Z., Sun, Y., Pan, S. Y., & Chiang, P. C. (2019). Integration of green energy and advanced energy-efficient technologies for municipal wastewater treatment plants. *International Journal of Environmental Research and Public Health, 16*(7), 1282.

Gupta, A., et al. (2016). Adsorption of fluoride from aqueous solutions using modified clay. *Journal of Environmental Management, 182*, 280–290. https://doi.org/10.1016/j.jenvman.2016.07.025

Hammer, M. J. (2000). *Water and wastewater technology*. Prentice Hall of India Pvt. Ltd.

Hassaan, M. A. A., & El Nemr, A. (2017). Advanced oxidation processes for textile wastewater treatment. *International Journal of Photochemistry and Photobiology, 2*(3), 85–93. https://doi.org/10.11648/j.ijpp.20170203.11

Henze, M., Gujer, W., Mino, T., & Van Loosedrecht, M. (2006). *Activated sludge models ASM1, ASM2, ASM2d and ASM3*. IWA Publishing.

Hiller, C. X., Hübner, U., Fajnorova, S., Schwartz, T., & Drewes, J. E. (2019). Antibiotic microbial resistance (AMR) removal efficiencies by conventional and advanced wastewater treatment processes: A review. *Science of the Total Environment, 685*, 596–608. https://doi.org/10.1016/j.scitotenv.2019.06.043

Hirom, K., & Devi, T. T. (2022). Application of computational fluid dynamics in sedimentation tank design and its recent developments: A review. *Water, Air, & Soil Pollution, 233*, 1–26. https://doi.org/10.1007/s11270-022-05847-8

Høibye, L., Clauson-Kaas, J., Wenzel, H., Larsen, H. F., Jacobsen, B. N., & Dalgaard, O. (2008). Sustainability assessment of advanced wastewater treatment technologies. *Water Science and Technology, 58*(5), 963–968. https://doi.org/10.2166/wst.2008.575

Hou, T., Du, H., Yang, Z., Tian, Z., Shen, S., Shi, Y., & Zhang, L. (2019). Flocculation of different types of combined contaminants of antibiotics and heavy metals by thermo-responsive flocculants with various architectures. *Separation and Purification Technology, 223*, 123–132. https://doi.org/10.1016/j.seppur.2019.05.012

Hu, E. (2019). Catalytic ozonation for textile dyeing wastewater treatment and reuse.

Huang, C. P., Dong, C., & Tang, Z. (1993). Advanced chemical oxidation: Its present role and potential future in hazardous waste treatment. *Waste Management, 13*(5–7), 361–377. https://doi.org/10.1016/0956-053X(93)90002-C

Huidobro-López, B., Martínez-Hernández, V., Barbero, L., Meffe, R., Nozal, L., & de Bustamante, I. (2023). Evaluation of contaminants of emerging concern attenuation through a vegetation filter managed using different operating conditions. *Journal of Hazardous Materials, 459*, 132217. https://doi.org/10.1016/j.jhazmat.2022.132217

Hussain, K., Khan, N. A., Vambol, V., Vambol, S., Yeremenko, S., & Sydorenko, V. (2022). Advancement in ozone-based wastewater treatment technologies: Brief review. *Ecological Questions, 33*(2), 7–19. https://doi.org/10.12775/EQ.2022.016

Ikehata, K., & Li, Y. (2018). Ozone-based processes. In *Advanced oxidation processes for wastewater treatment* (pp. 115–134). Academic Press. https://doi.org/10.1016/B978-0-12-812658-8.00006-9

Jahanshahi, M., & Taghizadeh, M. M. (2018). Pre-sedimentation tank effects on water treatment unit operation. *EQA-International Journal of Environmental Quality, 28*, 35–42. https://doi.org/10.6092/issn.2281-4485/8555

Jiao, Y., Liu, L., Zhang, Q., Zhou, M., & Zhang, Y. (2021). Treatment of reverse osmosis concentrate from industrial coal wastewater using an electro-peroxone process with a natural air diffusion electrode. *Separation and Purification Technology, 279*, 119667. https://doi.org/10.1016/j.seppur.2021.119667

Juraev, S., Akramov, A., Abdurazzokov, A., & Pathidinova, U. (2022, August). Increasing the efficiency of sedimentation tanks for drinking water treatment. *IOP Conference Series: Earth and Environmental Science, 1076*(1), 012049. IOP Publishing. https://doi.org/10.1088/1755-1315/1076/1/012049

Khadse, G. K., Patni, P. M., & Labhasetwar, P. K. (2015). Removal of iron and manganese from drinking water supply. *Sustainable Water Resources Management, 1*, 157–165. https://doi.org/10.1007/s40899-015-0015-7

Koutahzadeh, N., Esfahani, M. R., & Arce, P. E. (2016). Sequential use of UV/H_2O_2—(PSF/TiO_2/MWCNT) mixed matrix membranes for dye removal in water purification: Membrane permeation, fouling, rejection, and decolorization. *Environmental Engineering Science, 33*(6), 430–440. https://doi.org/10.1089/ees.2015.0228

Kumar, L., & Singh, S. K. (2015). Removal of chloride from groundwater by bioadsorption. *International Journal of Advanced Research, 3*, 140–154. https://doi.org/10.21474/IJAR01/1061

Kumar, V., Othman, N., & Ashanuddin, S. (2017). Applications of natural coagulants to treat wastewater: A review. *MATEC Web of Conferences, 103*, 06016. EDP Sciences. https://doi.org/10.1051/matecconf/201710306016

Lai, F. Y., Thai, P. K., O'Brien, J., Gartner, C., Bruno, R., Kele, B., ... & Mueller, J. F. (2013). Using quantitative wastewater analysis to measure daily usage of conventional and emerging illicit drugs at an annual music festival. *Drug and alcohol review, 32*(6), 594–602.

Lee, C. S., Robinson, J., & Chong, M. F. (2014). A review on the application of flocculants in wastewater treatment. *Process Safety and Environmental Protection, 92*(6), 489–508. https://doi.org/10.1016/j.psep.2014.05.006

Li, J., Pang, S. Y., Wang, Z., Guo, Q., Duan, J., Sun, S., & Jiang, J. (2021). Oxidative transformation of emerging organic contaminants by aqueous permanganate: Kinetics, products, toxicity changes, and effects of manganese products. *Water Research, 203*, 117513. https://doi.org/10.1016/j.watres.2021.117513

Lim, Y. J., Goh, K., Kurihara, M., & Wang, R. (2021). Seawater desalination by reverse osmosis: Current development and future challenges in membrane fabrication—A review. *Journal of Membrane Science, 629*, 119292. https://doi.org/10.1016/j.memsci.2021.119292

Liu, R., Zhang, H., & Lal, R. (2016). Effects of stabilized nanoparticles of copper, zinc, manganese, and iron oxides in low concentrations on lettuce (*Lactuca sativa*) seed germination: Nanotoxicants or nanonutrients? *Water, Air, & Soil Pollution, 227*, 1–14. https://doi.org/10.1007/s11270-016-2937-3

Mandoreba, C., Gwenzi, W., & Chaukura, N. (2021). Defluoridation of drinking water using a ceramic filter decorated with iron oxide-biochar composites. *International Journal of Applied Ceramic Technology, 18*(4), 1321–1329. https://doi.org/10.1111/ijac.13754

Masschelein, W. J., & Rice, R. G. (2016). *Ultraviolet light in water and wastewater sanitation*. CRC Press. https://doi.org/10.1201/9781315279742

Metcalf & Eddy. (2004). *Wastewater engineering*. Tata McGraw-Hill Publishing Company Ltd.

Modak, A., Osumi, Y. T. C., Pomeranz, S. F., & Mendelow, A. S. (2020). U.S. Patent Application No. 16/505,456. https://www.uspto.gov/patents/search

Mohapatra, T., Kumar, V., Sharma, M., & Ghosh, P. (2021). Hybrid Fenton oxidation processes with packed bed or fluidized bed reactor for the treatment of organic pollutants in wastewater: A review. *Environmental Engineering Science, 38*(6), 443–457. https://doi.org/10.1089/ees.2020.0212

Namasivayam, C., & Sangeetha, D. (2007). Removal of fluoride from drinking water by low-cost adsorbents. *Journal of Hazardous Materials, 147*(3), 1005–1010. https://doi.org/10.1016/j.jhazmat.2007.01.005

Neumann, S., & Fatula, P. (2009). Principles of ion exchange in wastewater treatment. *Asian Water, 19*, 14–19.

Permitsky, D. J., & Edzwald, J. K. (2006). Selection of alum and polyaluminum coagulants: Principles and applications. *Journal of Water Supply Research and Technology, AQUA, 55*(2), 121–141. https://doi.org/10.2166/aqua.2006.058

Peters, P. E. (2023). *Feasibility of sewer overflow treatment using chemical oxidation after rapid solids removal* [Master's thesis]. Marquette University. https://epublications.marquette.edu/etd/1204

Salmani, H. J., Karkhanechi, H., Jeon, S., & Matsuyama, H. (2022). Calculating osmotic pressure of liquid mixtures by association theory for sustainable separating of solvents by membrane processes. *Journal of Industrial and Engineering Chemistry, 109*, 137–146. https://doi.org/10.1016/j.jiec.2022.02.021

Satyavani, K., Saravanakumar, K., & Thangavel, K. (2016). Ion exchange for defluoridation of water: A review. *International Journal of Environmental Science and Technology, 13*(3), 765–776. https://doi.org/10.1007/s13762-016-0954-5

Saxena, K., Brighu, U., & Choudhary, A. (2018). Parameters affecting enhanced coagulation: A review. *Environmental Technology Reviews, 7*(1), 156–176. https://doi.org/10.1080/21622515.2018.1518893

Scaria, J., & Nidheesh, P. V. (2022). Comparison of hydroxyl-radical-based advanced oxidation processes with sulfate radical-based advanced oxidation processes. *Current Opinion in Chemical Engineering, 36*, 100830. https://doi.org/10.1016/j.coche.2021.100830

Sharma, A., Soni, R. S., Singh, K. P., & Shukla, M. (2017). Defluoridation of water using activated alumina: A review. *Water Science and Technology, 76*(7), 1760–1774. https://doi.org/10.2166/wst.2017.385

Shih, P. K., & Chang, W. L. (2015). The effect of water purification by oyster shell contact bed. *Ecological Engineering, 77*, 382–390. https://doi.org/10.1016/j.ecoleng.2015.02.022

Singh, L., & Mishra, P. K. (2020). Nano-membrane filtration: A novel application of nanotechnology for wastewater treatment. *Materials Today: Proceedings, 29*, 327–332. https://doi.org/10.1016/j.matpr.2020.01.367

Sivagami, K., Sakthivel, K. P., & Nambi, L. M. (2018). Advanced oxidation processes for the treatment of tannery wastewater. *Journal of Environmental Chemical Engineering, 6*(3), 3656–3663. https://doi.org/10.1016/j.jece.2018.05.016

Skipton, S., Dvorak, B. I., & Niemeyer, S. (2008). *G08-1491 drinking water treatment: Water softening (ion exchange)*. University of Nebraska–Lincoln.

Srivastav, A. L., & Kaur, T. (2020). Factors affecting the formation of disinfection by-products in drinking water: Human health risk. In *Disinfection by-products in drinking water* (pp. 433–450). Butterworth-Heinemann. https://doi.org/10.1016/B978-0-12-814650-0.00030-9

Sun, Y., Zhou, S., Chiang, P. C., & Shah, K. J. (2019). Evaluation and optimization of enhanced coagulation process: Water and energy nexus. *Water-Energy Nexus, 2*(1), 25–36. https://doi.org/10.1016/j.wen.2019.03.002

Suresh, A., Grygolowicz-Pawlak, E., Pathak, S., Poh, L. S., bin Abdul Majid, M., Dominiak, D., & Ng, W. J. (2018). Understanding and optimization of the flocculation process in biological wastewater treatment processes: A review. *Chemosphere, 210*, 401–416. https://doi.org/10.1016/j.chemosphere.2018.07.042

Thakur, G., Hebbi, V., Parida, S., & Rathore, A. S. (2020). Automation of dead-end filtration: An enabler for continuous processing of biotherapeutics. *Frontiers in Bioengineering and Biotechnology, 8*, 758. https://doi.org/10.3389/fbioe.2020.00758

Vaseghi, G., Ghassemi, A., & Loya, J. (2016). Characterization of reverse osmosis and nanofiltration membranes: Effects of operating conditions and specific ion rejection. *Desalination and Water Treatment, 57*(50), 23461–23472. https://doi.org/10.1080/19443994.2016.1185420

Vijayakumar, K., Radhakrishnan, S., & Sankaranarayanan, V. (2021). Microbial methods for fluoride removal: A review. *Environmental Biotechnology, 17*(1), 24–36. https://doi.org/10.3233/EB210010

Waldemer, R. H., Tratnyek, P. G., Johnson, R. L., & Nurmi, J. T. (2007). Oxidation of chlorinated ethenes by heat-activated persulfate: Kinetics and products. *Environmental Science & Technology, 41*(3), 1010–1015. https://doi.org/10.1021/es0610944

Wang, Q., Tang, X., Liang, H., Cheng, W., Li, G., Zhang, Q., & Wang, J. (2022). Effects of filtration mode on the performance of gravity-driven membrane (GDM) filtration: Cross-flow filtration and dead-end filtration. *Water, 14*(2), 190. https://doi.org/10.3390/w14020190

Wolf, C. A., Pavese, U., & von Gunten, U. (2019). Proxies to monitor the inactivation of viruses by ozone in surface water and wastewater effluent. *Water Research, 166*, 1150–1160. https://doi.org/10.1016/j.watres.2019.05.061

Wu, Q.-Y., Zhou, Y.-T., Li, W., Zhang, X., Du, Y., & Hu, H.-Y. (2019). Underestimated risk from ozonation of wastewater containing bromide: Both organic byproducts and bromate contributed to the toxicity increase. *Water Research, 162*, 38–50. https://doi.org/10.1016/j.watres.2019.06.054

Wu, Z., Dougherty, M., Chen, Z., Zhou, Y., Zuo, X., & He, J. (2022). Interaction between bioaccumulation and the efficiency of intermittent sand filters in wastewater treatment. *Journal of Cleaner Production, 335*, 130303. https://doi.org/10.1016/j.jclepro.2021.130303

Yang, Z., Liu, P., Wei, H., Li, H., Li, J., Qiu, X., & Guo, X. (2022). Alteration in microbial community and antibiotic resistance genes mediated by microplastics during wastewater ultraviolet disinfection. *Science of the Total Environment, 825*, 153918. https://doi.org/10.1016/j.scitotenv.2022.153918

Chapter 4
Wastewater Treatment Process

4.1 Introduction

Wastewater refers to water that has been contaminated by various effluents originating from human activities, including domestic, agricultural, and industrial wastes, as well as stormwater and surface runoff. It is alarming to note that half of the world's population is exposed to wastewater and its detrimental effects. Not only does it pose a danger to public health, but it also wreaks havoc on the environment. Shockingly, around 80% of untreated wastewater flows into the environment worldwide, polluting natural ecosystems. The nitrogen and phosphorus present in wastewater disrupts aquatic life and causes significant damage.

Wastewater can be defined as water that has undergone physical, chemical, or biological changes due to the introduction of certain substances, rendering it unsafe for certain purposes, such as drinking. Given that our daily activities heavily rely on water, we inevitably discharge various waste materials into it. These substances include body waste, hair products, food scraps, fat, laundry products, chemicals, detergents, household cleaners, dirt, and microorganisms that can cause illness and harm the environment.

It is crucial to recognize that a significant portion of the water supplied ultimately becomes wastewater, underscoring the importance of its proper treatment. Wastewater treatment involves the use of processes and technologies to remove contaminants and ensure a healthy environment and public health. Effective wastewater management is therefore essential to safeguard the environment, promote public health, and maintain economic, social, and political stability (Metcalf & Eddy, 1991; Templeton & Butler, 2011).

Overall, it is imperative to address the issue of wastewater in a professional and comprehensive manner, considering its far-reaching consequences for both humans and the environment. By understanding the sources and impacts of wastewater, we can work towards implementing effective wastewater management strategies to protect our surroundings and ensure the well-being of future generations.

S. A. Bandh and B. Mushtaq, *Wastewater Treatment Technology*, Springer Water,
https://doi.org/10.1007/978-3-031-86684-5_4

4.2 Principal Pollutants

The principal pollutants present in wastewater include:

4.2.1 *Organic Materials*

The presence of putrescible organic material in sewage is indicated by the biochemical oxygen demand (BOD). BOD represents the amount of oxygen required by microorganisms to decompose the organic substances in sewage. It is a critical parameter in the design and operation of a sewage treatment plant. Notably, industrial sewage may contain BOD levels significantly higher than those found in domestic sewage. The BOD of storm sewage becomes particularly concerning when it mixes with domestic sewage in combined sewerage systems (Sauer et al., 2006).

Dissolved oxygen plays a crucial role in assessing the water quality of lakes and rivers. The concentration of dissolved oxygen is directly linked to the overall water quality. When sewage infiltrates a lake or a stream, the decomposition process of organic materials begins. Microorganisms consume oxygen during their metabolic activities, leading to a rapid depletion of available oxygen in the water. Insufficient levels of dissolved oxygen can have severe consequences for trout and other aquatic species, ultimately resulting in their death. In fact, a complete absence of oxygen renders the water septic. The decomposition of organic compounds without oxygen is responsible for the unpleasant odors typically associated with septic or putrid conditions.

Understanding the levels of organic material and dissolved oxygen in sewage is crucial for effective sewage treatment plant design and operation. By monitoring and managing these factors, we can ensure the preservation of water quality and the well-being of aquatic ecosystems.

4.2.2 *Suspended Solids*

Suspended solids play a pivotal role in the analysis of sewage. The quantity of sludge produced in a treatment facility is directly correlated to the overall presence of suspended solids in the sewage. It is important to highlight that industrial and storm sewage often exhibit higher concentrations of suspended solids in comparison to domestic sewage. The effectiveness of the treatment process hinges on the treatment plant's ability to effectively eliminate both suspended solids and BOD (Biochemical Oxygen Demand).

4.2.3 Plant Nutrients

Plant nutrients including nitrogen and phosphorus, which are essential nutrients for plant growth, are present in domestic sewage. Their excessive amounts lead to the rapid growth of algae, a phenomenon known as eutrophication, often caused by sewage discharges and accelerating the natural aging process of lakes.

4.2.4 Microbes

Microbes are abundant in domestic sewage, with millions of microorganisms per gallon. The majority of these microorganisms are coliform bacteria originating from the human intestinal tract. Coliforms serve as indicators of sewage pollution, with a high coliform count typically indicating recent sewage contamination. Additionally, domestic sewage is likely to carry other types of microbes.

4.2.5 Pathogens

Pathogens, defined as organisms capable of causing diseases, are commonly found in sewage due to human fecal contamination. Among the various types of pathogens present, bacteria, viruses, and protozoa pose significant risks to human health.

4.2.6 Bacteria

Different types of bacteria present in sewage include *Salmonella*, C*ampylobacter*, *Shigella*, and *Vibrio cholerae*. While many bacteria are harmless or beneficial, pathogenic bacteria can cause infectious diseases. They enter cells and damage the body's immune system by producing endotoxins.

4.2.7 Viruses

Sewage may contain potential viruses such as hepatitis A, enteroviruses, and rotavirus. Diseases caused by viral pathogens include chickenpox, influenza, mumps, measles, Ebola, and HIV.

4.2.8 Protozoa

Protozoa found in sewage include *Entamoeba histolytica* and *Cryptosporidium parvum*. These single-celled eukaryotes feed on organic matter and exhibit animal-like behaviors.

4.2.9 Parasites

Parasites such as roundworms, hookworms, and whipworms are commonly found in sewage water. They harm their hosts by depleting nutrients and manipulating the host's immune system.

4.2.10 Micro-pollutants

Sewage contains a significant concentration of pharmaceutical compounds, with approximately 70% originating from households alone. Livestock farming contributes around 20%, hospital effluent accounts for 5%, and the remaining pollutants come from runoffs. Trihalomethanes, a byproduct of past disinfection processes, are also present in sewage. Furthermore, sewage contains high levels of nutrients that can lead to eutrophication, posing a threat to the aquatic ecosystem.

4.3 Classification of Wastewater

Wastewater can be broadly classified into two categories: sewage and non-sewage.

4.3.1 Sewage

Sewage refers to wastewater originating from domestic sources, including toilets, food waste, and laundry waste. These wastes primarily consist of human feces and urine. Additionally, commercial and industrial wastes contribute to sewage, with commercial wastes coming from local shops or service establishments, and industrial wastes stemming from manufacturing processes.

In terms of composition, domestic sewage is primarily composed of 99.9% water, with the remaining 0.1% containing suspended and dissolved impurities. These impurities include plant nutrients, putrescible organic materials, and disease-spreading microbes. Despite their small quantity, these impurities pose significant challenges in wastewater disposal.

4.3.2 Non-sewage

Non-sewage comprises storm or surface runoff water, carrying harmful elements and chemicals from the atmosphere, soil, and vegetation. Storm wastewater contains organic materials, dissolved and suspended solids, and other harmful substances. When properly treated and utilized, stormwater can be a valuable resource, especially in drought-prone regions.

Based on its source of generation wastewater can be further categorized into three subcategories.

4.3.3 Blackwater

Blackwater originates from toilets, bathrooms, dishwashers, kitchen sinks, and washing machines. It contains urine, feces, toilet paper, discarded food particles, and household cleaning agents. Due to its high contamination level and abundance of disease-causing pathogens, it is appropriately named Blackwater.

4.3.4 Greywater

Greywater refers to wastewater generated from washing machines and regular household cleaning activities, excluding bathroom and toilet water. It may also include water from industrial plants. While greywater may contain chemicals and other substances, it lacks human contaminants, making it suitable for recycling and treatment processes.

4.3.5 Yellow Water

Yellow water consists primarily of pure urine without fecal matter, chemicals, food particles, or other contaminants. It is collected from specific sources and is distinct from blackwater and greywater. Classifying wastewater based on its usage helps in understanding its composition and determining appropriate treatment methods for effective management and recycling, contributing to sustainability and environmental protection.

Based on the use of water, there are three types of wastewater, or sewage.

4.3.6 Domestic Sewage

Domestic sewage, also referred to as sanitary sewage, constitutes the wastewater discharged from residential dwellings, including houses and apartments. It predominantly comprises water, which accounts for slightly over 99.9% of its total weight. However, the remaining less than 0.1% consists of a diverse array of dissolved and suspended impurities. Despite their minimal proportion by weight, these impurities present significant challenges due to the large volumes in which they are carried, complicating the disposal of domestic wastewater (Kataki et al., 2021).

The principal impurities found in domestic sewage include putrescible organic materials and plant nutrients, along with the potential presence of disease-causing microbes. Domestic sewage is sourced from a variety of origins, encompassing households, hotels, restaurants, schools, and shopping centers. With rapid economic development and improved living standards, the volume of sewage discharge necessitating treatment has witnessed a steady rise. This type of sewage is characterized by a notable quantity of organic matter such as protein, starch, fat, and urea, as well as pathogenic microorganisms and suspended particles (Zhang et al., 2021a, 2021b, 2021c).

4.3.7 Industrial Sewage

Industrial sewage refers to the wastewater generated from manufacturing or chemical processes within industrial facilities. Unlike domestic sewage, industrial wastewater contains specific and often identifiable chemical compounds, the composition of which depends on the nature of the industrial activities involved. The quality and quantity of industrial wastewater can vary significantly depending on the industry from which it originates. It may range from highly biodegradable to completely resistant to decomposition, and it may contain compounds that are challenging to treat effectively.

Industrial sewage can contain a variety of pollutants, including synthetic organic substances, heavy metals, toxic chemicals, and other harmful compounds. The presence of these pollutants poses significant environmental and health risks, particularly if released untreated into the environment. In developing countries, the quantity and quality of industrial wastewater may differ substantially from those in developed nations, reflecting variations in industrial practices, regulations, and enforcement.

One of the primary concerns associated with industrial wastewater is the increasing prevalence and diversity of synthetic compounds released into the environment. Tracking and understanding the fate of these toxic compounds present significant challenges, as does the development and implementation of effective treatment methods to remove them from wastewater. To address these challenges, it is advisable to promote cleaner production methods in industries, which involve replacing toxic or resistant compounds with safer alternatives. Additionally, raising awareness among society about the importance of reducing the use of harmful compounds can contribute to mitigating the environmental impact of industrial activities (Jiménez, 2009).

4.3.8 Storm Sewage

Storm sewage, also known as stormwater, refers to the runoff generated from precipitation events such as rainfall or snowmelt. This runoff collects various pollutants and debris as it flows over impervious surfaces such as roads, parking lots, and rooftops, ultimately entering storm sewer systems. Storm sewers are designed to channel this runoff away from urban areas to prevent flooding and waterlogging.

The composition of storm sewage can vary depending on factors such as land use, weather conditions, and the presence of pollutants in the environment. It typically contains organic materials, suspended solids, dissolved substances, as well as pollutants such as heavy metals, oils, and chemicals washed off from streets and other surfaces. Storm sewage may also carry sediment, debris, and nutrients like phosphorus and nitrogen, which can contribute to water pollution and ecosystem degradation if not properly managed.

Storm sewer systems typically consist of a network of pipes, culverts, and drainage channels that collect and convey stormwater to discharge points such as rivers, lakes, or coastal waters. In urban areas, storm sewers are often separate from sanitary sewers to prevent contamination of water bodies with untreated sewage. However, in some older cities, combined sewer systems may exist, where stormwater and sanitary sewage are conveyed together to wastewater treatment plants or directly discharged into receiving waters during heavy rainfall events.

It is essential to differentiate between storm sewers and sanitary sewers to prevent the inadvertent discharge of sanitary waste into stormwater systems. Discharging sanitary waste into storm sewers can lead to contamination of surface waters, posing risks to public health and the environment. Therefore, proper waste management practices and public education efforts are crucial to ensure the integrity of water resources and protect human and ecological health.

4.4 Definitions of Common Terms Used in Sanitary Engineering

These definitions provide a comprehensive understanding of the terminology commonly used in the field of sanitary engineering:

- **Refuse**: A general term encompassing all types of waste, including rejects, sewage, and sullage.

- **Garbage**: Dry refuse such as waste papers, sweepings, and vegetable peelings, containing organic and putrefying matter.
- **Rubbish**: Various solid wastes found in residences, offices, and buildings, including broken furniture, paper, and rags.
- **Sullage**: Discharge from bathrooms, kitchens, and wash basins, excluding high-organic content sources like lavatories and hospitals.
- **Sewage**: A diluted mixture of waste from residential, public, and industrial areas, including foul discharge from water closets, urinals, and hospitals.
- **Storm Water**: Surface runoff from rainfall that enters sewers through inlets, devoid of foulness that can be disposed of in natural rivers.
- **Sanitary Sewage**: Sewage specifically from residential buildings.
- **Domestic Sewage**: Waste from lavatory basins, water closets, and urinals in residential, office, and institutional settings, often containing night soil and urine.
- **Industrial Sewage**: Wastewater generated by industries and commercial areas, with foulness levels varying based on industry processes.
- **Sewers**: Underground pipes designed to transport sewage to disposal points.
- **Sewerage**: The entire system responsible for collecting, carrying, and disposing of sewage through sewers.
- **Dry Weather Flow (DWF)**: Combined domestic and industrial sewage flow excluding stormwater, representing normal flow during dry seasons.
- **Bacteria**: Bacteria are microscopic organisms. They can be categorized into:

  - **Aerobic Bacteria**: These bacteria require oxygen and light for their survival.
  - **Anaerobic Bacteria**: These bacteria can survive without free oxygen and light.
  - **Facultative Bacteria**: These bacteria can exist in the presence or absence of oxygen, but they tend to thrive in oxygen-deprived environments.

- **Invert**: Invert refers to the lowest point inside the sewer at any cross-section.
- **Sludge**: Sludge is the organic matter that accumulates and settles in the sedimentation tank during the treatment process of any wastewater.
- **On-Site System**: The On-Site System is a method of wastewater disposal that takes place directly at the source of waste production, such as individual houses, eliminating the need for transportation. This system includes various approaches, such as dry methods like pit latrines and composting toilets, as well as water-saving methods like pour-flush latrines and aqua privy with soakage pits. Additionally, there are methods that require a higher water usage, such as flush toilets with septic tanks and soakage pits that are not regularly emptied.
- **Off-Site System**: In this system, wastewater is transported to a location other than its point of production. Off-site methods include bucket latrines, pour-flush toilets with vault and tanker removal, and conventional sewerage systems.
- **Conventional Sewerage Systems**: It can be either combined sewers, where wastewater is carried along with stormwater, or separated sewers.
- **Septic Tank**: A septic tank is an on-site system designed to hold black water for a sufficient period of time to allow for sedimentation. Typically, it is a watertight single-story tank.

- **Faccal Sludge**: Faccal sludge refers to all sludge collected and transported from on-site sanitation systems by vacuum trucks for disposal or treatment.
- **Unit Operation**: Unit Operation involves the removal of contaminants through physical forces.
- **Unit Process**: Unit Process involves the biological and/or chemical removal of contaminants.
- **Wastewater Treatment Plant**: A Wastewater Treatment Plant is a facility that incorporates a series of designed unit operations and processes aimed at reducing certain constituents of wastewater to acceptable levels.

4.5 Sewerage Systems

A sewerage system, also known as a wastewater collection system, is a sophisticated network comprising pipes, pumping stations, and various components that efficiently transport sewage from its source to a designated treatment and disposal point.

4.5.1 Combined Systems

Combined sewer systems are indeed a challenging aspect of urban wastewater management. These systems, prevalent in older cities, pose significant pollution risks, especially during wet weather events when stormwater overwhelms the capacity of treatment plants. As a result, untreated sewage is discharged directly into receiving waters, leading to water pollution issues.

To address combined sewer overflow problems, various strategies have been implemented. Some cities divert the initial flush of combined sewage into storage basins or underground tunnels for settling and disinfection treatment before discharge. Others opt for controlled treatment at nearby wastewater treatment plants to prevent overloading. Additionally, swirl concentrators are used to guide sewage through cylindrical structures, concentrating impurities for more efficient treatment.

Implementing these advanced techniques can help cities manage combined sewer overflows more effectively, reducing water pollution and promoting sustainable wastewater management practices.

4.5.2 Advantages of a Combined System

The combined system offers several notable advantages:

1. **Cost-Effective Maintenance**: By utilizing a single set of sewers, the combined system significantly reduces maintenance costs, making it an economically viable option.

2. **Enhanced Sewer Functionality**: The larger size of the sewers in this system minimizes the likelihood of blockages, while also facilitating easier cleaning processes.
3. **Dilution for Improved Treatment**: The presence of stormwater or rainwater dilutes the strength of the sewage, thereby simplifying and optimizing the treatment process. This not only enhances efficiency but also contributes to cost savings.
4. **Automatic Flushing**: The combined system incorporates automatic flushing mechanisms through the inclusion of stormwater or rainwater. This feature ensures a continuous flow and helps maintain the system's functionality.
5. **Simplicity**: The collection of sewage in this system is relatively straightforward, making it a simple and efficient solution.

4.5.3 Disadvantages of the Combined System

The combined system presents several drawbacks that need to be considered:

1. ***High Construction Costs***: The implementation of the combined system involves the construction of large-sized sewers at considerable depths below the ground surface, which requires extensive excavation. Consequently, the overall cost of construction is significantly higher.
2. ***Handling and Transportation Challenges***: The large size of the sewers poses difficulties in terms of their handling and transportation. This aspect adds complexity to the maintenance and operation of the system.
3. ***Increased Treatment Requirements***: Incorporating stormwater or rainwater treatment facilities of larger dimensions becomes necessary in the combined system. This results in additional costs and places a heavier load on the treatment units.
4. ***Economic Inefficiency***: If the lifting of sewage at the treatment works necessitates pumping the combined system proves to be uneconomical. Both the capital costs and the running costs are negatively impacted.
5. ***Unnecessary Pollution of Stormwater***: The inclusion of stormwater in the combined system leads to its unnecessary pollution. This aspect raises concerns about environmental sustainability and water quality.
6. ***Potential for Overflow and Pollution***: During periods of heavy rainfall, the sewers in the combined system may overflow, creating unhygienic conditions and causing pollution issues. This poses a significant challenge to maintaining a clean and healthy environment.
7. ***Silt Accumulation in Large Sewers***: Improperly designed large-sized sewers are prone to easy silt accumulation. Additionally, due to the small amount of dry weather flow compared to the total flow, these large sewers often become silted up during the dry part of the year, as the flow velocity decreases.

8. ***Ventilation Difficulties***: Ventilating large sewers is more challenging compared to smaller ones and this aspect adds complexity to the overall maintenance and operation of the combined system.

4.5.4 Conditions Favourable for Combined System

The combined system is advantageous in the following circumstances:

(i) ***Even Distribution of Rainfall***

When rainfall is evenly distributed throughout the year, the combined system can be implemented.

(ii) ***Pumping Requirements***

In cases where it is necessary to pump domestic or sanitary sewage stormwater (or rainwater), the combined system may be adopted.

(iii) ***Limited Space Availability***

When there are restrictions on the space available for laying sewers, it is advisable to opt for a combined sewer system.

(iv) ***Conversion of Existing Stormwater Sewer***

If it is feasible to convert an existing stormwater sewer into a combined sewer, then the combined system may be preferred. However, this conversion is only possible when the quantity of domestic or sanitary sewage and industrial sewage is relatively small compared to that of stormwater (or rainwater). By considering these favorable conditions, the implementation of a combined system can provide efficient and effective management of sewage and stormwater.

4.6 Separate Systems

In modern wastewater management systems, storm sewers and sanitary sewers are designed as separate entities, each serving distinct purposes. Storm sewers are dedicated to channeling surface runoff to disposal points like streams or rivers. They often incorporate small detention basins to temporarily hold stormwater and lessen peak flow rates during heavy rainfall events (Chupin et al., 2019). Reinforced concrete pipes are commonly used for constructing storm sewers, although corrugated metal pipes may be utilized in some cases. Strategic placement of stormwater inlets or catch basins along streets or private properties facilitates gravity flow towards disposal points, minimizing the need for pumping stations.

Sanitary sewers, on the other hand, transport domestic wastewater to sewage treatment plants. While pretreated industrial wastewater may be permitted into municipal sanitary sewerage systems, stormwater is strictly excluded. Sanitary sewer systems consist of laterals, submains, and interceptors. Laterals, the smallest sewers, rely on gravity to convey sewage into larger submains or collector sewers. Collector sewers merge into main interceptors or trunk lines, which transport sewage to treatment plants. Interceptors are typically constructed using precast sections of reinforced concrete pipe, with diameters reaching up to 5 m (15 ft.) (Imam & Elnakar, 2014).

These meticulous designs and dedicated systems ensure efficient wastewater management, minimizing environmental impacts and safeguarding public health.

4.6.1 Advantages of a Separate System

The advantages of a separate wastewater system are well-documented and contribute significantly to its preference in modern urban infrastructure planning:

1. **Efficient Treatment**: The smaller volume of sewage in a separate system allows for the use of smaller treatment facilities. This reduces the size and capacity requirements of treatment units, enhancing their efficiency by alleviating the load on them (Swain et al., 2010).
2. **Pollution Prevention**: By keeping stormwater separate from sewage, the separate system prevents contamination of natural water sources. This ensures that stormwater can be safely discharged into streams or rivers without additional treatment, preserving the quality of water bodies and reducing environmental pollution (Ferguson et al., 2008).
3. **Cost-Effectiveness**: In situations where pumping is necessary, the separate system offers economic advantages in terms of both initial capital costs and ongoing operational expenses. The smaller size of treatment facilities and sewers results in lower construction and maintenance costs, making the separate system a cost-effective solution.
4. **Economical Infrastructure**: The smaller diameter of sewers in a separate system contributes to its overall affordability. Additionally, stormwater can be conveyed through open or closed drains near the ground surface, further reducing installation costs while maintaining functionality (Eckenfelder, 2000).
5. **Enhanced Ventilation**: Smaller sewer sections in a separate system are easier to ventilate, improving air circulation and reducing the risk of odors and corrosion. This enhanced ventilation contributes to maintaining a healthier and more efficient sewer network (Liu et al., 2015).
6. **Streamlined Treatment Process**: The separate system ensures that sewage with relatively consistent characteristics flows through the sewers. This uniformity simplifies the treatment process, making it more streamlined and effective in removing contaminants (Baral, 2023).

By capitalizing on these advantages, communities can implement separate wastewater systems that are not only efficient and cost-effective but also environmentally sustainable.

4.6.2 Disadvantages of the Separate System

The separate wastewater system, while offering significant advantages, also presents several drawbacks that must be carefully considered:

1. **Cleaning Difficulties**: The small size of sewers in the separate system can make cleaning and maintenance challenging, requiring specialized equipment and techniques (Morrison et al., 2020).
2. **Choking Hazards**: Due to their smaller diameter, separate sewers are more susceptible to becoming clogged, leading to blockages that can disrupt the flow of wastewater (O'Connor et al., 2019).
3. **Insufficient Self-cleansing**: Without a steep gradient, separate sewers may lack sufficient velocity for self-cleansing, necessitating frequent flushing to prevent buildup and obstructions. This increases maintenance requirements and associated costs.
4. **Seasonal Use of Storm Water Drains**: Stormwater drains in separate systems are utilized primarily during the rainy season, leading to potential issues with garbage accumulation and blockages during periods of inactivity.
5. **Traffic Obstructions**: Maintenance or repair activities on separate sewers or drains can cause significant disruptions to traffic flow, especially when multiple systems are present in the same street (Rathi & Kumar, 2021).
6. **Foul Odors**: Limited air contact in smaller sewers of separate systems can contribute to the accumulation of sewage gases, resulting in foul odors that may affect nearby areas (Guzmán et al., 2016).
7. **Double House-Plumbing Requirements**: Implementing separate systems necessitates the installation of dual plumbing systems for households, increasing construction and maintenance costs (Ranasinghe et al., 2021).
8. **Risk of Incorrect Connections**: Incorrect connections between stormwater and sewage systems pose a risk of stormwater entering the sewage system, potentially causing overflow and environmental contamination (Sharma et al., 2020).

Careful consideration of these drawbacks is essential when evaluating the suitability and sustainability of separate wastewater systems for urban infrastructure planning.

4.6.3 Conditions Favorable for Separate System

(i) **Uneven Rainfall**

When precipitation is sporadic or confined to a brief period annually, adopting a separate system becomes economically viable.

(ii) **Separate Outlets for Sewage and Storm Water**

A separate system is warranted when there is a necessity to channel domestic, sanitary, and industrial sewage to a designated treatment facility, while utilizing a distinct outlet, such as a natural watercourse, for stormwater discharge.

(iii) **Pumping Requirement**

In scenarios necessitating the pumping of domestic, sanitary, and industrial sewage, a separate system alleviates the load on pumps.

(iv) **Limitations of Available Funds**

When initial funds are insufficient, constructing sewers exclusively for domestic, sanitary, and industrial sewage is feasible, with stormwater managed via open drains. These drains can be upgraded to standard sewers as additional funds become available.

(v) **Flat Topography**

In flat terrains, sewers need to be installed at a certain depth to achieve the necessary gradient. Separate sewers are more cost-effective in these situations, as building a larger, combined sewer at greater depth is costly.

(vi) **Gradient of Sewers**

When laying sewers at appropriate gradients is impractical, the risk of sewage backup into homes increases. Thus, a separate system is advisable.

(vii) **Subsoil Condition**

In cases of hard subsoil, installing a larger combined sewer is challenging and more expensive. Therefore, a separate system is preferable.

(viii) **Steep Topography**

In regions with steep slopes, managing stormwater or rainwater is more straightforward using open drains that lead directly to natural streams or rivers. Hence, implementing a separate system for this purpose is highly recommended.

(ix) **Optimal Timing for Sewer Installation**

To ensure maximum efficiency, installing sewers prior to area development is recommended. This approach facilitates the adoption of a separate system, offering numerous benefits.

(x) **Conversion of Existing Sewers**

Converting existing sewers to handle both domestic, sanitary, and industrial sewage within a combined system is undesirable due to high costs and inconveniences. It is more practical and cost-effective to implement a separate sewer system for stormwater or rainwater drainage.

4.7 Partially Separate System

In this system, domestic or sanitary sewage, industrial sewage, and stormwater from backyards and rooftops are conveyed through a unified set of sewers. Conversely, stormwater from house fronts, streets, and roads is collected and directed separately via distinct open drains (Imam & Elnakar, 2014). The combined sewage and stormwater transported by the sewers are typically routed to a sewage treatment plant for processing. Meanwhile, the stormwater collected by the open drains is channeled directly to a natural stream or river for disposal. This partially separate system facilitates efficient management of various wastewater types by integrating domestic and industrial sewage with specific stormwater sources. This integration optimizes the use of existing infrastructure.

The separation of stormwater from house fronts, streets, and roads ensures that potential contaminants are kept away from the main sewer system. This method balances functionality with environmental considerations, promoting the proper treatment and disposal of wastewater sustainably.

4.7.1 Advantages of a Partially Separate System

The partially separate system offers numerous advantages by combining the benefits of both separate and combined systems:

1. **Enhanced Efficiency**

By employing a partially separate system, sewers are designed to be of a manageable size, simplifying their cleaning and maintenance. This leads to a more efficient and effective wastewater management process.

2. **Prevention of Choking**

The inclusion of stormwater or rainwater in the system mitigates the risk of sewer blockages. Furthermore, sewers are thoroughly flushed during the rainy season, significantly reducing the likelihood of obstructions.

3. **Simplified Drainage**

The partially separate system streamlines the drainage of stormwater or rainwater from houses. This simplification ensures a more seamless and effective removal of excess water, enhancing overall system performance.

4.7.2 Disadvantages of a Partially Separate System

While the partially separate system offers numerous advantages, it is important to consider its potential drawbacks including:

1. **Increased Load on Infrastructure**

The admission of stormwater or rainwater into the sewers may place additional strain on pumping and treatment units, potentially impacting the overall efficiency and capacity of the system.

2. **Reduced Flow Velocity**

During dry weather conditions, the absence of rainwater can lead to a decrease in flow velocity within the system. This reduction in flow velocity can hinder the self-cleansing process, potentially resulting in a buildup of debris and sediment.

3. **Overflow Requirements**

To accommodate the influx of stormwater, it may be necessary to provide stormwater overflows. These overflows act as safety measures to prevent potential flooding or overwhelming of the system during periods of heavy rainfall.

In general, the partially separate system offers a balanced approach to wastewater management, combining the advantages of separate and combined systems. While it presents certain challenges, such as increased load on infrastructure and reduced flow velocity, these issues can be mitigated through proper planning and maintenance.

4.8 Alternative Systems

In certain situations, the cost of traditional gravity sewers can be excessively high due to factors such as low population densities or challenging site conditions like a high water table or bedrock. However, there are three alternative wastewater collection systems that can be utilized under these circumstances: small-diameter gravity sewers, pressure sewers, and vacuum sewers.

Small-diameter gravity systems involve the use of septic tanks to remove settleable and floating solids from the wastewater of each house before it enters a network of collector mains, typically measuring 100 mm (or 4 in.) in diameter. These systems are particularly suitable for small rural communities. Since these pipes do not transport grease, grit, and sewage solids, they can be of smaller diameter and placed at reduced slopes or gradients, thereby minimizing the costs associated with trench excavation (Imam & Elnakar, 2014).

Pressure sewers, on the other hand, are most effective in flat areas or locations where expensive rock excavation would be necessary. In this system, grinder pumps discharge wastewater from each home into the main pressure sewer, which can follow the natural slope of the ground.

A vacuum sewerage system operates by allowing sewage from one or more buildings to flow by gravity into a sump or tank. From there, vacuum pumps located at a central vacuum station pull the sewage out and direct it into a collection tank. Subsequently, the sewage is pumped to a treatment plant.

By considering these alternative systems, communities can overcome the challenges posed by high costs, low population densities, and difficult site conditions.

These solutions offer flexibility and efficiency, ensuring the effective collection and treatment of wastewater in various settings.

4.9 Pumps

Pumping stations are vital infrastructural components erected in scenarios where sewage requires elevation from lower to higher points or when natural downhill flow is impeded by the landscape. To facilitate the conveyance of raw sewage, specialized nonclogging pumps are deployed within structures known as lift stations, which can be broadly classified into two main types: wet well and dry well configurations.

In a wet-well setup, a singular chamber or tank serves as the receptacle for sewage accumulation before its expulsion. This chamber features submersible pumps and motors meticulously positioned at its base, fully immersed below the water surface.

Conversely, dry-well installations consist of distinct chambers: one for wastewater reception and another housing pumps and controls in a protected environment. The provision of a dedicated dry chamber facilitates convenient access for inspection and maintenance endeavors.

Regardless of wet-well or dry-well classification, it is imperative to incorporate a minimum of two pumps within lift stations. This redundancy ensures continuous operation, with one pump operational while the other undergoes maintenance or repair.

The deployment of these pumping stations enables the efficient and dependable conveyance of sewage, even in challenging topographical conditions, thereby bolstering wastewater management infrastructure.

4.10 Flow Rates

The fluctuating nature of sewage flow rates demands a sewerage system capable of accommodating these variations effectively. Urban areas, in particular, experience peak domestic sewage flows during morning and evening hours, with minimal levels recorded during late-night hours. Population density, water consumption patterns, and commercial or industrial activities significantly influence sewage flow quantities, closely correlating with community water usage averages.

In lateral sewers, short-term peak flow rates can surge to approximately four times the average, while trunk sewers experiencing peak rates up to two-and-a-half times the average flow. Importantly, sewage flows are not solely reliant on residential, commercial, and industrial connections but may also increase due to inflows and infiltration (I&I) into the sanitary sewer system.

Inflows occur when stormwater infiltrates sewers through improper connections like roof drains, storm drains, downspouts, and sump pumps. During precipitation or storm events, substantial rainwater runoff may inundate the sewer system, exacerbated by seasonal spring flooding from melting ice. Infiltration involves groundwater seepage into sewers through defective or damaged pipes, further contributing to hydraulic overload downstream at utilities and treatment plants.

To mitigate such strains, utilities may request residents to curtail usage of dishwashers, washing machines, showers, and, in extreme cases, limit toilet flushing. These measures aim to alleviate system burdens, particularly crucial in aging water infrastructures susceptible to issues related to inflows and infiltration.

4.11 Sewage Characteristics

Understanding the characteristics of sewage is pivotal for establishing an efficient and cost-effective waste management framework. It facilitates the selection of suitable treatment methodologies, determination of requisite treatment levels, assessment of potential waste applications, and orchestrated utilization of water bodies' natural purification capacities. Through comprehensive comprehension and ongoing monitoring of sewage traits, informed decisions pertaining to waste management and treatment processes can be made.

Domestic sewage constitutes utilized water from diverse sources such as kitchens, bathrooms, toilets, and others. Its attributes may vary contingent upon factors including daily water consumption per capita, water supply quality, type, condition, and extent of sewerage systems, as well as societal habits. Meanwhile, municipal sewage encompasses both domestic and industrial effluents, with variations observable across different locales owing to the presence and nature of industrial activities and establishments (Lynn et al., 2015).

The essential characteristics of sewage are:

- **Temperature**: Monitoring sewage temperature is critical for assessing oxygen solubility, influencing the efficiency of aeration equipment in aerobic systems, and regulating biological activity rates. Extremely low temperatures can detrimentally affect the performance of biological treatment systems and sedimentation efficiency. In Indian conditions, raw sewage temperatures typically range between 15 and 35 °C across various locations and seasons.
- **pH**: The pH level, indicating hydrogen ion concentration, plays a pivotal role in the operation of biological units. Fresh sewage generally exhibits a slightly higher pH compared to the community's water supply. However, organic matter decomposition may lower the pH, while industrial wastewater presence can lead to significant pH fluctuations. Raw sewage typically falls within the pH range of 5.5–8.0.
- **Color and Odor**: Fresh domestic sewage typically presents a slightly soapy and cloudy appearance, varying based on concentration. Over time, sewage becomes stale, resulting in darkened coloration and pronounced odors attributable to microbial activity.

- **Solids**: While sewage predominantly comprises water with less than 0.5% solids, the presence of these solids warrants attention due to their high degradability. Proper disposal of sewage solids is imperative. Categorized into dissolved, suspended, and volatile suspended forms, understanding the volatile or organic fraction's decomposition is crucial, representing the burden on biological treatment units or stream oxygen resources during disposal via dilution. Assessing suspended solids, both organic and inorganic, offers insight into sedimentation and grit removal system loads during sewage treatment. The dissolved inorganic fraction must be considered for land irrigation or any planned reuse of sewage (Chen et al., 2022).
- **Nitrogen and Phosphorus**: In domestic sewage, nitrogen compounds primarily comprise proteins, amines, amino acids, and urea. Bacterial decomposition of these organic constituents results in the presence of ammonia nitrogen. Nitrogen, a fundamental element of biological protoplasm, holds significant importance in ensuring the proper functioning of biological treatment systems. Typically, domestic sewage contains sufficient nitrogen levels to fulfill the requirements of biological treatment. However, in cases involving industrial wastewater with insufficient nitrogen content, external supplementation becomes necessary. Untreated sewage typically exhibits nitrogen concentrations ranging from 20 to 50 mg/L, measured as Total Kjeldahl Nitrogen (TKN).
- Phosphorus enters domestic sewage through food residues containing phosphorus and synthetic detergents, contributing significantly to its content. Phosphorus serves as an essential nutrient for biological processes. Domestic sewage typically contains phosphorus concentrations adequate for supporting aerobic biological wastewater treatment. However, phosphorus concentration becomes a concern when considering the reuse of treated effluent. The concentration of phosphate (PO_4) in raw sewage typically ranges from 5 to 10 mg/L (Chen et al., 2022). Nitrogen and phosphorus are integral components of domestic sewage. While nitrogen is generally present in sufficient quantities for biological treatment, additional nitrogen supplementation may be required for industrial wastewater. Phosphorus, on the other hand, is typically available adequately in domestic sewage for aerobic biological treatment, but its concentration becomes crucial when contemplating treated effluent reuse. Understanding nitrogen and phosphorus levels in sewage is imperative for effective wastewater management.
- **Chlorides**: The concentration of chlorides in sewage typically exceeds the normal chloride content found in water supplies, serving as an indicator of sewage strength. On average, each individual contributes approximately 8 g of chloride daily. Considering an average sewage flow of 150 L per capita per day (LPCD), this results in a chloride content in sewage approximately 50 mg per litre (mg/L) higher than that of the supplied water (Chen et al., 2022).
- Any abnormal increase in chloride levels warrants attention regarding the discharge of chloride-bearing wastes or infiltration of saline groundwater. Saline groundwater infiltration can also elevate sulphate levels, potentially leading to excessive hydrogen sulphide generation. Monitoring and controlling chloride concentrations in sewage are crucial for ensuring the proper functioning of

wastewater treatment systems and safeguarding the environment. Understanding chloride content provides insights into sewage strength and aids in identifying potential sources of contamination.

- **Organic Materials**: Organic compounds present in sewage hold significant importance for environmental engineering. A diverse range of microorganisms, found in both sewage and receiving water bodies, interact with these organic materials, utilizing them as sources of energy or materials in their metabolic processes. Within metabolism, catabolism involves the conversion of organic material to obtain energy, while anabolism entails the incorporation of organic material into cellular structures.
- To comprehend microbial metabolism and the oxidation of organic material, it is essential to quantitatively characterize the concentration of organic matter in various forms. Given the multitude of organic compounds in sewage, individually determining each one is impractical. Therefore, a parameter common to all these compounds must be utilized. In practice, two properties of organic compounds are typically employed: their oxidizability and organic carbon content.
- Environmental engineering employs two standard tests based on organic material oxidation: the Biochemical Oxygen Demand (BOD) test and the Chemical Oxygen Demand (COD) test. Both tests measure organic material concentration during the testing process. The key distinctions between the COD and BOD tests lie in the oxidant utilized and the operational conditions imposed, such as biochemical oxidation and chemical oxidation. An additional method for measuring organic material is the Total Organic Carbon (TOC) test, serving as an alternative for quantifying organic material concentration.
- By utilizing these standardized tests and characterizing organic matter concentration, environmental engineers acquire valuable insights into microbial behavior and organic compound oxidation.
- **Biological Oxygen Demand (BOD)**: Biochemical Oxygen Demand (BOD) quantifies the oxygen required for the biochemical breakdown of biodegradable organic matter under aerobic conditions, a critical parameter reflecting the amount of oxygen consumed during this process. This measurement directly correlates with the quantity of decomposable organic matter present. Typically, raw sewage exhibits a BOD range of 100–400 mg/L, with lower values commonly observed in average Indian cities (Chen et al., 2022; Hur et al., 2011; Metcalf & Eddy, 1991).
- **Chemical Oxygen Demand (COD)**: Chemical Oxygen Demand (COD), on the other hand, measures the oxygen needed for chemical oxidation, without distinguishing between biological oxidizable and nonoxidizable materials. Despite this, the COD to BOD ratio remains relatively constant for specific waste types, facilitating the assessment of treatment unit performance. Raw sewage COD levels typically range from 200 to 700 mg/L across different locations.
- In the COD test, organic matter oxidation typically completes within two hours, contrasting with the several weeks required for biochemical oxidation. Dealing with wastewater containing diverse organic compounds presents challenges for using BOD as a quantitative parameter due to variations in oxidation rates. Smaller molecules are readily metabolized by bacteria, while larger molecules

and colloidal or suspended matter necessitate hydrolysis as a preparatory step for metabolism. Establishing a general relationship between experimental five-day BOD and ultimate BOD (after several weeks) is impractical. For sewage (with a reaction constant (k) value of 0.23 d^{-1} at 20 °C), the BOD5 is approximately 0.68 times the ultimate BOD, with the ultimate BOD accounting for 87% of the COD. Thus, the COD/BOD ratio for sewage is approximately 1.7.

- **Toxic Metals and Compounds**: The infiltration of toxic heavy metals and compounds into municipal sewage presents a significant environmental concern. Substances such as chromium, copper, and cyanide, known for their harmful properties, can enter sewage systems through industrial discharges. The concentration of these compounds is crucial in determining suitable treatment methods for sewage or its safe disposal in streams or on land. While sanitary sewage typically maintains levels within acceptable toxic limits, the introduction of industrial discharges can exceed these limits, posing risks to municipal wastewater quality and environmental safety.

4.12 Wastewater Treatment

The treatment and safe disposal of wastewater are paramount for environmental protection and the conservation of natural resources. It is imperative that cities and towns properly treat wastewater before it is reintroduced into receiving waters or the land. To uphold stringent quality standards, specific guidelines are established for maximum allowable concentrations of solids (both suspended and dissolved), organic matter, nutrients, and pathogens (Droste & Gehr, 2018). The role of a design engineer in this process is to formulate a treatment plan that not only meets these standards but also considers various other factors such as construction and maintenance costs, availability of materials and equipment, and the need for specialized labor. By addressing these aspects, the design engineer ensures the technical feasibility of the treatment process (Droste & Gehr, 2018).

Primary treatment alone is insufficient to achieve effluent with an acceptable residual organic material concentration. Therefore, biological methods are commonly employed in treatment systems to achieve secondary treatment and remove organic material. Bacteria play a crucial role in metabolizing organic material in these biological treatment systems (Droste & Gehr, 2018; Liu & Lipták, 2020). Depending on the desired final effluent quality, tertiary treatment methods and/or pathogen removal may also be necessary.

Presently, most wastewater treatment plants utilize aerobic metabolism to eliminate organic matter. Common aerobic processes include the activated sludge process, oxidation ditch, trickling filter, and aerated lagoons. Stabilization ponds employ both aerobic and anaerobic mechanisms. However, there has been a growing interest in anaerobic treatment systems for wastewater and sewage treatment in recent years due to escalating costs associated with power and operation in aerobic processes (Crini & Lichtfouse, 2019).

4.13 Characteristics of a Treatment Plant

The quality of effluent generated by a treatment plant hinges upon the quality requirements of the receiving water body. These requirements are set forth either by legislation or via meticulous engineering analysis, considering the natural purification or self-purification mechanisms inherent in the receiving water. Regulation of these requirements can be accomplished through Stream Standards, which account for the water body's assimilative capacity, or through discharge standards uniformly enforced by governing authorities across jurisdictions, irrespective of the specific river water quality at individual locations.

4.14 Classification and Application of Wastewater Treatment Methods

Treatment methods are typically classified into three categories:

1. **Physical Unit Operations**: These methods rely on the application of physical forces to remove contaminants. Examples include screening, mixing, flocculation, sedimentation, flotation, and filtration (Liu & Lipták, 2020).
2. **Chemical Unit Processes**: These methods involve the removal or conversion of contaminants through the addition of chemicals or other chemical reactions. Examples include precipitation, gas transfer, adsorption, and disinfection.
3. **Biological Unit Processes**: These methods utilize biological activity to remove contaminants, primarily focusing on eliminating biodegradable organic substances from wastewater, whether in colloidal or dissolved form. Biological processes involve the conversion of organic matter into gases that can escape into the atmosphere and bacterial cells that can be removed through settling. Additionally, biological processes are utilized for nitrogen, phosphorus, and sulfate removal from wastewater.

Wastewater treatment plants employ various treatment methods, which can be categorized into three distinct categories:

1. **Primary Treatment**: This category encompasses physical unit operations aimed at removing solid materials and large particles from the wastewater.
2. **Secondary Treatment**: This category involves chemical and biological unit processes working together to further purify the wastewater by breaking down organic matter and removing additional contaminants.
3. **Tertiary Treatment**: This category refers to the utilization of one or a combination of physical unit operations, chemical processes, and biological processes with tertiary treatment employed after secondary treatment to achieve the highest level of purification.

By employing different treatment methods, wastewater treatment plants can effectively remove contaminants and ensure the safe disposal or reuse of wastewater.

4.15 Water Treatment

4.15.1 Pre-treatment

Pre-treatment plays a vital role in the water treatment process by focusing on the removal of materials that could potentially damage or clog the pumps and sewage lines of primary treatment clarifiers. These materials include plastics, tree limbs, leaves, branches, and other debris commonly found in raw sewage. The pre-treatment stage involves several processes, including:

4.15.1.1 Screening

During screening, influent sewage water passes through a bar screen designed to eliminate large objects such as cans, rags, sticks, and plastic packets carried in the sewage stream. In modern plants serving large populations, automated mechanically raked bar screens are commonly used, while smaller or less modern plants rely on manually cleaned screens. The raking action of mechanical bar screens is adjusted based on accumulation and flow rate. Collected solids are disposed of in a landfill or incinerated. Utilizing bar screens or mesh screens of varying sizes ensures optimal solid removal (Campos et al., 2016).

Types of Screens

Screens are categorized into three types based on the size of their openings:

(a) **Coarse Screens**: Coarse screens, also known as racks, are crucial for removing large floating objects from sewage. Typically, the spacing between bars is 50 mm or more. Material separated by these screens includes rags, wood, sticks, paper, and other decomposable debris, requiring proper disposal through incineration or burial. Coarse screens are constructed with parallel iron rods placed vertically or at a slight slope, usually inclined at approximately 45–60° to increase opening area and reduce flow velocity. Wastewater velocity passing through screens should not exceed 0.8–1 m/s. Material collected on the upstream side can be removed manually or mechanically (Lopez et al., 2022; Qasim, 2017).

(b) **Medium Screens**: Medium screens feature bar spacings ranging from 6 to 40 mm, collecting 30–90 L of material per million litres of sewage. Screenings often contain organic matter, necessitating proper disposal through incineration or burial. Rectangular-shaped coarse and medium screens made of steel are commonly used in sewage treatment plants, positioned at an angle of about 30–60° to increase opening area and reduce flow velocity (Qasim, 2017).

(c) **Fine Screens**: Fine screens have bar spacings ranging from 1.5 to 3 mm, removing approximately 20% of suspended solids from sewage. Made of fine wire, brass, or bronze plates with small openings, fine screens can be disc or drum type and are continuously operated by electric motors. They are commonly used for treating industrial wastewater or municipal waste containing industrial wastewater. Cleaning frequency is determined by head loss through the screen (Qasim, 2017).

4.15.1.2 Grit Channel or Chamber

In a sand or grit channel or chamber, the velocity of incoming sewage is regulated to facilitate the settlement of sand, grit, stones, and broken glass. Inorganic solids like pebbles, sand, silt, eggshells, glass, and metal fragments, along with heavier organics such as bone chips and seeds, combine to form grit. Removing this grit from wastewater prevents damage to pumps and the accumulation of grit in sludge digesters. Grit chambers function as sedimentation tanks, separating heavier inorganic materials through gravitational forces while allowing lighter organic material to pass through. They can be positioned either after or before the screen (Patziger, 2021).

The majority of substances found in grit are abrasive and can cause accelerated wear on pumps and sludge handling equipment. Grit deposited in low hydraulic shear areas like pipes, sumps, and clarifiers can absorb grease and solidify. These non-biodegradable materials take up valuable space in sludge digesters, underscoring the importance of separating them from organic suspended solids. Additionally, the quantity of grit present varies depending on the type, age, and condition of the pipes in the collection system, with infiltration being a major source of inorganic matter.

Grit removal facilities primarily consist of an enlarged channel area where reduced flow velocities allow grit to settle out. Two configurations of grit chambers are available: the channel type and the aerated rectangular basin.

4.15.1.3 Channel Type Basin

These are horizontal flow grit chambers where the horizontal velocity is maintained at approximately 0.3 m/s. Even a 25% increase in horizontal velocity may result in grit washout, while a 25% decrease may retain non-target organics. Hence, controlling horizontal velocity is crucial. This type of grit chamber is typically installed at small plants.

Grit from this chamber may contain a significant amount of biodegradable organics, necessitating their removal through washing or quick disposal to avoid nuisance problems. Grit containing organics should either be placed in a sanitary landfill or incinerated along with screenings to produce sterile ash for disposal.

4.15.1.4 Aerated Rectangular Basin

This type of basin is typically installed in large plants. By injecting compressed air, turbulence is generated, keeping lighter organic matter in suspension while allowing heavier grit to settle at the bottom. In this configuration, roll velocity holds more significance than horizontal velocity as it effectively separates non-target organics from grit, eliminating the need for artificial control of horizontal velocity. Adjusting the quantity of air enables control over settling.

4.15.1.5 Flow Equalization

Clarifiers and mechanized secondary treatment systems operate more efficiently under uniform flow conditions. Equalization basins serve as temporary storage for diurnal or wet-weather flow peaks, providing a location to hold incoming sewage temporarily during plant maintenance. They also serve to dilute and distribute batch discharges of toxic or high-strength waste, which could otherwise inhibit biological secondary treatment processes. Examples of waste requiring equalization include portable toilet waste, vehicle holding tanks, and septic tank pumpers.

Flow equalization basins necessitate variable discharge control and typically incorporate provisions for bypass and cleaning. They may also feature aerators. Locating the basin downstream of screening and grit removal processes facilitates cleaning.

4.15.1.6 Fat and Grease Removal

In larger plants, fat and grease removal involves passing sewage through a small tank. Skimmers are employed to collect floating fat on the surface. Additionally, air blowers positioned at the tank base may aid in recovering fat as froth. However, many plants prefer primary clarifiers equipped with mechanical surface skimmers for fat and grease removal.

4.15.2 Primary Treatment

During primary treatment, the removal of grit (fine, hard solids) and suspended solids occurs in two stages.

4.15.2.1 Pre-aeration

Initially, the wastewater undergoes aeration by introducing air through perforated pipes near the tank floor. This process reduces the water's density, facilitating the settling of grit. Air jets are strategically positioned to create swirling motion, preventing the settling of suspended solids. Additionally, the aeration supplies dissolved oxygen for later bacterial action. However, the wastewater does not remain in these tanks for sufficient bacterial action. The collected grit is washed and stored for on-site land reclamation and landscaping.

4.15.2.2 Sedimentation

Subsequently, the water flows slowly through sedimentation tanks. Within these tanks, suspended solids settle to the bottom, while scum rises to the surface. Scrapers remove solids from the tank bottom, and water jets wash off the scum. The scum and solids form sludge, sent for secondary treatment.

4.15.3 *Secondary Treatment*

The purpose of secondary treatment is to significantly degrade the organic matter content of sewage, originating from human waste, food waste, soaps, and detergents. Two processes are involved:

4.15.3.1 Activated Sludge

The Activated Sludge process is a widely used biological treatment method for wastewater management. It involves aerating sewage to promote the growth of microorganisms, which consume organic matter and nutrients. The resulting biomass is then separated from the treated water, allowing for effective reduction of pollutants and improved water quality before discharge.

4.15.3.2 Trickling Filters

Trickling filters are a biological wastewater treatment method that utilizes a bed of media, such as stones or plastic, where wastewater trickles over. Microorganisms grow on the media, breaking down organic pollutants. This process promotes natural aeration, enhancing treatment efficiency while reducing biochemical oxygen demand (BOD) and suspended solids in wastewater.

4.15.4 *Tertiary Treatment*

Tertiary treatment is the final stage of wastewater treatment, designed to remove remaining contaminants after primary and secondary processes. It typically involves advanced methods such as filtration, disinfection (using chlorine or UV light), and nutrient removal. This process significantly improves water quality, making it suitable for reuse or safe discharge into the environment.

4.15.4.1 Filtration

Filtration is a physical process used in water and wastewater treatment to remove suspended particles, such as sediments, pathogens, and organic matter (Metcalf & Eddy, 1991). It involves passing water through porous media like sand, gravel, or membranes. Filtration improves water clarity and quality, making it safer for consumption or further treatment (Hur et al., 2011).

4.15.4.2 Lagooning

Lagooning involves storing wastewater in large man-made ponds or lagoons, allowing settlement and further biological improvement. These highly aerobic lagoons promote the growth of native macrophytes, particularly reeds (Campos et al., 2016). Filter feeding invertebrates like Daphnia and certain species of Rotifera aid in treatment by removing fine particulates (Lopez et al., 2022).

4.15.4.3 Nutrient Removal

Excessive nitrogen and phosphorus levels in wastewater can harm the environment if released untreated, leading to eutrophication (Winkler & Straka, 2019). This process involves the rapid growth of weeds, algae, and cyanobacteria, causing algal blooms and oxygen depletion. Certain algal species produce toxins, further contaminating water supplies. Specific treatment processes are necessary to effectively remove nitrogen and phosphorus from wastewater.

These tertiary treatment methods collectively ensure that wastewater meets stringent quality standards before its release into the environment, safeguarding aquatic ecosystems and human health.

4.15.4.4 Nitrogen Removal

The removal of nitrogen from wastewater involves biological processes such as nitrification and denitrification. Nitrification converts ammonia to nitrate, while denitrification reduces nitrate to nitrogen gas, which is released into the atmosphere, effectively removing nitrogen from the water (Winkler & Straka, 2019).

Nitrification is a two-step aerobic process facilitated by different bacteria. *Nitrosomonas* spp. primarily convert ammonia (NH_3) to nitrite (NO_2), while *Nitrospira* spp. predominantly oxidize nitrite to nitrate (NO_3) in the environment.

Denitrification requires anoxic conditions to promote the formation of appropriate biological communities. Various bacteria participate in denitrification (Winkler & Straka, 2019).

While sand filters, lagooning, and reed beds can reduce nitrogen levels, the activated sludge process, if well-designed, offers an efficient solution for nitrogen removal.

4.15.4.5 Phosphorus Removal

Phosphorus removal is essential due to its role as a limiting nutrient for algal growth in freshwater systems, particularly in water reuse systems where high phosphorus concentrations can lead to equipment fouling (Winkler & Straka, 2019).

One effective method for phosphorus removal is Enhanced Biological Phosphorus Removal (EBPR), which involves selectively enriching polyphosphate accumulating organisms (PAOs). These bacteria have the ability to accumulate substantial amounts of phosphorus within their cells, up to 20% of their mass. When the biomass enriched with PAOs is separated from the treated water, the resulting bio-solids have significant fertilizer value (Winkler & Straka, 2019).

Alternatively, phosphorus removal can be achieved through chemical precipitation using salts of iron, aluminum, or lime. However, this method may lead to excessive sludge production and can be costly. The removed phosphorus can be stored in a landfill or sold for use in fertilizers as phosphate-rich sludge.

Overall, EBPR is an efficient method for phosphorus removal, as it selectively enriches PAOs, which accumulate substantial phosphorus within their cells, providing bio-solids with fertilizer value (Winkler & Straka, 2019).

In wastewater treatment plant design, understanding the following key terms and concepts is essential for effective planning and operation (CPHEEO, 1993).

1. **Flow Sheet**: A flow sheet is a graphical representation of the unit operations and processes used in the treatment plant. It provides a visual overview of how wastewater flows through the various treatment stages, aiding in understanding the treatment process.
2. **Process Loading Criteria**: Also known as designed criteria, these are foundational guidelines used to determine the size of each unit operation or process within the treatment plant. Adhering to these criteria ensures optimal and efficient plant operation.
3. **Solid Balance**: The solid balance involves quantifying the amounts of solids that enter and exit each unit operation or process in the treatment plant. This assessment helps maintain control over solid waste within the plant, ensuring proper management of solids throughout the treatment process.
4. **Hydraulic Profile**: The hydraulic profile is a tool used to determine the elevation of the wastewater's free surface as it flows through different treatment units. This profile aids in understanding flow dynamics and ensures smooth operation of the treatment process by identifying potential flow issues.

5. **Plant Layout**: Plant layout refers to the spatial arrangement of physical facilities within the treatment plant, as outlined in the flow sheet. Optimizing the layout is crucial for maximizing workflow efficiency and ensuring effective operation of the treatment plant.

By understanding these terms and their significance, engineers and operators can better plan, design, and operate wastewater treatment plants to provide clean and safe water for communities.

4.16 Order of Reaction

The reactions that occur during wastewater treatment are characterized by their slow nature, making it crucial to consider kinetics when designing treatment processes (Qasim, 2017). To relate the rate of change of concentration to time, a general equation is used:

$$\frac{dS}{dt} = K \times Sn$$

where,

S represents the concentration of the reacting substance,
K is the reaction rate constant per unit time, and
n denotes the order of the reaction.

The order of the reaction can be 1 for a first-order reaction, 2 for a second-order reaction, and so on.

The value of K is influenced by various environmental conditions within the reactor, including temperature, the presence of toxicity, the presence of catalysts, and the availability of nutrients and growth factors.

Zero-order reactions (n = 0) are independent of the concentration of the substance, resulting in a constant rate of change (dS/dt). Certain catalytic reactions and even some biological reactions can follow this pattern.

In first-order reactions, the rate of change of substance concentration is directly proportional to the concentration of that substance. Over time, both the concentration and rate will decrease. The decomposition of a single substrate is a prime example of a true first-order reaction.

A typical example of a pseudo-first-order reaction is the biological stabilization of organic matter in a batch reactor. In this case, the rate of reaction is proportional to the concentration of a single component, organic matter, assuming favorable conditions for other reaction-controlling parameters. If the substrate concentration (organic matter) is maintained within a narrow range, as seen in continuous flowing, completely mixed reactors, the rate of reaction remains practically constant, resembling a pseudo-zero-order reaction (Qasim, 2017). Some biological treatment systems exhibit this behavior.

By considering the order of reaction and its implications, wastewater treatment processes can be designed more effectively and efficiently. There are several intricate processes that exhibit an overall rate that is approximately first order in nature. When dealing with complex substrates such as sewage or industrial wastewaters, the overall reaction rate may resemble that of a first order reaction. However, it is important to note that individual substrates within this complex mixture may actually follow a zero order reaction. This occurs because the initial rate of reaction is higher, primarily due to the more readily biodegradable substrate being utilized. As time progresses, the rate of reaction slows down due to the presence of more challenging substrates remaining in the reactor.

4.17 Types of Reactors Used

a. **Batch Reactor**: A batch reactor is a closed system used in chemical processes where reactants are added, mixed, and allowed to react for a specific time. This type of reactor is commonly used in research and small-scale production, enabling precise control over reaction conditions, such as temperature and pressure, for optimized yields (Qasim, 2017).
b. **Plug-Flow Reactor**: A plug-flow reactor (PFR) is a continuous-flow system where reactants move through a cylindrical pipe or channel without back-mixing. In a PFR, reactants progress as "plugs," allowing for uniform reaction conditions. This design is efficient for chemical reactions, particularly in large-scale production, ensuring optimal conversion rates and product quality (Qasim, 2017).
c. **Continuous-Flow Stirred Tank Reactor**: A Continuous-flow Stirred Tank Reactor (CSTR) is a widely used system in chemical engineering where reactants are continuously fed into a well-mixed tank. The constant stirring ensures uniformity in temperature and concentration, facilitating efficient reactions. CSTRs are ideal for large-scale production and processes requiring steady-state conditions for optimal yields (Qasim, 2017).
d. **Arbitrary Flow Reactor**: An Arbitrary Flow Reactor is a versatile chemical reactor that allows for complex flow patterns and mixing configurations, enabling tailored reaction conditions. This type of reactor can accommodate varying feed compositions and operational parameters, making it suitable for a wide range of chemical processes, including batch and continuous operations in research and industry (Qasim, 2017).
e. **Packed Bed Reactors**: Packed Bed Reactors are commonly used in chemical processes, consisting of a cylindrical vessel filled with solid catalyst particles. Reactants flow through the packed bed, allowing for efficient mass and heat transfer. This design enhances reaction rates and is widely employed in catalysis, gas absorption, and wastewater treatment applications (Qasim, 2017).

f. **Fluidized Bed Reactor**: A Fluidized Bed Reactor utilizes a bed of solid particles that are suspended or "fluidized" by upward-flowing gas or liquid. This design enhances mixing and heat transfer, allowing for uniform reaction conditions. Fluidized bed reactors are widely used in various applications, including catalysis, combustion, and wastewater treatment, due to their efficiency (Qasim, 2017).

4.18 Flow Patterns of Reactors

The flow patterns within reactors are influenced by mixing conditions, which are determined by factors like reactor shape, energy input per unit volume, reactor size and scale, liquid up-flow velocity, and gas generation or supply rate. These factors collectively affect treatment exposure time and substrate distribution within the reactor (Qasim, 2017).

Reactor classification is based on the flow pattern observed, and can be categorized as follows:

(a) Batch reactors
(b) Ideal plug flow reactors
(c) Ideal completely-mixed flow reactors
(d) Non-ideal, dispersed flow reactors, and
(e) Series or parallel combinations of reactors.

The hydraulic regime within the reactor can be defined by the "Dispersion number", which characterizes the mixing conditions present (Kalbar et al., 2013).

$$Dispersion\,Number = \frac{D}{UL}$$

where,

D Axial or longitudinal dispersion coefficient, measured in m^2/s,
U Mean flow velocity along the reactor, measured in L/s, and
L Length of axial travel path, measured in L.

For ideal plug flow, D/UL = 0, indicates zero dispersion, D/UL $\leq$ 0.2 indicates a regime approaching plug flow conditions, while D/UL > 3.0–4.0 indicates a regime approaching completely mixed conditions. Understanding the flow patterns and hydraulic regimes within reactors is crucial for optimizing their performance and achieving desired treatment outcomes.

4.19 Advancements in Wastewater Treatment

The American Institute of Chemical Engineering defines advanced wastewater treatment as any process capable of reducing impurities in wastewater to levels beyond what can be achieved through conventional secondary or biological treatment methods (Hong et al., 2022). These innovative treatment solutions have garnered global attention as individuals, communities, and industries seek ways to preserve and maintain essential resources for use. The need for advanced wastewater treatment has become increasingly urgent as society grapples with the consequences of population growth, urbanization, industrialization, and the depletion of potable water sources. Inefficient wastewater treatment can give rise to a host of concerns, including unpleasant odors and potential health hazards. To address these challenges, new techniques are being implemented to enhance wastewater treatment processes. By harnessing these cutting-edge technologies, it becomes possible to elevate the quality of wastewater beyond the limitations of conventional methods, ultimately achieving the goal of resource recovery or conservation (Hong et al., 2022). The effluents from advanced wastewater treatment plants can be recycled either directly or indirectly, thereby bolstering the available domestic water supply.

4.20 The Importance of Advanced Wastewater Treatment

The necessity for advanced wastewater treatment arises from various factors that must be taken into consideration. These factors include:

1. **Meeting Stringent Discharge and Reuse Requirements**: Conventional secondary treatment processes may not be sufficient to eliminate organic matter and total suspended solids to the extent required. Advanced treatment methods are necessary to ensure compliance with more stringent regulations for discharge and reuse (Hong et al., 2022).
2. **Enhancing Disinfection Efficiency**: To effectively disinfect treated wastewater, it is crucial to remove residual suspended solids. Advanced treatment techniques aid in conditioning the wastewater for more efficient disinfection processes (Hong et al., 2022).
3. **Preventing Eutrophication**: Conventional secondary treatment processes may not adequately remove nutrients, leading to the eutrophication of sensitive water bodies. Advanced treatments are essential to limit this phenomenon and protect the health of aquatic ecosystems (Hong et al., 2022).
4. **Meeting Stringent Requirements for Specific Contaminants**: Advanced treatment methods are indispensable for the removal of specific inorganic (such as heavy metals) and organic constituents. These treatments are necessary to comply with stricter discharge and reuse regulations for both surface water and land-based effluent disposal. Additionally, they are crucial for indirect portable reuse applications, such as groundwater recharge (Hong et al., 2022).

4.21 Effective Removal of Organic and Inorganic Colloidal and Suspended Solids

The elimination of organic and inorganic colloidal and suspended solids is typically achieved through filtration processes including:

4.21.1 Depth Filtration

Depth filtration, a method widely used in wastewater treatment and various industries, involves the use of filters that trap particles within their porous structure, facilitating the removal of solids from the wastewater. This process is highly effective in eliminating a wide array of undesirable particles, including dirt, grit, sand, organic solids, and rust particles (Nejatishahidein & Zydney, 2021). Depth filtration employs meticulously chosen materials such as polypropylene, cotton, and glass fiber to construct cartridges that physically intercept these particles, preventing their presence in the final filtered liquid.

Within the realm of depth filtration, two primary types of filtration media prevail:

1. **Rolled Depth**: These media consist of bands specifically designed to cater to various applications, each possessing known micron ratings. This versatility allows for their utilization in diverse scenarios.
2. **Graded Density Media**: This type of media boasts a graded density of microns throughout its structure, rendering it suitable for a wide range of particle sizes.

By employing depth filtration, industries can ensure the removal of unwanted contaminants, thereby enhancing the quality and purity of their liquid products (Nejatishahidein & Zydney, 2021).

4.21.1.1 Working of Depth Filtration

Depth filtration is a highly efficient method for separating suspended particles or liquid droplets from their carrying fluid within the thickness of the filter medium. Unlike other filtration technologies, depth filters capture unwanted particles throughout the entire medium, rather than just on the surface. This characteristic makes them particularly effective in situations where water contains a high concentration of particles.

During depth filtration, large particulate matter is initially trapped as the water passes through the cartridge media, following the path of least resistance. Simultaneously, the water undergoes a twisting and turning motion, maximizing particle capture by the densely packed fibers. This sieving mechanism, known as mechanical retention, allows the fibers to accumulate smaller particles as they become increasingly compacted.

The outer layer of the cartridge captures larger particles, while the inner layer, tightly packed, targets smaller particles. This comprehensive approach ensures that particulate matter is captured at every level of the depth media. Additionally, particle bridging occurs at various points along the filtration media, further enhancing buildup and maximizing cartridge efficiency.

In addition to physical particle capture, cartridge fibers utilize electrostatic charges to attract particles through a process called adsorptive retention. The combination of these processes results in a highly efficient filtration mechanism suitable for a wide range of applications.

Overall, depth filtration offers a reliable and effective solution for separating suspended particles or liquid droplets from their carrying fluid. Its ability to capture and retain a large volume of particles before clogging sets it apart from other filtration methods. With its comprehensive design and efficient mechanisms, depth filtration is a valuable tool in various industries and applications.

4.21.1.2 Applications

Depth filters, deep bed filters, and thick media filters play crucial roles in various industries by efficiently removing unwanted materials from fluids. These filters are employed in a range of applications, including clarification of drinking and process water, polishing of wastewater before discharge, and cleaning of industrial liquids.

Depth filters are designed to remove small amounts of suspended material from fluid flows. They are commonly used to clean waste streams before discharge into the environment or as part of industrial processes. The primary function of depth filters is to remove solid waste material, and their cleaning process aims to restore their filtration characteristics rather than retrieve accumulated solids.

Deep bed filters are widely utilized in water treatment processes for both municipal treatment works and industrial sites. They are particularly effective in the clarification of drinking and process water and in polishing wastewater before discharge. While there is potential for cleaning hot exhaust gases, this application is less prevalent compared to water treatment.

Thick media filters are extensively used in industrial processes for liquid cleaning purposes. They play a critical role in maintaining the integrity of hydraulic and pneumatic systems and cleaning machine tool coolant. Thick media filters ensure the required clarity of liquids for membrane processes and find applications in industries such as food and beverage, fine chemicals, and toiletries.

Overall, depth filters, deep bed filters, and thick media filters serve vital roles in various industries, contributing to the efficient removal of unwanted materials and ensuring the quality of fluids used in processes or discharged into the environment.

4.21.2 Surface Filtration

Surface filtration is a critical process used in water filtration systems to effectively remove particles from the liquid's surface. This method involves the formation of a layer of material, known as the "cake layer," which enhances filtration efficiency. As suspended particles pass through the filter, they accumulate on the surface, gradually forming the cake layer. Surface filters are often employed before chemical or biological treatments, such as chlorine disinfection, as their surfaces may not withstand harsh chemicals. However, surface filters are essential for eliminating large particles that could obstruct subsequent filter stages.

The process of surface filtration involves the mechanical sieving of suspended particulate material in a liquid by passing it through a thin septum, also known as the filter material. Various materials can serve as filter septum, including woven metal fabrics, cloth fabrics with different weaves, and synthetic materials. Cloth-medium surface filters, with openings typically ranging from 10 to 30 μm or larger, are commonly used in this filtration method.

There are two primary types of cloth-medium surface filtration devices:

a. ***Discfilter***: The Discfilter (DF) consists of a series of vertically mounted parallel disks that support the filter cloth. Each disk is connected to a central feed tube. The two-dimensional cloth screen material used in the DF can be either polyester or type 316 stainless steel.
b. ***Cloth-Media Disk Filter***: The cloth-media disk filter (CMDF) comprises several vertically mounted disks within a tank. Two different types of cloth can be utilized in the CMDE:

   i. **A needle felt cloth made of polyester**
   ii. **A synthetic pile fabric cloth**.

The needle-felt cloth features a unique three-dimensional weave that effectively removes particles. In addition to regular backwashing, it is necessary to periodically clean the felt cloth using a high-pressure spray. On the other hand, the pile fabric does not require a high-pressure spray wash and can be thoroughly cleaned using backwash alone.

4.21.3 Membrane Filtration Process

Membrane filtration is a critical process in wastewater treatment, aimed at effectively separating particulate and colloidal matter, as well as dissolved constituents, from a liquid. Membrane filtration utilizes semi-permeable membranes as selective barriers, allowing the passage of certain compounds while retaining others based on their physical and/or chemical properties.

This method serves as a physical separation technique, leveraging the unique characteristics and varying sizes of molecules in the liquid. By acting as a physical obstruction, the membrane permits the passage of desired constituents while preventing the passage of unwanted ones. Membrane filtration encompasses a wide range of filters, each with different pore sizes and configurations, catering to diverse filtration needs.

The field of membrane filtration is rapidly expanding within the realm of water treatment, finding numerous applications in the removal of relatively large particulate material to the elimination of dissolved compounds. Its effectiveness and versatility make membrane filtration a pivotal tool in wastewater treatment, contributing to the overall efficiency and quality of treated water.

4.21.3.1 Membrane Materials

Most membranes utilized in filtration processes are synthetic organic polymers, which are tailored to meet specific pore size requirements. Microfiltration and ultrafiltration membranes are often made from similar materials but undergo different formation conditions to achieve varying pore sizes. These membranes play crucial roles in separating particles and molecules based on size and other properties.

In addition to organic polymers, membranes can also be fabricated from inorganic materials such as ceramics or metals. Ceramic membranes, for instance, are microporous, thermally stable, and chemically resistant, making them well-suited for microfiltration applications. However, their widespread adoption has been impeded by factors such as high cost and mechanical fragility.

Metallic membranes, typically composed of materials like stainless steel, exhibit finely porous structures. Although their primary application lies in gas separations, they can also be employed for high-temperature water filtration or as membrane supports in various processes. While each type of membrane has its advantages and limitations, the choice of membrane material depends on factors such as pore size requirements, chemical compatibility, and operating conditions.

4.21.3.2 Mechanism of Membrane Filtration

Membrane filtration is acknowledged as a highly efficient pretreatment method in comparison to conventional approaches due to its ability to achieve superior water quality outcomes while utilizing less space and fewer chemicals. As the cost of membrane materials continues to decrease, the economic feasibility of operating membrane filtration plants becomes increasingly attractive.

The mechanism underlying membrane filtration technology for capturing microorganisms involves two primary phenomena. First, there are physiochemical interactions between the membrane and microorganisms including adhesion, surface charge interactions, and hydrophobic or hydrophilic interactions, depending on the properties of both the membrane material and the microorganisms present in the water. Second, there is the sieving effect, where microorganisms larger than the pore size of the membrane are physically retained. Additionally, the negatively charged membrane surface can repel microorganisms through electrostatic forces, further enhancing the capture efficiency (Rezakazemi et al., 2018).

Overall, membrane filtration stands out as a highly effective and versatile technology for water treatment, offering advantages in terms of both performance and cost-effectiveness compared to traditional methods.

4.22 Removal of Dissolved Organic and Inorganic Constituents

Activated Carbon Adsorption is a widely utilized method for the removal of dissolved organic constituents from wastewater due to its effectiveness in capturing a wide range of organic compounds. This process involves the accumulation of substances from the liquid phase onto a solid surface, typically activated carbon, through physical adsorption.

Activated carbon is produced from organic materials such as almond, coconut, or walnut hulls, as well as wood, bone, or coal. The production process involves heating the base material to red heat in a retort to create char, which is then activated by exposure to oxidizing gases like steam and CO_2 at high temperatures. This activation process generates a porous structure within the carbon, increasing its surface area and adsorption capacity.

The activated carbon is available in two forms: Powdered Activated Carbon (PAC) and Granular Activated Carbon (GAC). PAC consists of finely ground particles with a high surface area, allowing for rapid adsorption kinetics. GAC, on the other hand, comprises larger granules that offer a slower adsorption rate but provide a longer contact time with the wastewater.

The choice between PAC and GAC depends on various factors such as the specific contaminants present in the wastewater, the required treatment efficiency, and the operational constraints of the treatment system (Table 4.1). Additionally, activated carbon can be regenerated and reused through thermal reactivation or chemical regeneration processes, further enhancing its cost-effectiveness and sustainability.

Overall, activated carbon adsorption stands as a versatile and efficient treatment method for the removal of dissolved organic constituents from wastewater, offering customizable options to meet the diverse needs of wastewater treatment plants (Guillossou et al., 2019).

Granular activated carbon (GAC) filters play a crucial role in water treatment, offering highly effective technology for the removal of dissolved organic compounds and other contaminants. Derived from organic sources such as wood, coconut shells, coal, or peat, GAC serves as a versatile filtration media in both fixed bed applications and pump and treat procedures for the purification of drinking water and the decontamination of polluted water.

Table 4.1 Comparison of granular and powdered activated carbon

Parameter	Unit	Types of activated carbon	
		GAC	PAC
Total surface area	m^2/g	700–1300	800–1800
Bulk density	kg/m^3	400–500	360–740
Particle density, wetted in water	kg/L	1.0–1.5	1.3–1.4
Particle size range	mm (μm)	0.1–2.36	(5–50)
Effective size	mm	0.6–0.9	na
Uniformity coefficient	UC	≤ 1.9	na
Mean pore radius		16–30	20–40
Iodine number		600–1100	800–1200
Abrasion number	Minimum	75–85	70–80
Ash	%	≤ 8	≤ 6
Moisture as packed	%	2–8	3–10

The inclusion of GAC in water treatment systems enables the targeted removal of specific chemicals, particularly organic contaminants, as well as substances responsible for unpleasant odors or tastes in water. Through the activation process, which involves heating, the surface area of the carbon is increased, enhancing its adsorption capacity for dissolved contaminants as water passes through the filter containing granular activated carbon.

Activated carbons, including both GAC and powdered activated carbons (PAC), are widely recognized for their ability to remove undesirable odor, color, taste, and other organic and inorganic impurities from both domestic and industrial wastewater. This efficacy is attributed to their extensive surface area, micro-porous structure, and nonpolar characteristics, making them economically viable options for water treatment processes.

The primary constituent of activated carbon is carbon itself, which accounts for a significant portion of its weight. Additionally, activated carbons may contain other heteroatoms such as hydrogen, nitrogen, sulfur, and oxygen, further enhancing their adsorption capabilities.

Overall, the utilization of granular activated carbon filters in water treatment processes offers a highly efficient and reliable means of removing dissolved organic compounds and improving water quality. With their unique properties and versatile applications, GAC filters play a critical role in ensuring the safety and purity of drinking water and the protection of aquatic environments from pollution (Ahmadpour, 2022).

4.23 Activated Carbon Treatment Processes Applications

Activated carbon treatment processes are highly versatile and find applications across various industries. One primary application is the removal of persistent organic compounds and residual inorganic compounds from wastewater. Activated carbon adsorption is particularly effective in eliminating stubborn organic pollutants and trace amounts of inorganic substances such as nitrogen, sulfides, and heavy metals.

Additionally, activated carbon treatment is extensively used for the removal of taste and odor compounds from wastewater, especially when water reuse is desired. Both powdered and granular activated carbon are employed in these processes due to their low adsorption affinity for low molecular weight polar organic species. However, challenges may arise in removing these compounds if the biological activity in the carbon contactor or other biological unit processes is low.

Typically, after undergoing carbon treatment, the effluent biochemical oxygen demand (BOD) ranges from 2 to 7 mg/L, while the effluent chemical oxygen demand (COD) ranges from 10 to 20 mg/L under normal conditions. However, under optimum conditions, it is possible to achieve even lower effluent COD levels, often less than 10 mg/L, highlighting the effectiveness of activated carbon treatment in reducing organic pollutants and improving water quality (Ahmadpour, 2022).

4.24 Treatment with Granular Activated Carbon

Granular Activated Carbon (GAC) treatment involves passing a liquid through a reactor containing activated carbon, which effectively removes various contaminants from the water. Different types of activated carbon reactors are used for advanced wastewater treatment, including pressure or gravity-based systems with downflow or upflow fixed-bed units arranged in two or three columns in series. Alternatively, an expanded bed upflow-countercurrent type can be employed. Figure 4.1a–d provides schematic diagrams illustrating these carbon reactor configurations (Ahmadpour, 2022).

Activated carbon treatment processes offer effective solutions for removing refractory organic compounds, inorganic compounds, taste, and odor compounds from wastewater. By utilizing granular activated carbon and selecting appropriate system configurations, significant reductions in effluent COD levels can be achieved.

4.24.1 Fixed-Bed Columns

Fixed-bed columns are commonly used for contacting wastewater with granular activated carbon (GAC). These columns can be operated individually, in series, or in parallel. To remove organic compounds associated with suspended solids in

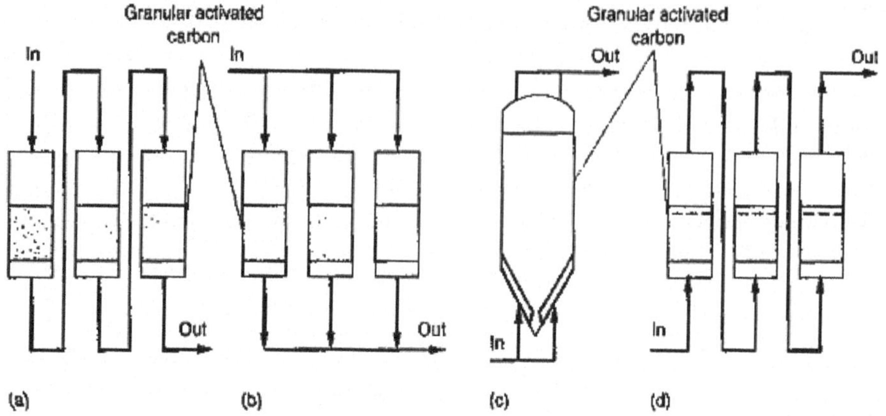

Fig. 4.1 a–d Types of activated carbon contractors: **a** downflow in series; **b** downflow in parallel; **c** moving bed; **d** upflow expanded in series

secondary effluent, granular-medium filters are often employed upstream of the activated carbon contactors. In this setup, water to be treated is applied to the top of the column and withdrawn from the bottom. An underdrain system is used at the bottom of the column to hold the carbon in place. Backwashing and surface washing are typically provided in wastewater applications to prevent headloss buildup caused by the removal of particulate suspended solids within the carbon column (Ahmadpour, 2022). In addition to fixed-bed columns, expanded-bed, moving-bed, and pulsed-bed carbon contactors have been developed to address headloss buildup challenges. In the expanded-bed system, the influent is introduced at the bottom of the column, allowing the activated carbon to expand, similar to how a filter bed expands during backwash. Once the adsorptive capacity of the carbon at the bottom of the column is depleted, the bottom portion of carbon is removed, and an equivalent amount of regenerated or virgin carbon is added to the top of the column.

4.25 Treatment with Powdered Activated Carbon (PAC)

An alternative method for achieving adsorption is by utilizing powdered activated carbon (PAC). PAC can be employed in various ways: it can be applied to the effluent from biological treatment processes, directly integrated into different biological treatment processes, or incorporated into physical–chemical treatment processes.

4.25.1 Biological Treatment Processes

In biological treatment plants, PAC is often introduced into the effluent within a contacting basin. Here, sufficient contact time is allowed for the carbon particles to settle to the bottom of the tank, and the treated water is then extracted. This process enhances the removal of soluble refractory organics, which are compounds resistant to conventional biological treatment methods. The addition of PAC directly into the aeration basin of an activated sludge treatment process has proven highly effective in removing a significant number of these organics. The presence of PAC in the aeration basin facilitates adsorption of soluble contaminants, which are then biodegraded by the activated sludge, thus enhancing overall treatment efficiency (Altmann et al., 2016).

4.25.2 Physical–Chemical Treatment Processes

In physical and chemical treatment processes, PAC is often combined with chemicals used for the precipitation of specific constituents. This combination can significantly enhance the overall efficiency of the treatment process. For instance, PAC can be used alongside coagulants and flocculants to improve the removal of targeted substances. The combined action of adsorption (by PAC) and precipitation (by chemicals) leads to a more effective removal of a wide range of contaminants, including organic compounds, heavy metals, and other refractory materials.

4.25.3 Advantages of PAC

Employing PAC in wastewater treatment offers several advantages:

1. **Versatility**: PAC can be tailored to various treatment processes, ensuring optimal removal of contaminants.
2. **Enhanced Adsorption**: The high surface area and adsorption capacity of PAC make it particularly effective for removing soluble refractory organics and other specific contaminants.
3. **Improved Treatment Efficiency**: By integrating PAC with biological and physical–chemical treatment processes, the overall efficiency and effectiveness of wastewater treatment are significantly improved.

The utilization of PAC presents a valuable solution for wastewater treatment, providing an efficient and adaptable means of achieving superior water quality. This method offers a versatile approach that can be tailored to different treatment processes, ensuring the optimal removal of contaminants and enhancing the overall treatment effectiveness. The integration of PAC into both biological treatment plants and physical–chemical treatment processes has demonstrated promising results in improving the quality of treated water, particularly in the removal of soluble refractory organics and targeted constituents (Altmann et al., 2016).

4.26 Reverse Osmosis

4.26.1 Basic Mechanism

Reverse osmosis (RO) is a process that produces demineralized water by exerting high pressure on water, forcing it through semi-permeable membranes. In ordinary osmosis, water naturally diffuses through a semi-permeable membrane from a less concentrated solution to a more concentrated one until osmotic pressure equalizes the levels. In reverse osmosis, high pressure is applied to the concentrated solution, reversing the natural flow and allowing pure water to pass through the membrane while retaining dissolved solids and impurities.

4.26.2 Applications in Industrial Effluent Treatment

Reverse osmosis can be applied to various industrial sectors for treating effluent, recovering resources, and reusing water:

- **Treating Colored Effluents**: RO can treat outflows containing dyes, potentially enabling the recovery of these dyes for reuse.
- **Oily Emulsions and Paints**: Effluents containing oily emulsions, latex, and electro-phoretic paints can be effectively treated using RO, separating and recovering valuable materials.
- **Metal-Finishing Industry**: RO can treat outflows to recover concentrated solutions of metal salts, allowing for the reuse of water in cleaning processes.
- **Chemical and Pharmaceutical Industries**: RO can handle wastewater from organic, inorganic chemical, and pharmaceutical industries, which often contain complex and hazardous contaminants (Garud et al., 2011).

4.26.3 Challenges in Treating Industrial Wastewater

Wastewater treatment using RO differs from general water purification due to the higher levels and variability of contaminants. Industrial wastewater's diverse nature poses several challenges, including:

- **Organic Fouling**: Organic substances, including surfactants and bacteria, can foul the membranes, reducing their efficiency.

- **Mineral Scaling**: Inorganic salts like calcium and iron can precipitate on the membrane surface, causing scaling.
- **Chemical Degradation**: Certain chemicals in wastewater can degrade membrane materials, reducing their lifespan and effectiveness.

4.26.4 Pre-treatment Requirements

Effective RO treatment of industrial wastewater requires comprehensive pretreatment to mitigate fouling and scaling with the key pre-treatment steps including:

- **Cation/Anion Balance**: Establishing a balance of ions in the wastewater to predict and control potential scaling.
- **Flocculation**: Identifying and using appropriate flocculants to aggregate and remove particulate matter.
- **Removing Specific Contaminants**: Pre-treatment technologies vary depending on the industry and contaminants e.g.,
- **Metal Finishing and Electronics**: Effluents often contain heavy metals which are removed before discharge. RO can then reduce total dissolved solids (TDS), making the water suitable for reuse. Ion exchange can further polish RO product water for high-purity applications.
- **Microelectronics**: Specialized pre-treatment technologies are required to handle the specific contaminants found in rinse waters (Wenten, 2016).

4.26.5 Pre-treatment Technologies

A wide range of pretreatment technologies can be employed to address specific fouling and scaling issues e.g.,

- **Coagulation and Flocculation**: These processes aggregate fine particles into larger clusters that can be more easily filtered out.
- **Media Filtration**: Removes larger suspended solids that could clog membranes.
- **Softening**: Reduces hardness ions such as calcium and magnesium that contribute to scaling.
- **Activated Carbon Adsorption**: Removes organic contaminants that could foul the membranes.
- **Chemical Oxidation**: Breaks down organic foulants and neutralizes biological activity.

Designing an Effective RO System

Designing a successful RO system requires a detailed evaluation of the wastewater characteristics, including pH, oxidizing potential, and soluble salt concentrations. The goal is to determine the optimum preconditioning chemistry and select the most suitable pretreatment technology for the specific application. Each industrial effluent presents unique challenges that must be addressed to ensure the cost-effectiveness and efficiency of the RO treatment process.

4.27 Chemical Precipitation

4.27.1 Basic Mechanism

Chemical precipitation is an essential process in water and wastewater treatment that transforms dissolved materials into solid particles by introducing counter-ions. This process primarily targets metallic cations but is also effective in removing anions such as fluoride, cyanide, and phosphate, and certain organic molecules, including phenols and aromatic amines, through enzymatic and detergent-induced precipitation (Mbamba et al., 2015).

4.27.2 Precipitation Process

The process of chemical precipitation involves several key steps:

1. **Addition of Chemicals**: Chemicals, known as precipitants, are added to the wastewater, which react with dissolved ions to form insoluble compounds.
2. **Formation of Precipitates**: The reactions result in the formation of solid particles (precipitates) that are less soluble in water.
3. **Aggregation of Particles**: Coagulants, such as polymers, are often used to aggregate smaller particles into larger flocs that are easier to remove.
4. **Separation**: The solid precipitates are separated from the liquid through filtration, centrifugation, or sedimentation.

4.27.3 Common Precipitants

Different precipitants are used depending on the target contaminants. Some common precipitants include:

- **Calcium Hydroxide (Lime)**: Used for the precipitation of heavy metals.
- **Ferric Chloride and Aluminum Sulfate**: Used for the removal of phosphates.
- **Barium Chloride**: Effective for removing oily emulsions and sulfates.
- **Sodium Sulfide**: Used for precipitating heavy metals like mercury and cadmium.

4.27.4 *Coagulants*

Coagulants enhance the precipitation process by facilitating the aggregation of smaller particles. Some common coagulants include:

- **Polymers**: Long-chain molecules that can be cationic, anionic, or nonionic. Their charge properties enable them to act as bridges between suspended particles or to neutralize charges, enhancing the formation of larger aggregates (Jacangelo et al., 1989).

4.27.5 *Applications*

Chemical precipitation is widely used to remove a variety of contaminants from wastewater:

- **Metallic Cations**: Heavy metals such as lead, mercury, cadmium, and chromium.
- **Anions**: Phosphate, fluoride, and cyanide.
- **Organic Compounds**: Phenols, aromatic amines, and organophosphates.
- **Suspended Solids**: Fats, oils, greases, and colloidal particles.

4.27.6 *Effectiveness and Considerations*

The effectiveness of chemical precipitation depends on several factors:

- **Dosage**: The amount of precipitant and coagulant used must be carefully controlled to achieve optimal results.
- **pH Control**: The pH of the wastewater significantly influences the solubility of the precipitates.
- **Mixing and Contact Time**: Adequate mixing and sufficient contact time are essential for complete precipitation and floc formation.
- **Post-treatment**: The settled precipitate needs to be efficiently separated from the treated water, which can be achieved through various methods such as sedimentation, filtration, and centrifugation.

Chemical precipitation is a versatile and proven technology for removing a wide range of contaminants from wastewater. By transforming dissolved materials into solid precipitates, which can then be separated, this method ensures a clear effluent that is substantially free from suspended matter or colloidal substances. The use of coagulants further enhances the aggregation of particles, making the process more efficient. Chemical precipitation continues to be a cornerstone of water and wastewater treatment, offering reliable solutions for the removal of metals, inorganics, suspended solids, and certain organic substances.

4.28 Metals Removal

Water hardness is primarily caused by the dissolution of compounds such as calcium and magnesium carbonate and bicarbonate in water. Additionally, sulfates, chlorides, and silicates of these metals contribute to water hardness to a lesser extent. The process of removing these dissolved compounds, known as water softening, often involves chemical precipitation. When lime (calcium oxide) is added to hard water, it reacts with the water to form calcium carbonate. This calcium carbonate can act as a coagulant, effectively removing ions from the water and allowing them to settle. However, using lime alone requires a significant amount of lime to be effective. To address this, lime is often combined with ferrous sulfate, resulting in the formation of insoluble ferric hydroxide. It is important to note that the combination of lime and ferrous sulfate only works in the presence of dissolved oxygen. Another method of water softening involves the use of alum. When alum is added to water containing calcium and magnesium bicarbonate alkalinity, it reacts with the alkalinity to form an insoluble aluminum hydroxide precipitate. Additionally, hydroxide compounds can be added to convert soluble heavy metal ions into insoluble metal hydroxides or carbonates. Ferrous sulfate and lime can also be used to form insoluble metal sulfides. Once these compounds become insoluble, they tend to precipitate and settle. The solubility of these metal compounds is pH dependent, with most being least soluble in alkaline solutions. Therefore, the optimal pH for precipitation depends on the specific metal and counter ion used (hydroxide, carbonate, or sulfide). It is crucial to determine the best treatment procedure on a case-by-case basis, considering factors such as metal solubility data available in various sources (Benefield & Morgan, 1990). To expedite the settling process, a polymer coagulant that gathers the particles of insoluble metal compounds into larger compounds can be added.

4.29 Removal of Fats, Oils, and Greases

Fats, oils, and greases are organic substances that have a tendency to clump together or form slick layers on the surface of water-based solutions. This behavior is due to their insolubility in water and non-polar nature. Additionally, these substances are less dense than water, causing them to float rather than sink to the bottom. When the oily substance is present in slicks, skimming the surface of the solution is the most effective method for removing the majority of the material. However, fats, oils, and greases can also become emulsified in water, meaning that small droplets of the oily product become dispersed throughout the solution. These droplets are formed by the aggregation of similarly charged molecules that are hydrophobic, or repel water. In some cases, other substances present in the solution, such as soaps and detergents, can aid in making hydrophobic substances soluble in water.

To remove emulsified oils and greases, it is necessary to break up the emulsion by destabilizing the electrical charge attractions that keep the clusters of oily molecules stable in the solution. This can be achieved by introducing a specially designed polymer that neutralizes the charges. By doing so, the attractive forces between the oily particles are disrupted, allowing them to separate from the water-based solution. This method ensures that the emulsified oils and greases can be efficiently removed from the water, thus improving the overall quality and purity of the treated water.

4.30 Phosphorus Removal

Metal salts, typically ferric chloride or aluminum sulfate (also known as alum), as well as lime, have been commonly used for removing phosphate compounds from water. When lime is employed, a sufficient amount must be added to raise the pH of the solution to at least 10. This creates an environment where excess calcium ions can react with the phosphate, resulting in the formation of an insoluble precipitate called hydroxylapatite. Lime is a highly efficient agent for removing phosphates, although it does result in a significant increase in sludge volume.

On the other hand, when ferric chloride or alum is employed, the iron or aluminum ions present in the solution react with phosphates, forming insoluble metal phosphates. The degree of insolubility of these compounds is contingent upon the pH level. Ferric chloride or alum introduces iron or aluminum ions into the solution, which then react with phosphates to create insoluble metal phosphates. It is worth mentioning that the solubility of these compounds is influenced by the pH level.

In the realm of phosphate removal, lime stands out as a remarkably effective agent. However, it does come with the drawback of generating a substantial amount of sludge. Alternatively, the utilization of ferric chloride or alum introduces iron or aluminum ions into the solution, which then react with phosphates to create insoluble metal phosphates with the solubility of these compounds influenced by the pH levels.

4.31 Suspended Solids

Finely divided particles suspended in a solution can evade filtration and other similar removal processes due to their small size, which enables them to remain suspended for extended periods. Typically, the particles present in wastewater carry a negative charge. To address this, cationic polymers which serve two purposes are commonly introduced into the solution reducing the surface charge of the particles and creating bridges between them. As a result, particle coagulation occurs, leading to settling (Tchobanoglous et al., 1991).

Alternatively, lime can be employed as a clarifying agent to eliminate particulate matter. When calcium hydroxide reacts with the wastewater solution, it forms calcium carbonate, which acts as a coagulant, effectively removing particles from the solution. This dual approach—using cationic polymers to neutralize charge and induce bridging, or employing lime to form calcium carbonate—facilitates the aggregation and subsequent removal of fine particles, ensuring more effective wastewater treatment.

4.32 Applicability

Chemical precipitation is a versatile and effective technique for removing a variety of contaminants from both municipal and industrial wastewaters applicable in several areas:

1. **Water Softening**: Chemical precipitation effectively removes compounds such as calcium and magnesium that contribute to water hardness. By introducing chemicals like lime (calcium oxide), these compounds form insoluble precipitates that can be easily removed from the water.
2. **Heavy Metal Removal**: This method is particularly valuable for eliminating heavy metals from industrial effluents, such as metal plating wastes. By adding appropriate precipitants, metals like lead, copper, and zinc are converted into insoluble compounds, facilitating their removal from the wastewater.
3. **Oil and Grease Removal**: Chemical precipitation can effectively break emulsified solutions, allowing for the removal of oils and greases. This is especially useful in industries where emulsified oils are prevalent, such as in food processing and metalworking.
4. **Phosphate Removal**: Phosphates, commonly found in wash-waters and other industrial effluents, can be precipitated using chemicals like lime, ferric chloride, or aluminum sulfate (alum). This process helps in reducing the nutrient load in wastewater, which is crucial for preventing eutrophication in receiving water bodies.
5. **Wastewater Polishing**: Beyond primary treatment, chemical precipitation serves as a polishing step to remove remaining particulate matter and further clarify the effluent. This enhances the overall quality of treated water, making it suitable for various reuse applications or safe discharge into the environment.

Chemical precipitation is a highly applicable and effective method for the removal of a broad range of contaminants from wastewater. Its ability to target specific pollutants makes it a valuable tool in both municipal and industrial wastewater treatment processes.

4.33 Advantages and Disadvantages

4.33.1 Advantages

Chemical precipitation offers several distinct advantages, making it a widely used and reliable method for wastewater treatment:

1. **Well-Established Technique**: Chemical precipitation is a proven and established method, with extensive knowledge and experience available regarding its application and effectiveness. This historical foundation ensures reliability and predictability in treatment outcomes.
2. **Cost-Effectiveness**: Many treatment chemicals used in chemical precipitation, particularly lime, are highly cost-effective. Lime is readily available and inexpensive, making it an economically viable option for large-scale operations.
3. **Autonomous Operation**: Enclosed systems used in chemical precipitation are often designed to operate autonomously with minimal human intervention. These systems typically require only the periodic replenishment of chemicals, reducing the need for continuous monitoring and manual operation.
4. **Low Maintenance**: Chemical precipitation systems generally require minimal maintenance. The main operational requirement is the replenishment of treatment chemicals, which can often be managed without the need for highly skilled operators. This reduces labor costs and simplifies the overall management of the treatment process.

These advantages make chemical precipitation an attractive option for various wastewater treatment applications, combining cost-efficiency, ease of operation, and reliability to effectively remove contaminants from wastewater.

4.33.2 Disadvantages

Chemical precipitation, while effective, comes with several notable disadvantages that must be carefully considered:

1. **Challenging Dosage Calculations**: Determining the proper dosage of chemicals for precipitation can be complex due to competing reactions, fluctuating alkalinity levels, and other factors. Achieving the right balance is crucial for optimal treatment outcomes, but it often requires frequent testing and adjustments, adding complexity to the process.
2. **Operator Safety Concerns**: Handling corrosive chemicals, such as those used in chemical precipitation, poses risks to operator safety. Accidental exposure or mishandling of these chemicals can lead to injuries or health hazards. Strict safety protocols and protective equipment are necessary to minimize these risks.
3. **Increase in Waste Sludge**: The addition of treatment chemicals, particularly lime, can significantly increase the volume of waste sludge generated during the

treatment process. This excess sludge must be managed and disposed of properly, which adds complexity and cost to the overall treatment process.

4. **Transportation Risks**: Transporting large quantities of chemicals to treatment facilities can pose logistical challenges and safety risks. Accidents during transportation can result in spills or environmental contamination, requiring careful planning and mitigation strategies.
5. **Cost of Polymers**: The use of polymers, often employed as coagulants or flocculants in chemical precipitation, can contribute to overall treatment costs. These chemicals can be expensive, particularly for large-scale treatment operations, adding to the financial burden of the treatment process.

Despite these disadvantages, chemical precipitation remains a valuable and widely used method for water treatment, particularly in situations where other treatment options may not be feasible or cost-effective. Effective management of these drawbacks through proper planning, monitoring, and safety protocols is essential to ensure the successful implementation of chemical precipitation processes.

4.34 Advanced Chemical Oxidation

The emergence of persistent and harmful compounds in natural waters, resistant to conventional treatment methods, has driven the development of advanced oxidation processes (AOPs) as a solution. AOPs involve chemical oxidation processes that modify the structure and properties of organic substances, breaking them down into smaller, more easily biodegradable compounds. This approach typically results in the formation of oxidized compounds such as alcohols and carboxylic acids, which are more amenable to degradation (Andreozzi et al., 1999; Rasalingam et al., 2014).

Advanced Oxidation Technologies (AOTs) encompass a range of physicochemical conversion methods, including AOPs, non-thermal plasmas (NTP) for air and wastewater treatment, electrohydraulic cavitation and sonolysis for wastewater treatment, catalytic oxidation for air treatment, wet air oxidation for wastewater treatment, and supercritical water oxidation treatment.

By employing these advanced oxidation technologies, the challenges posed by persistent and harmful compounds in natural waters can be effectively addressed. These processes offer promising solutions for treating water contaminated with toxic, mutagenic, and carcinogenic pollutants. Through the utilization of AOTs, efficient and sustainable water treatment can be achieved, ensuring environmental protection and the well-being of all living organisms.

4.34.1 Advanced Oxidation Processes (AOPs)

Advanced Oxidation Processes (AOPs) represent a cutting-edge approach to water purification, leveraging the potent reactivity of hydroxyl radicals. These processes utilize various oxidation technologies, including UV/O_3, O_3/H_2O_2, UV/H_2O_2, and the photo Fenton reaction ($UV/H_2O_2/Fe^{++}$ or Fe^{+++}), to achieve water treatment objectives (Andreozzi et al., 1999).

AOPs revolutionize water treatment by generating significant quantities of hydroxyl radicals, which efficiently target and eliminate contaminants present in water sources. Through oxidation technologies like UV/O_3, O_3/H_2O_2, UV/H_2O_2, and the photo Fenton reaction, AOPs offer comprehensive and effective solutions for water purification.

The UV/O_3 method combines ultraviolet (UV) light and ozone (O_3) to produce hydroxyl radicals, effectively breaking down organic pollutants and disinfecting water. Similarly, the O_3/H_2O_2 technique involves the combination of ozone and hydrogen peroxide (H_2O_2) to generate hydroxyl radicals, ensuring the removal of harmful substances from water sources.

In the UV/H_2O_2 approach, UV light and hydrogen peroxide synergistically generate hydroxyl radicals, facilitating the degradation of organic compounds and the elimination of pathogens. The photo Fenton reaction utilizes UV light, hydrogen peroxide, and iron ions (Fe^{++} or Fe^{+++}), to produce hydroxyl radicals that oxidize and remove contaminants from water.

The application of AOPs in water treatment represents a significant advancement, offering highly efficient and environmentally friendly solutions. By harnessing the reactivity of hydroxyl radicals, AOPs ensure thorough purification of water sources, addressing a wide range of contaminants and impurities.

4.34.2 Theory of Advanced Oxidation

The theory underlying advanced oxidation processes (AOPs) centers on the generation of hydroxyl free radicals (HO*), which serve as potent non-selective chemical oxidants. These radicals are employed to eliminate organic compounds that resist oxidation by conventional oxidants like ozone, oxygen, and chlorine (Munter, 2001).

Hydroxyl radicals exhibit exceptional efficacy in breaking down organic chemicals due to their highly reactive electrophilic nature. They react swiftly and non-selectively with nearly all electron-rich organic compounds, owing to their oxidation potential of 2.80 V, facilitating faster oxidation rates compared to conventional oxidants (Deng & Zhao, 2015).

Once hydroxyl radicals are generated, they initiate attacks on organic chemicals through processes such as electron transfer, hydrogen abstraction, and radical combination (Metcalf et al., 1991). The ensuing reactions can be summarized as follows:

(i) $R + HO\cdot \rightarrow ROH$
(ii) $R + HO\cdot \rightarrow R + H_2O$
(iii) $Rn + HO\cdot \rightarrow Rn^{-1} + OH$.

By harnessing the power of hydroxyl free radicals, advanced oxidation processes offer a highly effective means of eliminating recalcitrant organic compounds. This innovative approach holds great promise for various applications in water treatment, wastewater management, and environmental remediation.

It is important to note that complete oxidation of a compound or group of compounds is not always necessary. In many cases, partial oxidation is sufficient to reduce the toxicity of specific compounds or make them more suitable for subsequent biological treatment. To better understand the extent of degradation achievable, it is helpful to classify it into four groups:

1. Primary degradation involves a structural change in the parent compound, without necessarily reducing its toxicity.
2. Acceptable degradation refers to a structural change in the parent compound that results in reduced toxicity, making the compound less harmful.
3. During mineralization, organic carbon compounds are converted into inorganic carbon dioxide, representing the ultimate degradation by breaking down the compound into its basic components.
4. Unacceptable degradation involves a structural change in the parent compound that increases its toxicity, which is undesirable and should be avoided.

By comprehending the different levels of degradation, we can effectively manage and mitigate the environmental impact of these compounds (Covinich et al., 2014; Diya'uddeen et al., 2011).

4.34.3 Mechanism of Advanced Oxidation Processes

Advanced Oxidation Processes (AOPs) encompass two distinct types: homogeneous processes and heterogeneous processes. Homogeneous processes involve chemical alterations resulting solely from the interactions between the chemical reagents and the target substances (Aramyan & Moussavi, 2017).

On the other hand, heterogeneous processes involve the adsorption and desorption of reactants and products on the active sites of a catalyst. As the reaction progresses, the desorption of products and the absorption of new species occur on these active sites. Consequently, the efficiency of the process is influenced by the surface characteristics and pore structure of the catalyst (Ribeiro et al., 2015).

Various methods exist for the production of hydroxyl radicals, crucial in AOPs. These methods include the Fenton process (Fe^{2+}/H_2O_2), solar photo-Fenton process (solar/Fe^{2+}/H_2O_2), solar photocatalysis (solar/TiO_2/H_2O_2), peroxone (Ozone/H_2O_2), the combination of peroxone with ultraviolet (Ozone/H_2O_2), Ozone/UV, and photolysis (H_2O_2/UV). The formation mechanisms of oxidation by-products in these diverse AOPs are thoroughly discussed.

Homogeneous Advanced Oxidation Processes (AOPs) that utilize UV radiation are typically employed to degrade compounds that absorb UV radiation within the corresponding range of the spectrum. These compounds, which absorb UV light at lower wavelengths, are particularly suitable for this type of photodegradation process.

Photolytic Ozonation (O/UV): O/UV represents an advanced water treatment method that effectively addresses the oxidation and elimination of toxic organic compounds present in wastewater. This technique involves subjecting ozone-saturated aqueous systems to UV light with a wavelength of 254 nm. Extensive research conducted by Lin and Lin (2021) and Gualda-Alonso et al. (2022) has consistently demonstrated the remarkable efficacy of O/UV in the destruction of chlorophenols (CPs), surpassing the results achieved through UV photolysis and ozonation alone.

4.35 Hydrogen Peroxide and Ultraviolet Radiation (H_2O_2/UV)

The H_2O_2/UV advanced oxidation process involves the formation and generation of hydroxyl radicals through the photolysis of hydrogen peroxide and various propagation reactions. The photolysis of hydrogen peroxide occurs when UV radiation is applied, as shown in the following equation:

$$\left[H_2O_2 + h\upsilon \rightarrow 2HO^{\cdot}\right]$$

To achieve effective results, this method requires a longer exposure to UV radiation and/or a relatively high dose of H_2O_2. Numerous studies have demonstrated that the rate of photolysis of hydrogen peroxide is highly dependent on the pH level, with alkaline conditions leading to an increase in the reaction rate. To enhance the molar absorptivity of hydrogen peroxide, UV lamps emitting lower wavelengths can be utilized, as the absorptivity is low at 253.7 nm. A drawback of this method is that when the water to be treated has a higher absorbance, it tends to compete with hydrogen peroxide for radiation (Fotiou et al., 2015).

The H_2O_2/UV system has the capability to completely mineralize any organic compound, resulting in the production of water and CO_2. The technique offers several key advantages. Hydrogen peroxide is highly soluble, allowing it to be added to the source water at high concentrations. Additionally, H_2O_2/UV processes generate larger amounts of hydroxyl radicals compared to O_3/UV, using the same amount of energy.

However, there are major disadvantages to consider. The method is expensive due to the additional costs of necessary devices and energy requirements. Furthermore, the presence of residual hydrogen peroxide in the treated effluent can promote biological re-growth within the distribution system (Rashid & Barakat, 2015).

4.36 The Fenton Process

The Fenton process, pioneered by Henry J. H. Fenton, utilizes a combination of iron (II) salt and hydrogen peroxide, collectively termed the Fenton reagent. Fenton's research demonstrated that hydrogen peroxide can oxidize a variety of organic molecules, generating hydroxyl radicals (OH) in the presence of Fe(II) as a catalyst (Bautista et al., 2008). This mechanism is heralded as a highly effective method for the remediation of severely contaminated water due to its ability to degrade persistent compounds with ease. Despite its efficacy, the primary limitation of the Fenton process is the production of iron sludge waste.

4.36.1 The Solar Photo-Fenton Method

Among the spectrum of advanced oxidation processes (AOPs), the solar photo-Fenton process emerges as a compelling method for the degradation of a variety of hazardous pollutants. This technique is lauded for its environmental compatibility. By harnessing sunlight as an energy source, the solar photo-Fenton process has attracted significant interest, especially in regions with abundant sunlight such as India. This makes it a cost-effective AOP option (Barwal & Chaudhary, 2016). Figure 4.2 depicts the schematic diagram of the SPF. Both the Fenton process and the solar photo-Fenton process present robust solutions for water remediation, with the latter being particularly beneficial due to its dependence on renewable solar energy.

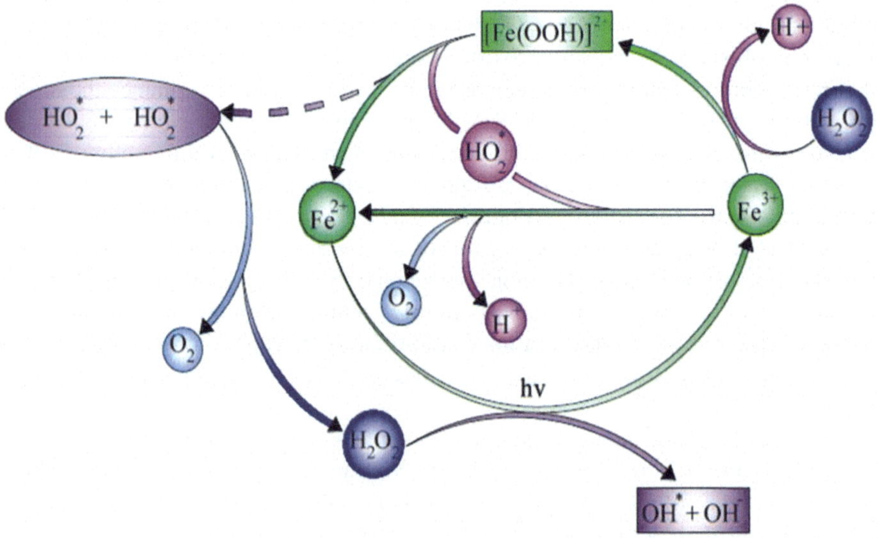

Fig. 4.2 Schematic diagram of a solar photo-Fenton process

4.36.2 Reaction Rates and Influencing Factors in Fenton and Photo-Fenton Processes

The reaction rates and stoichiometry of both the Fenton and Photo-Fenton processes are significantly influenced by the conditions under which they are performed and the composition of the water being treated. These effects stem from the solution chemistry, with pH being a critical factor. The efficiency of the reactive system formed is highly dependent on pH levels, which profoundly affect the decomposition of organic compounds (Zhang et al., 2005).

Research indicates that the Fenton and Photo-Fenton processes achieve optimal efficiency at a pH of 3. At higher pH levels, the decomposition rates diminish due to the precipitation of $Fe(OH)_3$. Conversely, at pH values below this optimal range, the reaction rate decreases because of a lower light absorption coefficient associated with complex formation.

Another essential factor influencing these processes is the concentration of hydrogen peroxide (H_2O_2) (Maezono et al., 2011). This concentration directly impacts the rate and extent of organic compound degradation. An increase in hydrogen peroxide concentration generally results in an increased reaction rate. It is crucial to add sufficient hydrogen peroxide to drive the reaction effectively, especially during the pretreatment of complex organic wastewater to reduce toxicity.

Increasing the hydrogen peroxide dose can lead to a steady reduction in chemical oxygen demand (COD) with minimal or no change in toxicity until a threshold is reached. However, exceeding the optimal hydrogen peroxide limit does not enhance degradation efficiency. This is due to the auto-decomposition of hydrogen peroxide into water and oxygen, as well as the recombination of hydroxyl radicals (OH) (Maezono et al., 2011; Pesakhov et al., 2007). Understanding these optimal conditions and the constituents of the water is crucial for selecting the most suitable method for wastewater treatment.

4.36.3 Electro-Fenton Method

An alternative advanced oxidation process is the Electro-Fenton (EF) method. This technology operates by continuously generating H_2O_2 at a suitable cathode, supplied with O_2 or air. Additionally, an iron catalyst is added to the treated solution to facilitate the production of hydroxyl radicals ($OH\cdot$) via a Fenton reaction (Nidheesh & Gandhimathi, 2012). The Electro-Fenton method offers several advantages, including on-site production of H_2O_2 and a higher degradation rate of organic pollutants. This efficiency is achieved through the continuous regeneration of Fe^{2+} at the cathode, which also helps minimize sludge production (Kavitha & Palanivelu, 2004).

4.37 Anodic Oxidation

Another advanced oxidation process suitable for wastewater treatment is anodic oxidation (Candia-Onfray et al., 2018). This is an accelerated electrochemical process that utilizes the natural oxide skin of aluminum. Over time, the transparent oxide layer formed becomes significantly thicker compared to the natural oxide layer. One notable advantage of this process is its long-term protection and strong resistance to corrosion (Nidheesh & Gandhimathi, 2012).

4.38 Heterogeneous Advanced Oxidation Processes

In heterogeneous advanced oxidation processes (AOPs), catalysts are employed to degrade contaminants. The term "heterogeneous" signifies that the contaminants are in the aqueous phase, while the catalyst is usually in the solid phase. This configuration enhances chemical reactions by creating electron–hole pairs. The photogenerated electrons and holes facilitate oxidation and reduction processes, respectively. When the catalyst absorbs water molecules in aqueous solutions, hydroxyl radicals (OH·) are produced. This process is typically conducted under aerobic conditions, with oxygen acting as the electron acceptor, resulting in the formation of superoxide radicals.

4.39 Role of Photocatalysts in Heterogeneous Advanced Oxidation Processes

Photocatalysts are integral to heterogeneous advanced oxidation processes, where they play a crucial role in facilitating the degradation of compounds. For a photocatalyst to be considered effective, it must possess different characteristics like: biological and chemical inertness, photoactivity, capability to utilize visible or near UV light, cost-effectiveness, photostability, and non-toxicity (Chong et al., 2010).

Examples of photocatalysts include Si, ZnO, WO_3, TiO_2, CdS, $SrTiO_3$, ZnS, SnO_2, Fe_2O_3, and WSe. Among these, Titanium dioxide (TiO_2) stands out as the most commonly used and has proven to be an excellent catalyst for the photo-oxidation of various inorganic and organic compounds.

In heterogeneous advanced oxidation processes, catalysts play a crucial role in driving oxidation and reduction reactions by generating electron–hole pairs in the solid phase. Photocatalysts, such as Titanium dioxide, are essential components of these processes due to their ability to efficiently drive the desired chemical reactions.

4.40 Titanium Oxide: Photocatalytic Perspective

Titanium, ranking ninth among the most abundant elements in the Earth's crust, exists in three crystalline forms: rutile, anatase, and brookite, with its most stable form being an oxide. From a photocatalytic viewpoint, only anatase and rutile are pertinent, with rutile demonstrating the highest catalytic activity.

The photocatalytic process initiates as TiO_2 absorbs UV radiation, light, or sunlight, generating electron–hole pairs. Upon illumination, excited electrons migrate to the conduction band of titanium dioxide, producing positive holes and negative electrons.

4.41 Operating Parameters

The efficacy of the photocatalytic process is influenced by several key operating parameters. These include the quantity of catalyst utilized, the initial concentration of the reactant, and the pH level. A comprehensive understanding and precise control of these factors are imperative for optimizing the photocatalytic reaction (Bagheri et al., 2014).

4.42 Applications of Advanced Oxidation Processes (AOPs)

Advanced Oxidation Processes (AOPs) have garnered significant attention due to their superior efficacy compared to individual agents such as hydrogen peroxide, UV, or ozone. Studies have consistently shown that AOPs outperform these agents, especially in treating wastewaters with low Chemical Oxygen Demand (COD), where the cost-effectiveness of H_2O_2 and/or ozone is advantageous for generating hydroxyl radicals. These radical-based processes exhibit remarkable capability in transforming previously resistant materials into compounds that are more amenable to further biological treatment.

4.43 Disinfection and Treatment of Refractory Organic Compounds

One of the primary applications of AOPs lies in the disinfection of treated wastewater and the treatment of refractory organic compounds (Merayo et al., 2013). AOPs offer enhanced disinfection potential by harnessing the superior oxidizing power of free radicals generated from ozone compared to ozone alone. This heightened oxidizing capability effectively neutralizes microorganisms and refractory organic materials present in wastewater. However, it is important to note that the relatively short half-life of hydroxyl free radicals makes achieving high concentrations impractical. Consequently, prolonged detention times are required for disinfection of microorganisms, necessitating the application of the CT (concentration × time) concept for determination. As a result, hydroxyl radicals are rarely used for disinfection purposes but are more commonly employed in oxidizing trace elements of refractory organic compounds found in extensively treated effluents.

4.44 Versatile Applications

AOPs offer a wide range of applications that surpass the capabilities of individual agents. Their potential for disinfection and treatment of refractory organic compounds makes them a valuable tool in various wastewater treatment processes.

4.45 Electrodialysis: A Highly Efficient Water Treatment Process

Electrodialysis stands as a highly effective water treatment method utilizing semipermeable reverse osmosis membranes to transport ions based on their charge. Through the application of electrical current, it significantly reduces the ionic content of water. This electrochemical process entails the migration of ions through ion-selective, semipermeable membranes.

4.45.1 Principle and Operation

Unlike reverse osmosis, which relies on external pressure, electrodialysis employs an electric current to facilitate water passage across the membrane. This technique excels in salt removal and the principle underlying electrodialysis is that dissolved salts, as impurities, carry either positive or negative charges. Upon application of an electric field, these salts migrate towards electrodes of opposite charge. Electrodialysis utilizes two types of ion-selective membranes: cation membranes, permitting only cations transfer, and anion membranes, allowing only anions transfer.

4.45.2 Working Mechanism

The functioning of an electrodialysis cell is depicted in Fig. 4.3. A typical system comprises an electrolytic cell divided into three compartments, utilizing two semipermeable membranes. Alternating cationic and anionic membranes between electrodes create parallel channels. Upon passing an electric current through the waste or saline water, negatively charged ions migrate towards the anode, and positively charged ions towards the cathode (Gurreri et al., 2020). As a result, solutions in alternate channels become more concentrated, while others become more dilute, yielding ion-free, pure water.

4.45.3 Effectiveness of Electrodialysis in Dissolved Solids Removal

The electrodialysis method has demonstrated high effectiveness in removing a substantial portion of dissolved solids from wastewater. Research conducted by Valero et al. (2011) reveals that this method can eliminate anywhere from 50 to 94% of total dissolved solids in wastewater with an initial concentration of 12,000 mg/L.

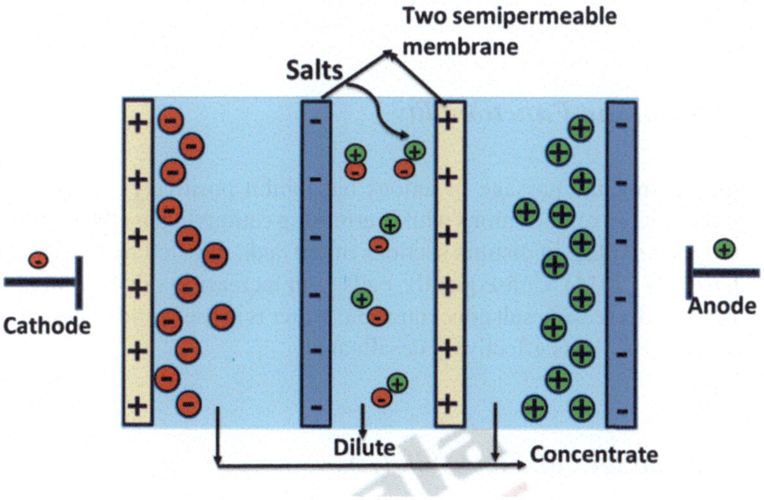

Fig. 4.3 A phenomena of electrodialysis

4.45.4 Limitations and Drawbacks

However, it is essential to acknowledge the limitations of electrodialysis, with its inability to effectively remove organic molecules being a major drawback. This limitation arises due to the larger size of organic molecules compared to inorganic ones, leading to their entrapment in the small pores of the semi-permeable membrane used in the process.

Additionally, electrodialysis presents a challenge in terms of its high energy requirement for desalinating concentrated feed solutions. This energy-intensive nature poses significant challenges, both economically and environmentally, making it imperative to explore alternative approaches to mitigate these drawbacks.

4.45.5 Operation of Electrodialysis

The electrodialysis process involves immersing electrodes in a salt solution and connecting them to a suitable direct current supply, facilitating the flow of current carried by ions present. Cations, positively charged ions, are attracted to the negative cathode, while anions, negatively charged ions, move towards the positive anode. In electrodialysis, filters or membranes selectively allowing the passage of cations or anions are alternately positioned between the electrodes.

4.45.6 Membrane Functionality

Cation filters permit the passage of anions but inhibit positively charged cations, whereas anion filters exclude anions while permitting cations to pass through. As ions migrate, they accumulate in distinct sections of the tank, directed by the appropriate filter (Valero et al., 2011). Consequently, cells with increasing salt content alternate with cells having decreasing salt concentration. Water is subsequently extracted from the relevant compartments effectively desalinated.

4.45.7 Advantages of Electrodialysis

a. The electrodialysis (ED) system separates substances without altering their phases, resulting in reduced energy consumption.
b. When desalting brackish water using an ED system, minimal pre-treatment is required.
c. The ED system is particularly effective in distinguishing between ionized and non-ionized components, as it exclusively eliminates ionized species.

d. Osmotic pressure is not a concern in the ED system, allowing it to concentrate salt solutions to 20% or higher.

4.45.8 Disadvantages of the Electrodialysis (ED)

In the realm of drawbacks, the Electrodialysis (ED) system presents a few noteworthy challenges:

a. The ED system falls short in effectively eliminating organic debris, colloids and SiO_2 from the water.
b. To prevent the ED stacks from succumbing to clogging, it becomes imperative to subject the feed water to pre-treatment.
c. The ED system demands meticulous control measures, and the task of maintaining them in optimal condition can prove to be quite arduous.
d. The selection of building materials for membranes and stacks must be approached with utmost care to guarantee compatibility with the feed stream.

4.46 Distillation: A Method for Water Purification

Distillation stands as an effective technique for removing various inorganic compounds such as metals (e.g., iron and lead), nitrate, hardness (e.g., calcium and magnesium), and particulates from contaminated water sources. Moreover, the boiling process efficiently eradicates microorganisms like bacteria and certain viruses. The efficacy of distillation in eliminating organic compounds depends on their solubility and boiling points. Organic compounds with boiling points higher than that of water, including certain pesticides, can be effectively removed. Conversely, organic compounds with lower boiling points, such as benzene and toluene, will volatilize alongside the water. Hence, it is imperative to eliminate these harmful compounds before the condensation process (Yan et al., 2021).

4.46.1 Principle of Operation

The distillation process operates on the principle of evaporation to purify water. Contaminated water is heated, causing it to transition into steam. Inorganic compounds and large non-volatile organic molecules do not vaporize and are left behind. Subsequently, the steam cools and condenses, yielding purified water.

4.46.2 Distillation Units

Distillation water treatment encompasses two widely used methods: solar distillation and multistage flash (MSF) distillation.

4.46.2.1 Solar Distillation for Water Purification

Solar energy can be effectively utilized to heat water by implementing a mirror system that tracks the sun's path, concentrating sunlight onto water sheets within a solar still. In these stills, salt water is contained in basins with glass panes serving as roofs, positioned at an angle of 10–18° (Fig. 4.4). These roofs are tightly sealed to the holding frame, ensuring a perfect fit between the still cover and the vertical walls.

4.46.2.2 Operation

When exposed to solar heat, water molecules undergo conversion into steam, leaving behind the salts in the basin. The steam then condenses at the bottom of the glass roof, and the resulting water flows down the slopes of the glass panel, ultimately

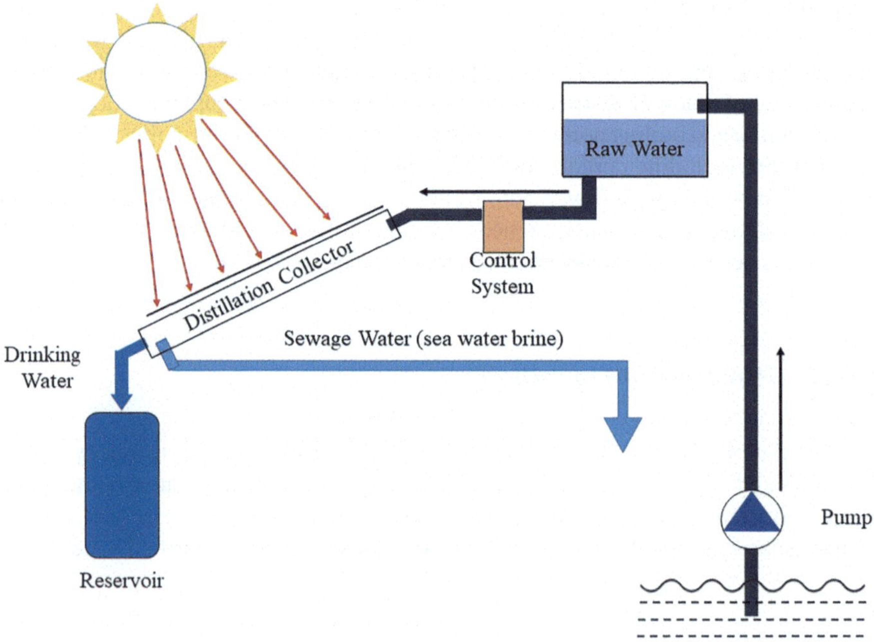

Fig. 4.4 Solar distillation

collecting in troughs designed for this purpose. These collecting troughs, located at the base of the glass pane cover, are thoughtfully constructed to facilitate the free drainage of water into a pipe that transports the distillate to a fresh water tank (Ge et al., 2023).

4.46.2.3 Sustainable Solution

By employing solar distillation, the abundant energy from the sun can be harnessed to purify water, separating it from impurities and salts. This method offers a sustainable and environmentally friendly solution for water treatment, ensuring the availability of clean and safe drinking water.

4.47 Multi-stage Flash Distillation Water Treatment

Multi-stage flash (MSF) distillation is a highly effective water treatment method. It involves heating the feed water and reducing the pressure, causing the water to rapidly transform into steam. This process is repeated in a series of stages, with each stage operating at a lower pressure level.

4.47.1 Distillation Process

Over the past few decades, numerous distillation processes have been extensively studied and implemented. These processes utilize various types of evaporators and employ different techniques for heat energy utilization and transfer. The key distillation processes include:

1. Boiling with submerged-tube heating surface
2. Boiling with long-tube vertical evaporator
3. Flash evaporation
4. Forced circulation with vapor compression
5. Solar evaporation
6. Rotating surface evaporation
7. Wiped-surface evaporation
8. Vapor reheating process
9. Direct heat transfer using an immiscible liquid
10. Condensing-vapor-heat transfer vapor other than steam.

Among these distillation methods, multistage flash evaporation, multiple effect evaporation and vapor-compression distillation have emerged as the most viable options for the reclamation of municipal wastewater.

By employing the MSF distillation technique, we can efficiently treat water by harnessing the power of steam generation. This process not only removes impurities but also ensures the production of high-quality water suitable for various applications.

4.48 Multiple Effect Evaporation (MEE)

Multiple Effect Evaporation (MEE) is a highly efficient process that involves a well-structured sequencing of evaporators. Typically, the number of evaporators or effects is limited to four, unless a specific design necessitates otherwise (Metcalf & Eddy, 1991; Nandy et al., 2002). This method boasts several salient features that make it a preferred choice in various industries. Firstly, a multiple effect evaporator is a sophisticated apparatus that maximizes the utilization of steam heat to evaporate water (Fig. 4.5). By utilizing a sequence of vessels, each operating at a progressively lower pressure, this process ensures optimal efficiency. The key principle behind this technique lies in the fact that the boiling point of water decreases as the pressure decreases. Consequently, the vapor produced in one vessel can be effectively utilized to heat the subsequent vessel, resulting in a cascading effect. Remarkably, only the initial vessel in the sequence requires an external heat source. Once the process is initiated, the subsequent vessels can rely on the vapor generated by the previous stages, significantly reducing the need for additional energy input. It is worth noting

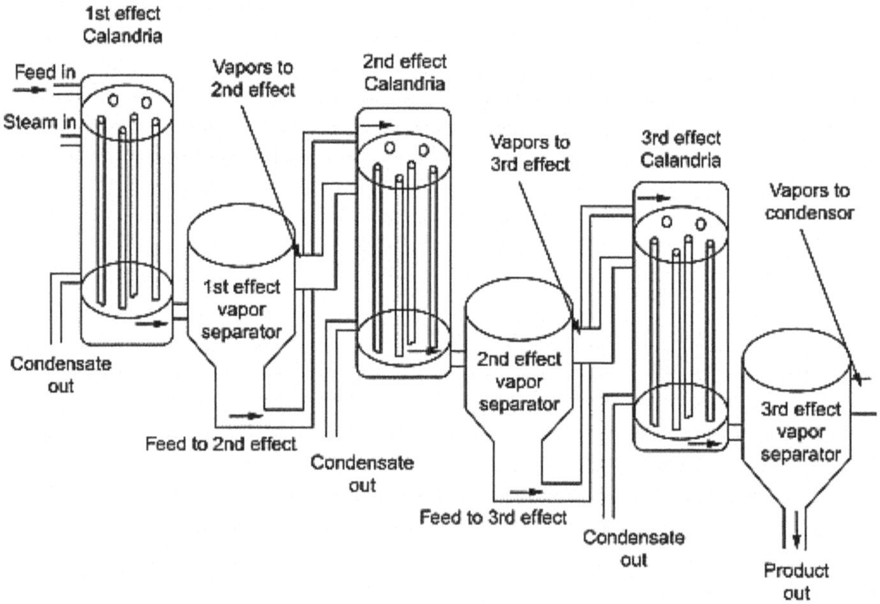

Fig. 4.5 Schematic of multiple effect evaporator

that evaporators with more than four to five effects are rarely employed due to practical considerations. However, in specific cases where unique design requirements arise, the number of effects can be adjusted accordingly. Multiple effect evaporation is a highly efficient, resourceful and well-structured sequencing of evaporators that ensure optimal utilization of energy, making it a preferred choice in various industries.

4.49 Multistage Flash Evaporation (MFE)

Multistage flash evaporation (MFE) systems have long been utilized in the desalination field. In this process, incoming wastewater undergoes initial treatment to remove Total Suspended Solids (TSS) and is then subjected to a deaeration process. Subsequently, the treated wastewater is pumped through heat transfer units within multiple stages of the distillation system, each operating at low pressure. The reduction in pressure within these stages induces vapor generation or boiling, known as flashing. As water enters each stage through a pressure-reducing nozzle, a portion undergoes flashing, transforming into vapor, which condenses on the outer surface of condenser tubes and is collected in trays (Fig. 4.6). During condensation, the latent heat of the vapor preheats the wastewater returned to the main heater. This preheated wastewater receives additional heat before entering the first flashing stage. As wastewater concentration increases, it reaches the lowest pressure stage and is pumped out of the system. The MFE system efficiently separates freshwater from saline water using flashing and condensation principles. By optimizing heat transfer and employing multiple stages, this process maximizes freshwater extraction while minimizing energy consumption.

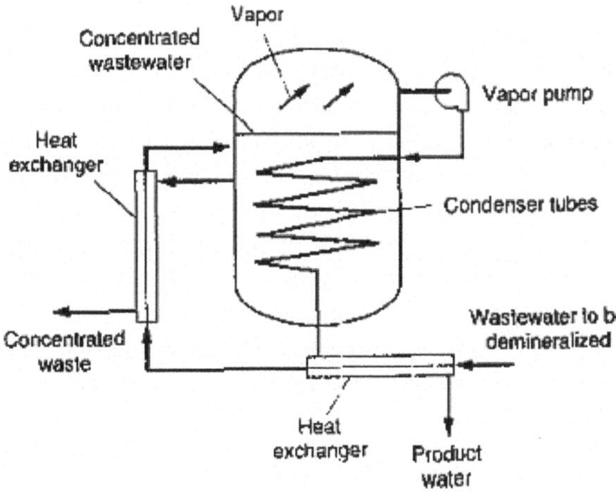

Fig. 4.6 Schematic of multistage flash evaporation

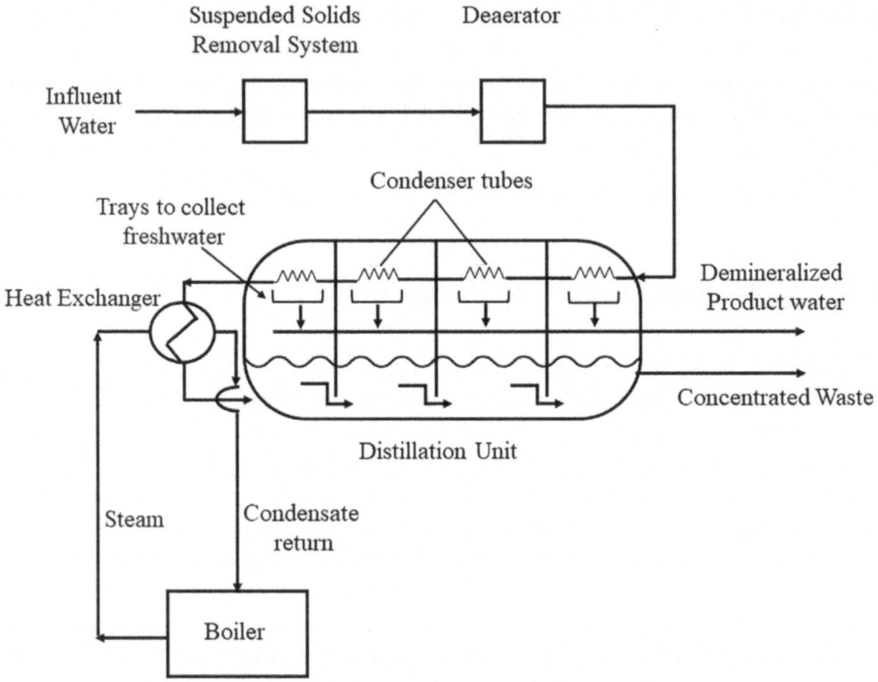

Fig. 4.7 Schematic of vapor compression distillation

4.50 Vapor-Compression Distillation

Vapor-compression distillation stands as a highly efficient process employing an increase in vapor pressure to establish a temperature difference for effective heat transfer. Initially, the wastewater is heated, followed by the operation of a vapor pump to condense the vapor at higher pressure in the condenser tubes. Simultaneously, this process releases an equivalent amount of vapor from the concentrated solution. Heat exchangers play a crucial role in conserving heat from both the condensate and the waste brine. One notable advantage of this process is its reliance solely on mechanical energy for the operation of the vapor pump, rendering it an energy-efficient solution. However, regular discharge of the hot concentrated wastewater is essential to prevent the accumulation of excessive salt concentrations in the boiler. By employing vapor-compression distillation, optimal heat transfer can be achieved, effectively treating wastewater. This process not only ensures the removal of impurities but also conserves energy, making it a sustainable and environmentally friendly solution (Fig. 4.7).

References

Ahmadpour, A. (2022). Using activated carbon adsorption in wastewater industries. *Journal of Chemistry Letters, 3*(1), 2–9. https://doi.org/10.22034/jcl.2022.160448.1010

Altmann, J., Massa, L., Sperlich, A., Gnirss, R., & Jekel, M. (2016). UV_{254} absorbance as real-time monitoring and control parameter for micropollutant removal in advanced wastewater treatment with powdered activated carbon. *Water Research, 94*, 240–245. https://doi.org/10.1016/j.watres.2016.02.008

Andreozzi, R., Caprio, V., Insola, A., & Marotta, R. (1999). Advanced oxidation processes (AOP) for water purification and recovery. *Catalysis Today, 53*, 51–59. https://doi.org/10.1016/S0920-5861(99)00137-9

Aramyan, S. M., & Moussavi, M. (2017). Advances in Fenton and Fenton-based oxidation processes for industrial effluent contaminants control—A review. *International Journal of Environmental Science and Natural Resources, 2*, 555594. https://doi.org/10.19080/IJESNR.2017.02.555594

Bagheri, S., Muhd Julkapli, N., & Bee Abd Hamid, S. (2014). Titanium dioxide as a catalyst support in heterogeneous catalysis. *The Scientific World Journal, 2014*, 727496. https://doi.org/10.1155/2014/727496

Baral, A. (2023). Biosorption and bioaccumulation of geosmin from recirculating aquaculture system: Fate of geosmin in aquaculture water. *Water Research* (in press).

Barwal, A., & Chaudhary, R. (2016). Effectiveness of solar photo-Fenton process for simultaneous detoxification of heavy metals and disinfection in municipal wastewater by using response surface method. *Environmental Progress & Sustainable Energy, 36*, 448–459. https://doi.org/10.1002/ep.12306

Bautista, P., Mohedano, A. F., Casas, J. A., Zazo, J. A., & Rodriguez, J. J. (2008). An overview of the application of Fenton oxidation to industrial wastewater treatment. *Journal of Chemical Technology & Biotechnology: International Research in Process, Environmental & Clean Technology, 83*(10), 1323–1338. https://doi.org/10.1002/jctb.1845

Benefield, L. D., & Morgan, J. M. (1990). Chemical precipitation. In R. D. Letterman (Ed.), *Water quality and treatment: A handbook of community water supplies* (4th ed., pp. 235–287). McGraw-Hill Inc.

Campos, C. J., Avant, J., Lowther, J., Till, D., & Lees, D. N. (2016). Human norovirus in untreated sewage and effluents from primary, secondary, and tertiary treatment processes. *Water Research, 103*, 224–232. https://doi.org/10.1016/j.watres.2016.07.007

Candia-Onfray, C., Espinoza, N., Sabino da Silva, E. B., Toledo-Neira, C., Espinoza, L. C., Santander, R., García, V., & Salazar, R. (2018). Treatment of winery wastewater by anodic oxidation using BDD electrode. *Chemosphere, 206*, 709–717. https://doi.org/10.1016/j.chemosphere.2018.05.043

Chen, Y., Lin, M., & Zhuang, D. (2022). Wastewater treatment and emerging contaminants: Bibliometric analysis. *Chemosphere, 297*, 133932. https://doi.org/10.1016/j.chemosphere.2022.133932

Chong, M. N., Jin, B., Chow, C. W., & Saint, C. (2010). Recent developments in photocatalytic water treatment technology: A review. *Water Research, 44*(10), 2997–3027. https://doi.org/10.1016/j.watres.2010.02.015

Chupin, R., Pham, N. M., & Chupin, V. (2019, November). Optimization of the sewerage systems scheme of cities and populated areas. *IOP Conference Series: Materials Science and Engineering, 667*(1), 012018. IOP Publishing. https://doi.org/10.1088/1757-899X/667/1/012018

Covinich, L. G., Bengoechea, D. I., Fenoglio, R. J., & Area, M. C. (2014). Advanced oxidation processes for wastewater treatment in the pulp and paper industry: A review. *American Journal of Environmental Engineering, 4*(2), 56–70. https://doi.org/10.5923/j.ajee.20140402.03

CPHEEO, Central Public Health and Environmental Engineering Organization. (1993). *Manual on sewerage and sewage treatment* (2nd ed.). Ministry of Urban Development.

Crini, G., & Lichtfouse, E. (2019). Advantages and disadvantages of techniques used for wastewater treatment. *Environmental Chemistry Letters, 17*, 145–155. https://doi.org/10.1007/s10311-018-0084-2

Deng, Y., & Zhao, R. (2015). Advanced oxidation processes (AOPs) in wastewater treatment. *Current Pollution Reports, 1*, 167–176. https://doi.org/10.1007/s40726-015-0015-z

Diya'uddeen, B. H., Daud, W. M. A. W., & Aziz, A. A. (2011). Treatment technologies for petroleum refinery effluents: A review. *Process Safety and Environmental Protection, 89*(2), 95–105. https://doi.org/10.1016/j.psep.2010.09.005

Droste, R. L., & Gehr, R. L. (2018). *Theory and practice of water and wastewater treatment*. Wiley.

Eckenfelder, Jr., W. W. (2000). Wastewater treatment. In *Kirk-Othmer encyclopedia of chemical technology*. https://doi.org/10.1002/0471238961.1205202110210210.a01

Ferguson, C. M., Charles, K., & Deere, D. A. (2008). Quantification of microbial sources in drinking-water catchments. *Critical Reviews in Environmental Science and Technology, 39*(1), 1–40. https://doi.org/10.1080/10643380701430457

Fotiou, T., Triantis, T., Kaloudis, T., & Hiskia, A. (2015). Photocatalytic degradation of cylindrospermopsin under UV-A, solar, and visible light using TiO_2: Mineralization and intermediate products. *Chemosphere, 119*, 589–594. https://doi.org/10.1016/j.chemosphere.2014.07.040

Garud, R. M., Kore, S. V., Kore, V. S., & Kulkarni, G. S. (2011). A short review on process and applications of reverse osmosis. *Universal Journal of Environmental Research & Technology, 1*(3).

Ge, Z., Ma, Z., Zou, J., Zhang, Y., Li, Y., Zhang, L., & Zhang, J. (2023). Purification of aquaculture wastewater by macrophytes and biofilm systems: Efficient removal of trace antibiotics and enrichment of antibiotic resistance genes. *Science of the Total Environment, 901*, 165943. https://doi.org/10.1016/j.scitotenv.2023.165943

Gualda-Alonso, E., Soriano-Molina, P., Lopez, J. L. C., Sanchez, J. L. G., Plaza-Bolaños, P., Agüera, A., & Perez, J. A. S. (2022). Large-scale raceway pond reactor for CEC removal from municipal WWTP effluents by solar photo-Fenton. *Applied Catalysis B: Environmental, 319*, 121908. https://doi.org/10.1016/j.apcatb.2022.121908

Guillossou, R., Le Roux, J., Mailler, R., Vulliet, E., Morlay, C., Nauleau, F., … Rocher, V. (2019). Organic micropollutants in a large wastewater treatment plant: What are the benefits of an advanced treatment by activated carbon adsorption in comparison to conventional treatment? *Chemosphere, 218*, 1050–1060. https://doi.org/10.1016/j.chemosphere.2018.11.071

Gurreri, L., Tamburini, A., Cipollina, A., & Micale, G. (2020). Electrodialysis applications in wastewater treatment for environmental protection and resources recovery: A systematic review on progress and perspectives. *Membranes, 10*(7), 146. https://doi.org/10.3390/membranes10070146

Guzmán, J., Mosteo, R., Sarasa, J., Alba, J. A., & Ovelleiro, J. L. (2016). Evaluation of solar photo-Fenton and ozone-based processes as citrus wastewater pre-treatments. *Separation and Purification Technology, 164*, 155–162. https://doi.org/10.1016/j.seppur.2016.01.022

Hong, J., Cho, K. H., Presser, V., & Su, X. (2022). Recent advances in wastewater treatment using semiconductor photocatalysts. *Current Opinion in Green and Sustainable Chemistry*, 100644. https://doi.org/10.1016/j.cogsc.2022.100644

Hur, J., Lee, T. H., & Lee, B. M. (2011). Estimating the removal efficiency of refractory dissolved organic matter in wastewater treatment plants using a fluorescence technique. *Environmental Technology, 32*(16), 1843–1850. https://doi.org/10.1080/09593330.2011.570367

Imam, E. H., & Elnakar, H. Y. (2014). Design flow factors for sewerage systems in small arid communities. *Journal of Advanced Research, 5*(5), 537–542. https://doi.org/10.1016/j.jare.2014.01.005

Jacangelo, J. G., Patania, N. L., Reagan, K. M., Aieta, E. M., Krasner, S. W., & McGuire, M. J. (1989). Ozonation: Assessing its role in the formation and control of disinfection by-products. *Journal—American Water Works Association, 81*(8), 74–84. https://doi.org/10.1002/j.1551-8833.1989.tb05114.x

Jiménez, B. (2009). Wastewater risks in the urban water cycle. In *Urban water security: Managing risks* (pp. 151–200). CRC Press. https://doi.org/10.1201/9781315240989-12

Kalbar, P. P., Karmakar, S., & Asolekar, S. R. (2013). Assessment of wastewater treatment technologies: Life cycle approach. *Water and Environment Journal, 27*(2), 261–268. https://doi.org/10.1111/wej.12002

Kataki, S., Chatterjee, S., Vairale, M. G., Dwivedi, S. K., & Gupta, D. K. (2021). Constructed wetland, an eco-technology for wastewater treatment: A review on types of wastewater treated and components of the technology (macrophyte, biofilm, and substrate). *Journal of Environmental Management, 283*, 111986. https://doi.org/10.1016/j.jenvman.2021.111986

Kavitha, V., & Palanivelu, K. (2004). The role of ferrous ion in Fenton and photo-Fenton processes for the degradation of phenol. *Chemosphere, 55*(9), 1235–1243.

Lin, H. H. H., & Lin, A. Y. C. (2021). Solar photo-Fenton oxidation of cytostatic drugs via Fe(III)-EDDS at circumneutral pH in an aqueous environment. *Journal of Water Process Engineering, 41*, 102066. https://doi.org/10.1016/j.jwpe.2021.102066

Liu, D. H., & Lipták, B. G. (Eds.). (2020). *Wastewater treatment*. CRC Press.

Liu, Z. H., Lu, G. N., Yin, H., Dang, Z., & Rittmann, B. (2015). Removal of natural estrogens and their conjugates in municipal wastewater treatment plants: A critical review. *Environmental Science & Technology, 49*(9), 5288–5300. https://doi.org/10.1021/acs.est.5b00044

Lopez, F. J., Pitarch, E., Botero-Coy, A. M., Fabregat-Safont, D., Ibáñez, M., Marin, J. M., … Hernandez, F. (2022). Removal efficiency for emerging contaminants in a WWTP from Madrid (Spain) after secondary and tertiary treatment and environmental impact on the Manzanares River. *Science of the Total Environment, 812*, 152567. https://doi.org/10.1016/j.scitotenv.2021.152567

Lynn, C. J., Dhir, R. K., Ghataora, G. S., & West, R. P. (2015). Sewage sludge ash characteristics and potential for use in concrete. *Construction and Building Materials, 98*, 767–779. https://doi.org/10.1016/j.conbuildmat.2015.07.020

Maezono, T., Tokumura, M., Sekine, M., & Kawase, Y. (2011). Hydroxyl radical concentration profile in the photo-Fenton oxidation process: Generation and consumption of hydroxyl radicals during the discoloration of azo-dye Orange II. *Chemosphere, 82*(10), 1422–1430. https://doi.org/10.1016/j.chemosphere.2010.10.045

Mbamba, C. K., Batstone, D. J., Flores-Alsina, X., & Tait, S. (2015). A generalised chemical precipitation modelling approach in wastewater treatment applied to calcite. *Water Research, 68*, 342–353. https://doi.org/10.1016/j.watres.2014.10.014

Merayo, N., Hermosilla, L., Blanco, L., Cortijo, L., & Blanco, Á. (2013). Assessing the application of advanced oxidation processes, and their combination with biological treatment, to effluents from the pulp and paper industry. *Journal of Hazardous Materials, 262*, 420–427. https://doi.org/10.1016/j.jhazmat.2013.08.020

Metcalf and Eddy, Inc. (1991). *Wastewater engineering: Treatment, disposal, and reuse* (3rd ed.). McGraw-Hill, Inc.

Metcalf, L., Eddy, H. P., & Tchobanoglous, G. (1991). *Wastewater engineering: Treatment, disposal, and reuse* (Vol. 4). McGraw-Hill.

Morrison, C. M., Betancourt, W. Q., Quintanar, D. R., Lopez, G. U., Pepper, I. L., & Gerba, C. P. (2020). Potential indicators of virus transport and removal during soil aquifer treatment of treated wastewater effluent. *Water Research, 177*, 115812. https://doi.org/10.1016/j.watres.2020.115812

Munter, R. (2001). Advanced oxidation processes—Current status and prospects. *Proceedings of the Estonian Academy of Sciences: Chemistry, 50*(2), 59–80. https://doi.org/10.3176/chem.2001.2.02

Nandy, T., Shastry, S., & Kaul, S. N. (2002). Wastewater management in a cane molasses distillery involving bioresource recovery. *Journal of Environmental Management, 65*(1), 25–38. https://doi.org/10.1006/jema.2001.0506

Nejatishahidein, N., & Zydney, A. L. (2021). Depth filtration in bioprocessing—New opportunities for an old technology. *Current Opinion in Chemical Engineering, 34*, 100746. https://doi.org/10.1016/j.coche.2021.100746

Nidheesh, P. V., & Gandhimathi, R. (2012). Trends in electro-Fenton process for water and wastewater treatment: An overview. *Desalination, 299*, 1–15. https://doi.org/10.1016/j.desal.2012.04.022

O'Connor, M., Helal, S. R., Latch, D. E., & Arnold, W. A. (2019). Quantifying photo-production of triplet excited states and singlet oxygen from effluent organic matter. *Water Research, 156*, 23–33. https://doi.org/10.1016/j.watres.2019.02.033

Patziger, M. (2021). Improving wastewater treatment plant performance by applying CFD models for design and operation: Selected case studies. *Water Science and Technology, 84*(2), 323–332. https://doi.org/10.2166/wst.2021.304

Pesakhov, S., Benisty, R., Sikron, N., Cohen, Z., Gomelsky, P., Khozin-Goldberg, I., Dagan, R., & Porat, N. (2007). Effect of hydrogen peroxide production and the Fenton reaction on membrane composition of *Streptococcus pneumoniae*. *Journal of Hazardous Materials, 176*(1–3), 590–597. https://doi.org/10.1016/j.jhazmat.2009.01.031

Qasim, S. R. (2017). *Wastewater treatment plants: Planning, design, and operation*. Routledge.

Ranasinghe, R. P. L. N., Dissanayaka, D. M. S. H., Rathnayaka, R. A. A. S., Nirmanee, K. G. S., & Jayaneththi, J. P. H. U. (2021). Water quality in different inlet and outlet canals connected to Nuwarawewa, Anuradhapura, Sri Lanka. *International Journal of Latest Technology in Engineering, Management & Applied Science, 10*(2), 01–06.

Rasalingam, S., Peng, R., & Koodali, R. (2014). Removal of hazardous pollutants from wastewaters: Applications of TiO_2–SiO_2 mixed oxide materials. *Journal of Nanomaterials, 2014*, 1–42. https://doi.org/10.1155/2014/312789

Rashid, J., & Barakat, M. A. (2015). Ag_3PO_4 enhanced TiO_2 for visible light photocatalysis of 2-chlorophenol in wastewater. *International Journal of Environmental Engineering, 2*, 93–97.

Rathi, B. S., & Kumar, P. S. (2021). Application of adsorption process for effective removal of emerging contaminants from water and wastewater. *Environmental Pollution, 280*, 116995. https://doi.org/10.1016/j.envpol.2021.116995

Rezakazemi, M., Khajeh, A., & Mesbah, M. (2018). Membrane filtration of wastewater from gas and oil production. *Environmental Chemistry Letters, 16*, 367–388. https://doi.org/10.1007/s10311-018-0706-0

Ribeiro, A. R., Nunes, O. C., Pereira, M. F. R., & Silva, A. M. T. (2015). An overview on the advanced oxidation processes applied for the treatment of water pollutants defined in the recently launched Directive 2013/39/EU. *Environmental International, 75*, 33–51. https://doi.org/10.1016/j.envint.2014.10.007

Sauer, T. P., Casaril, L., Oberziner, A. L. B., José, H. J., & Moreira, R. D. F. P. M. (2006). Advanced oxidation processes applied to tannery wastewater containing Direct Black 38—Elimination and degradation kinetics. *Journal of Hazardous Materials, 135*(1–3), 274–279. https://doi.org/10.1016/j.jhazmat.2005.11.035

Sharma, J., Kumar, V., Kumar, S. S., Malyan, S. K., Mathimani, T., Bishnoi, N. R., & Pugazhendhi, A. (2020). Microalgal consortia for municipal wastewater treatment—Lipid augmentation and fatty acid profiling for biodiesel production. *Journal of Photochemistry and Photobiology B: Biology, 202*, 111638. https://doi.org/10.1016/j.jphotobiol.2019.111638

Swain, S. K., Baral, P., Hutin, Y. J., Rao, T. V., Murhekar, M., & Gupte, M. D. (2010). A hepatitis E outbreak caused by a temporary interruption in a municipal water treatment system, Baripada, Orissa, India, 2004. *Transactions of the Royal Society of Tropical Medicine and Hygiene, 104*(1), 66–69. https://doi.org/10.1016/j.trstmh.2009.08.002

Tchobanoglous, G., Burton, F. L., & Stensel, H. D. (1991). *Wastewater engineering: Management* (7th ed.). McGraw-Hill.

Templeton, M. R., & Butler, D. (2011). *Introduction to wastewater treatment*. Bookboon.

Valero, F., Barceló, A., & Arbós, R. (2011). Electrodialysis technology: Theory and applications. In *Desalination, trends and technologies* (Vol. 28, pp. 3–20). https://doi.org/10.1016/j.desal.2011.02.002

Wenten, I. G. (2016). Reverse osmosis applications: Prospect and challenges. *Desalination, 391*, 112–125. https://doi.org/10.1016/j.desal.2016.02.013

Winkler, M. K., & Straka, L. (2019). New directions in biological nitrogen removal and recovery from wastewater. *Current Opinion in Biotechnology, 57*, 50–55. https://doi.org/10.1016/j.copbio.2018.09.008

Yan, Z., Jiang, Y., Liu, L., Li, Z., Chen, X., Xia, M., … Ding, A. (2021). Membrane distillation for wastewater treatment: A mini review. *Water, 13*(24), 3480. https://doi.org/10.3390/w13243480

Zhang, F., Peng, Y., Wang, Z., Jiang, H., Ren, S., & Qiu, J. (2021a). Achieving synergetic treatment of sludge supernatant, waste activated sludge and secondary effluent for wastewater treatment plants (WWTPs) sustainable development. *Bioresource Technology, 337*, 125416. https://doi.org/10.1016/j.biortech.2021.125416

Zhang, H., Choi, H. J., & Huang, C. B. (2005). Optimization of Fenton process for the treatment of landfill leachate. *Journal of Hazardous Materials, 125*(1–3), 166–174. https://doi.org/10.1016/j.jhazmat.2005.06.005

Zhang, Q., He, C., Ren, J., & Goodsite, M. E. (Eds.). (2021b). *Waste to renewable biohydrogen: Volume 1: Advances in theory and experiments*. Academic Press. https://doi.org/10.1016/B978-0-12-819173-9.00001-8

Zhang, S., Wang, F., Mei, Z., Lv, L., & Chi, Y. (2021c). Status and development of sludge incineration in China. *Waste and Biomass Valorization, 12*, 3541–3574. https://doi.org/10.1007/s12649-021-01424-5

Chapter 5
Sludge Treatment and Disposal

5.1 Sludge Management

The management of sludge originating from wastewater treatment plants is a highly complex and costly activity, crucial for maintaining environmental and sanitary advantages expected from treatment systems. Recognizing its significance, Agenda 21 addressed environmentally sound management of solid wastes and sewage-related issues, emphasizing reduction in production, maximizing reuse and recycling, and promoting environmentally sound treatment and disposal practices (Cieślik et al., 2015).

Sludge, comprising solids that settle during the wastewater treatment process, necessitates proper treatment or disposal at various treatment stages, including primary sedimentation, secondary sedimentation, and advanced (tertiary) treatment. The quantity of sludge generated correlates with the level of treatment or quality of treated effluent required. As treatment intensifies, so does the volume of sludge, necessitating effective handling to minimize operational costs.

Given stringent regulations governing sludge handling and disposal, reducing sludge volume becomes imperative to manage overall plant costs efficiently. Therefore, a properly designed and efficiently operated sludge processing and disposal system is essential for the wastewater treatment plant's success. By managing and treating various types of sludge effectively, plants can ensure regulatory compliance, cost reduction, and operational success.

The sludge generated during wastewater treatment can be categorized into three main types.

S. A. Bandh and B. Mushtaq, *Wastewater Treatment Technology*, Springer Water,
https://doi.org/10.1007/978-3-031-86684-5_5

5.2 Primary Sludge

Primary sludge plays a crucial role in wastewater treatment processes, precipitating during the initial phase of sedimentation. This semi-solid residue, typically containing 60–75% moisture, 20–30% volatile solids, and 5–10% inorganic matter (Wang et al., 2020), effectively captures a significant portion of suspended solids and contaminants from influent wastewater. Characterized by a gray color, slimy texture, coarse nature, and unpleasant odors, primary sludge production varies with influent characteristics and plant capacity, with large-scale facilities generating approximately 120–160 g of primary sludge per liter of wastewater treated. Effective management of primary sludge, including dewatering and proper disposal or utilization, is indispensable for ensuring the operational efficiency and environmental sustainability of wastewater treatment plants.

5.3 Secondary Sludge

Secondary sludge is a vital byproduct of biological wastewater treatment, formed during the secondary clarification phase. Typically comprising 75–85% moisture content, 10–20% volatile solids, and 5–10% inorganic matter (Zhang et al., 2021a, 2021b, 2021c), it exhibits a brownish color, flocculent appearance, and earthy odor. Scientific data indicates that secondary sludge production varies widely, influenced by factors like influent characteristics and the efficiency of biological processes. On average, wastewater treatment plants generate approximately 35–50 g of secondary sludge per liter of wastewater treated. Effective management of secondary sludge, including dewatering, treatment, and disposal strategies, is essential for optimizing the overall sustainability and performance of wastewater treatment facilities.

5.4 Tertiary Sludge

Tertiary sludge is a vital aspect of advanced wastewater treatment, originating from the tertiary treatment stage aimed at refining effluent quality. Typically composed of 80–90% moisture content, 5–10% volatile solids, and 5–10% inorganic matter (Smith et al., 2023), its production varies based on the extent of advanced treatment processes, typically ranging from 10 to 30 g per liter of treated wastewater. Effective management of tertiary sludge is crucial for treatment plant efficiency and environmental sustainability, necessitating strategies for dewatering, treatment, and responsible disposal or beneficial use.

Sludge management encompasses various stages, from generation through treatment (thickening, stabilization, conditioning, and dewatering) to eventual use and disposal. Inadequate handling can lead to environmental issues such as odors, health hazards, and water contamination. Despite being costly, sludge management is vital, accounting for 50% of municipal treatment works expenses but resulting in 90% of operational problems (Cieślik et al., 2015).

Sludge in wastewater treatment plants exists in two forms: settleable solids in primary settlement tanks and biological cell mass generated during secondary treatment, settling in secondary tanks. With high water content, further treatment is necessary to reduce water content and oxygen demand before disposal. Stabilizing the sludge aims to achieve objectives like reducing pathogens, eliminating odors, and improving dewatering characteristics.

- Reduce pathogens
- Eliminate odors
- Inhibit, reduce, or eliminate the potential for decomposition
- Improve dewatering characteristics of the sludge to reduce volume for disposal.

There are four effective methods to eliminate this troublesome condition through stabilization.

1. Biological reduction of volatile solids
2. Chemical oxidation of volatile solids
3. Addition of chemicals to create unfavorable conditions for bacterial growth
4. Application of heat to disinfect or sterilize the sludge.

Implementing stabilization techniques is crucial for addressing issues such as pathogen reduction, odor elimination, prevention of decomposition, and improved dewatering characteristics in sludge management. However, disposing of sludge presents significant challenges due to its predominantly organic nature, leading to decomposition. Additionally, the volume of sludge far exceeds that of the solid constituents, necessitating further treatment to manage effectively. Methods like thickening, dewatering, and drying are employed to reduce sludge volume, while stabilization of organic matter can be achieved through various processes such as digestion (aerobic or anaerobic), incineration, composting, heat treatment, chlorine oxidation, or lime stabilization.

Organic sludge originates from different sources within wastewater treatment plants, including primary settlement tanks (PST), secondary settlement tanks (SST) of the activated sludge process (ASP), and SST of trickling filters (TF) (Ahammad et al., 2020). PST and SST sludge typically contain 65–75% volatile solids, while sludge from TF or SST ranges from 45 to 70% volatile solids. After thickening, the dry solids content of primary and secondary sludge is approximately 4–6%, with a specific gravity of 1.01. Separating water from undigested sludge is challenging, and the fuel value of undigested sludge is around 6500–8000 BTU/lb (15,100–18,600 kJ/kg of dry solids). Following digestion, the sludge contains volatile solids of about 32–48%, dry solids of 8–13%, and a specific gravity of 1.03–1.05. The fuel value increases to approximately 8100–9300 kJ/kg of dry solids. Digested sludge is more easily dewatered, and it effectively destroys around 99.8% of coliform in sewage sludge.

5.5 Sludge Sources, Characteristics and Quantities

Conventional water and wastewater treatment processes inevitably yield significant quantities of end products, namely supernatant liquid (effluent) and dilute solid mixtures known as sludge. The composition of these sludges is contingent upon various factors, including the sources and characteristics of the raw wastewater, the treatment processes utilized, and the extent of treatment applied. Typically, sludge resulting from treatment processes consists predominantly of water, constituting as much as 97% of its composition. Consequently, sludge treatment processes primarily aim to separate this substantial amount of water from the solid residues, which are subsequently subjected to further processing at the water or wastewater treatment plant (Gawdzik et al., 2015).

As previously mentioned, the characteristics and quantities of sludge are influenced by three key factors:

(i) The initial composition of water and wastewater
(ii) The specific water and wastewater treatment processes employed
(iii) The degree of treatment applied.

By considering these factors, water and wastewater treatment facilities can effectively manage and optimize the treatment of sludge, ensuring efficient and environmentally responsible handling of these byproducts.

The characteristics and quantity of sludge generated can fluctuate on different time scales, including annually, seasonally, or even daily, owing to changes in the incoming wastewater flow and composition. These variations can be especially pronounced in wastewater systems receiving substantial industrial waste discharges. The main sources of solids and sludge and the types generated are:

In the wastewater treatment process, various types of solid materials are separated and managed through different treatment stages (Gupta et al., 2012):

Screening: Coarse solids are initially removed through bar screens, which are then manually cleaned. The removed solids are typically disposed of in onsite or offsite landfills.

Grit Chamber: The grit chamber collects dense materials such as sand, broken glass, nuts, and bolts. Since grit is mostly non-biodegradable, it is usually disposed of onsite or offsite through landfilling.

Primary or Raw Sludge: Sludge collected from the bottom of primary clarifiers typically contains 3–8% solids, with around 70% being organic. This sludge is highly putrescible.

Secondary Sludge: This sludge comprises materials, including scum and fixed solids, removed from the secondary treatment process. It is approximately 90% organic and emits strong odors. Additionally, it may contain significant amounts of chemicals used for phosphorus removal through chemical precipitation in aeration basins.

Tertiary Sludge/Water Treatment Sludge: The characteristics of sludge from tertiary treatment vary depending on the specific processes employed. For example, if phosphorus removal is conducted in the activated sludge process, chemically precipitated sludge is combined with biological sludge. Sludges generated in water treatment processes are primarily inorganic and can be recovered through acidification, recalcination, and other methods.

Septage: Septage is a mixture of sludge, scum, and liquid extracted from septic tanks or cesspools. It is important to note that septage is only partially treated and requires further treatment before proper disposal.

5.6 Characteristics of Sludge

In order to effectively treat and dispose of sludge generated from wastewater treatment plants, it is crucial to have a comprehensive understanding of the characteristics of the solids and sludge that will undergo processing (Ghangrekar et al., 2005). These characteristics can vary depending on the source of the solids and sludge, the extent of aging they have undergone, and the specific processing methods they have been subjected to. Some of the key characteristics of the sludges are:

1. Various chemical constituents, including nutrients, play a significant role in determining the ultimate disposal of processed sludge and the liquid extracted from it during processing. Parameters such as pH, alkalinity, and organic acid content are particularly important for process control in anaerobic digestion. When considering methods like incineration and land application, it becomes crucial to assess the presence of heavy metals and hydrocarbons. Additionally, the energy content of sludge becomes a vital factor to consider when contemplating thermal reduction processes like incineration.
2. The specific constituents present in sludge can either restrict or enhance the available options for sludge utilization or disposal. Among these constituents, five are typically important:

    (i) ***Organic Content (Typically Measured as Volatile Solids)***: This parameter provides insights into the amount of organic matter present in the sludge, which can influence the selection of appropriate sludge use or disposal methods.
    (ii) ***Nutrients***: The presence and concentration of nutrients in sludge are crucial considerations, as they can impact the potential applications of the sludge and the overall environmental impact of its disposal.
    (iii) ***Pathogens***: The presence of pathogens in sludge poses a significant concern, as it can affect public health and necessitate specific treatment or disposal measures to ensure safety.

(iv) ***Metals***: The identification and quantification of metals in sludge are essential, as they can influence the suitability of various sludge management options and the potential risks associated with their use or disposal.

(v) ***Toxic Organic Chemicals***: The presence of toxic organic chemicals.

The organic content of sludge is a crucial parameter, measured as a percentage of the total solids (TS) represented by volatile solids (VS). Volatile solids are organic compounds that are eliminated when sludge undergoes heating to 500 °C under oxidizing conditions. This organic content influences various aspects such as thermal value, potential for odor issues during storage and land application, and the likelihood of gas generation.

Municipal sludges are rich in three essential nutrients for plant growth: nitrogen (3.3%), phosphorus (2.3%), and potassium (0.3%). However, these nutrient levels are considerably lower compared to commercial fertilizers, which typically contain nitrogen (5%), phosphorus (10%), and potassium (10%). Potassium is usually found in sludge in minimal concentrations (Di Costanzo et al., 2021).

During the wastewater treatment process, a significant concentration of bacteria, viruses, protozoa, and parasitic worm eggs accumulates in sludge. While only a small percentage of these organisms may be pathogenic, their presence raises concerns. Sludge treatment methods like anaerobic digestion can effectively reduce pathogen levels by approximately 99.4–99.99%.

In addition to organic matter and pathogens, sludges may also contain varying amounts of heavy metals and inorganic ions such as boron, cadmium, chromium, copper, lead, nickel, mercury, silver, and zinc. At low concentrations, these metals can act as micronutrients for plants and animals, but at high concentrations, they pose risks of toxicity to humans, animals, and plants. The concentration of metals in sludge is critical when considering its application to land, as it can potentially harm crops and, in the case of cadmium, enter the human food chain.

The management of sludge requires careful consideration of its organic content, nutrient levels, pathogen reduction, and metal concentrations. While sludges may contain synthetic organic chemicals from industrial waste, household chemicals, and pesticides, most sludges have minimal levels of these substances, posing no significant threats to human health or the environment. Evaluating sludge quantity involves considering both the volume of wet sludge, accounting for water and solid content, and the mass of dry sludge solids, which are essential for determining the feasibility of various disposal or utilization options.

5.7 Mass Balance

The daily influx of solids into wastewater treatment systems can vary substantially. To guarantee that the tank capacity can adequately manage these fluctuations, it is imperative for the designer of sludge processing and disposal facilities to consider two primary factors: (i) the average and maximum rates of sludge production, and

(ii) the potential storage capacity of treatment units within the plant. By meticulously assessing these aspects, the designer can ensure the facility operates efficiently and effectively, accommodating the demands of varying sludge production rates while maintaining optimal storage capabilities.

5.8 Sludge Treatment

A variety of sludge treatment processes have been developed and applied to wastewater treatment operations. These processes are crucial in effectively managing sludge and ensuring its safe disposal or beneficial reuse. The typical sludge processing stages are summarized in Fig. 5.1.

Sludge treatment processes typically encompass five main stages, each fulfilling a distinct purpose:

A. **Sludge Thickening or Concentration**: This initial stage aims to separate as much water as possible from the sludge. Various methods such as centrifugation, gravity settling, flotation, and clarifiers are employed to achieve this objective.
B. **Sludge Stabilization or Digestion**: During this phase, the organic solids present in the sludge are conditioned and transformed into more inert forms. This process is crucial for minimizing any potential nuisance or health hazards associated with the sludge. Additionally, the stabilized sludge can be utilized as a soil conditioner or safely disposed of.
C. **Dewatering**: In this stage, the sludge undergoes treatment with chemicals or heat to facilitate the separation of water from the solids. This step significantly reduces the moisture content of the sludge, making it easier to handle and transport.
D. **Reduction or Drying**: The primary goal of this stage is to convert the solids into a stable form. This can be achieved through methods such as wet oxidation or incineration. By reducing the volume and stabilizing the sludge, it becomes more suitable for long-term storage or disposal.

These sludge treatment processes are essential for ensuring the efficient and environmentally responsible management of wastewater treatment operations. By following these stages, wastewater treatment facilities can effectively handle and dispose of sludge, minimizing any potential negative impacts on public health and the environment.

To optimize the utilization of resources and meet treatment requirements, it is crucial to consider wastewater and sludge treatment and disposal systems as a whole (Raheem et al., 2018). Making an informed decision on the most resource-efficient and environmentally sound system requires a methodological selection process. With numerous treatment and disposal options available, it is important to consider various technical and non-technical factors that influence and determine the suitability and choice of a system. These factors include:

(i) **Effectiveness**: The ability of the system to achieve desired treatment goals and standards.

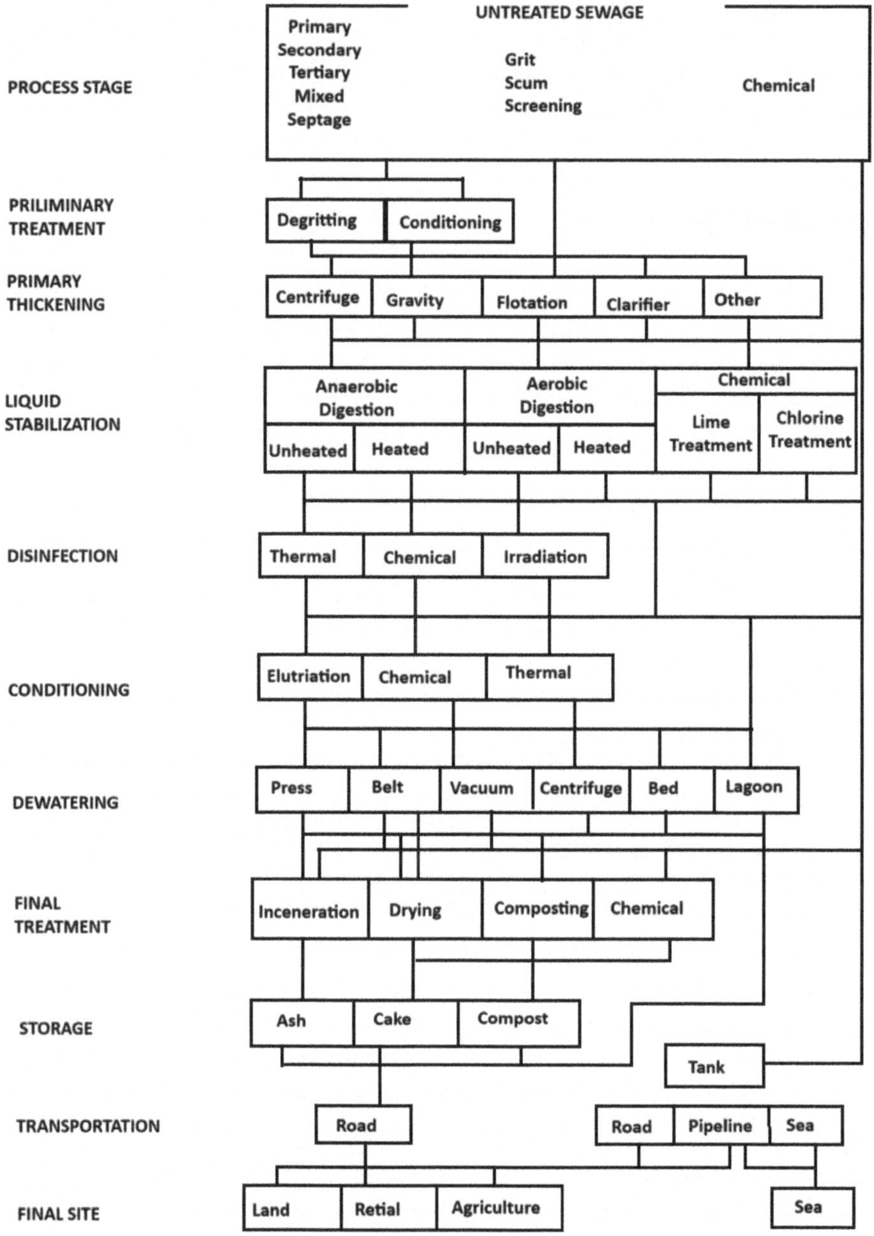

Fig. 5.1 Sludge treatment processes. *Source* EPA (1978)

(ii) **Implementability**: Feasibility of implementing and maintaining the system considering available resources and expertise.
(iii) **Comparability**: Comparison of different treatment options based on their performance and outcomes.
(iv) **Economic Aspects**: Consideration of costs associated with installation, operation, and maintenance of the system.
(v) **Environmental Impacts**: Assessment of potential environmental consequences, including air and water pollution, habitat disruption, and resource depletion.
(vi) **Energy**: Evaluation of energy requirements and efficiency of the system, including potential for energy recovery.
(vii) **Administrative Burdens**: Examination of regulatory compliance requirements and administrative complexities associated with the system implementation and operation.

5.9 Sludge Thickening and Dewatering

5.9.1 Preliminary Considerations

Thickening is a critical process employed to increase the solid content of sludge by removing a portion of the liquid fraction. The main aim is to minimize the capital and operating costs associated with subsequent sludge processing steps by significantly decreasing the sludge volume. For example, thickening from one to two percent solids concentration can effectively halve the sludge volume, while further concentration to five percent solids can reduce the volume to one-fifth of its original size.

In addition to volume reduction, thickening offers several other benefits, including sludge blending, sludge flow equalization, sludge storage, grit removal, gas stripping, and clarification. Various physical methods are employed to achieve thickening, such as primary and secondary sedimentation, gravity thickening, flotation, and centrifugation.

Considering that the primary objective of sludge thickening and dewatering (Knocke et al., 1987) is to decrease the water content in the sludge and subsequently reduce its volume, it is essential to select the most suitable thickening method based on factors such as the characteristics of the sludge, required degree of thickening, and available resources.

5.9.2 *Sludge Thickening*

5.9.2.1 Gravity Thickeners

Gravity thickening is a simple and economical process commonly employed for primary sludge treatment. It operates similarly to sedimentation in settling tanks. The sludge is directed into the central well of the tank, as depicted in Fig. 5.2, allowing the solids to settle at the bottom. A sturdy mechanized scraper then transfers the sludge into the hopper, from which the settled solids are removed.

The performance of the gravity thickening process is significantly influenced by the type of sludge being processed, with optimal results typically achieved when dealing exclusively with primary sludge, which generally ranges from 3 to 10% concentration. Gravity thickeners are structurally similar to sedimentation tanks, usually circular in shape, center-fed, and equipped with bottom sludge withdrawal and supernatant removal over their perimeter. The thickened sludge is then directed to the subsequent stage, often digestion, while the supernatant returns to the plant headworks. Figure 5.2 illustrates the schematics of a gravity thickener, offering a visual representation of its design.

The behavior of the sludge within the thickener conforms to the principles of zone settling and the solids flux theory. Tank sizing can be determined based on these principles or by considering solids and hydraulic loading rates. Hydraulic loading is vital in managing excessive detention times, which can lead to the generation of unpleasant odors. To ensure optimal performance, it is advisable to maintain loading rates within the range of 20–30 $m^3/m^2/d$. However, achieving these values may not always be feasible with the influent sludge alone. In such cases, it is common practice to recycle the final effluent back to the thickener. This recycling process helps increase influent flow, thereby reducing hydraulic detention time. Importantly, this incremental flow does not adversely affect the thickener's performance.

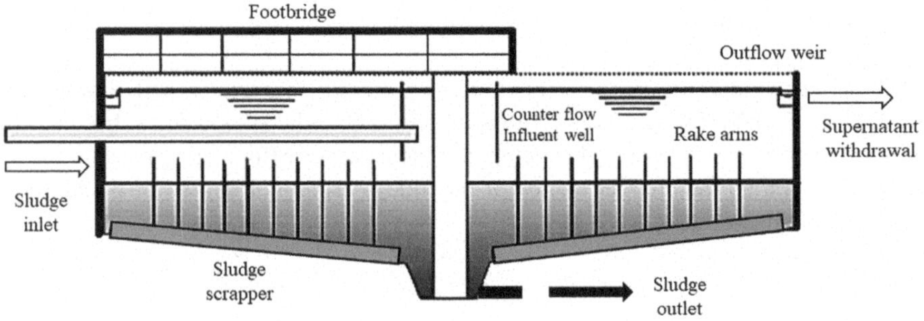

Fig. 5.2 Schematic cross section of a gravity thickener

5.9.2.2 Dissolve Air Flow Thickners

The dissolved air flotation (DAF) process is a highly effective method utilized in wastewater treatment, involving the introduction of air into a solution maintained under high pressure. Within this pressurized environment, the air remains dissolved within the solution; however, upon release of the pressure, the dissolved air forms small bubbles. As these bubbles ascend to the surface, they carry with them sludge particles, effectively separating the sludge from the liquid and facilitating easy removal. The separated sludge is subsequently skimmed off the surface, leaving behind a clarified liquid (Butler et al., 1997).

Flotation thickening, employing the DAF process, is particularly beneficial for excess activated sludge that does not adequately thicken in gravity thickeners. Through dissolved air flotation, sludge particles can be efficiently removed, resulting in a more concentrated sludge.

Moreover, dissolved air flotation finds application in treatment plants practicing biological phosphorus removal, where maintaining aerobic conditions for the sludge is crucial to prevent the release of particulate phosphorus back into the liquid phase as dissolved phosphorus. By utilizing the DAF process, sludge can be effectively treated while minimizing the risk of phosphorus contamination.

The dissolved air flotation process emerges as a versatile and effective method for wastewater treatment, offering the capability to separate sludge particles and maintain aerobic conditions. Its utilization presents treatment plants with the opportunity to achieve optimal results concerning sludge concentration and phosphorus removal.

5.9.2.3 Centrifuging

Centrifuges are essential for thickening and dewatering sludge, primarily waste activated sludge (Ginisty et al., 2021). Various types, including disc nozzle, sludge treatment, imperforate basket, and solid bowl decanter centrifuges are used. Among these, baskets and solid bowls are prevalent, with solid bowl decanters offering continuous operation. However, their use requires careful consideration due to higher power and maintenance costs, best suited for constrained spaces with skilled operators and challenging sludge thickening needs. Centrifuges are essential for thickening and dewatering sludge, primarily waste activated sludge. Various types are used, including disc nozzle, sludge treatment, imperforate basket, and solid bowl decanter centrifuges. Among these, baskets and solid bowls are prevalent, with solid bowl decanters offering continuous operation. However, their use requires careful consideration due to higher power and maintenance costs, best suited for constrained spaces with skilled operators and challenging sludge thickening needs.

5.9.3 *Sludge Conditioning*

Sludge conditioning is vital for efficient dewatering, altering particle size, distribution, surface charges, and interactions. Chemicals like inorganic and organic agents or thermal treatment achieve this. Hydration, chemical demand, and dewatering resistance rise with particle surface area. Sewage sludges often contain colloids and thin particles captured during biological treatment, complicating dewatering and increasing chemical use (Zhou et al., 2014).

Conditioning aims to agglomerate particles into larger flocs via coagulation and flocculation. Coagulation destabilizes particles by reducing electrostatic repulsion, while flocculation promotes gentle mixing for agglomeration. The amount of conditioner needed varies based on sludge and dewatering equipment.

Factors like sludge transportation method and storage duration impact water content and fine solids levels, affecting conditioning needs. Considering these factors is crucial for effective sludge management.

5.9.3.1 Organic Polymers

Organic polymers play a vital role in sludge conditioning, offering a wide range of products with diverse chemical compositions, performance characteristics, and cost-effectiveness (EPA, 1978). Compared to chemical conditioners, organic polymers offer several advantages, including a reduction in sludge mass by 15–30%, contributing to a more efficient and sustainable sludge management process (EPA, 1978). Additionally, the handling of organic polymers is cleaner and simpler, leading to improved operational efficiency and reduced maintenance issues (EPA, 1978).

Another significant advantage of organic polymers is that they do not diminish the calorific value of dewatered sludge, allowing it to be utilized as a fuel for incineration, thereby minimizing waste and providing an additional energy source (EPA, 1978). When dissolved in water, organic polymers form solutions with varying viscosities, influenced by factors such as molecular weight, ionization charge, and water salt content (EPA, 1978).

In solution, organic polymers attach themselves to sludge particles, initiating a sequence of events that include desorption of surface water, neutralization of charges, and agglomeration of small particulate matter through bridging (EPA, 1978). Selecting the most suitable polymers requires routine and continuous testing involving both the operational staff of the treatment plant and polymer suppliers, with on-site testing advisable due to the variability of sludge characteristics (EPA, 1978).

5.9.3.2 Composition and Surface Charges

Polymers are composed of lengthy chains of specialized chemical elements that are soluble in water, synthesized through a series of consecutive reactions called polymerization. These polymers can be derived from individual monomers, which form a sub-unit or repeated unit within the molecular structure, or by adding monomers or functional groups to natural polymers. Acrylamide is the most commonly used monomer for producing organic synthetic polymers (Beckett & Le, 1990).

Polymers can be classified based on their surface charges as neutral or non-ionic, cationic, or anionic. Anionic flocculants, containing polyacrylamides, introduce negative charges into aqueous solutions, while cationic polyacrylamides carry positive charges. Considering that most sludges have predominantly negative electric charges, cationic polymers are typically preferred for sludge conditioning. The choice of cationic polymer depends on the sludge characteristics and dewatering equipment, needing to be both productive and cost-effective. For instance, sludge particles with fine texture, high water content, and increased relative surface charges may require polymers with higher electric charges.

Polymers are available in powder or liquid form, with liquid polymers often commercialized as aqueous solutions or water-in-oil emulsions. It is crucial to protect polymers from drastic temperature changes during storage. Dry polymer powders can be stored for varying periods, typically from one to several years, while liquid products generally have storage periods of 6–12 months. Polymers come in different molecular weights and charge densities, significantly impacting their performance in sludge conditioning.

5.9.3.3 Dry Polymers

A diverse array of polymers is available, each characterized by a unique chemical composition that dictates its performance properties. Dry polymers can be synthesized as either granular powder or flocs, contingent on the specific manufacturing processes employed (Haleem, 2024). These dry polymers are typically highly concentrated, with active solids comprising approximately 90–95% of their total composition. For optimal preservation, dry polymers must be stored in environments free from moisture, as exposure to humidity can induce clumping, rendering them ineffective. Furthermore, meticulous attention is required during the dissolution of dry polymers.

Effective dissolution necessitates a system incorporating an ejector or an alternative polymer moistening device. This apparatus pre-humidifies the polymer powder before it is gradually introduced into a mixing tank, ensuring thorough dissolution. A subsequent mixing period of around 60 s is essential to confirm complete polymer dissolution. Inadequate dissolution can precipitate several operational problems, including the clogging of pumps and pipes, as well as scaling in filter presses and belt presses. This additional mixing period is also critical as it allows polymer molecules to elongate and assume a configuration conducive to sludge flocculation.

Insufficient mixing time can significantly impair the performance of the polymeric solution.

The choice of polymers for various applications is extensive, with chemical composition playing a pivotal role in their functional performance. Dry polymers, whether in granular powder or floc form, require precise storage conditions to prevent moisture-induced clumping. Dissolving these polymers correctly demands the use of a polymer moistening device and an adequate mixing period to achieve full dissolution and peak performance. Failure to adhere to these protocols can lead to operational inefficiencies and diminished efficacy of the polymeric solution.

5.9.3.4 Liquid Polymers

Liquid polymers are marketed in various concentrations and types of polymeric materials, functioning as the active agents for sludge conditioning. The dissolution process of these polymers hinges on the viscosity of the final solution (Quintero-García et al., 2023). Unlike other solutions, concentrated polymeric solutions do not necessitate stirring, as the polymers inherently form true solutions. These liquid polymer solutions are readily available in 208 L containers, 1040 L vessels, or bulk quantities, contingent on demand.

Nevertheless, caution is imperative when storing liquid polymers in cold climates, as variations in density can impede pumping. Preparing the dosing solution necessitates a mixing tank and a storage tank for the diluted polymer. Typically, a 0.1% polymer solution is achieved by mixing a concentrated polymer solution with water for at least 30 min. It is crucial to acknowledge that this solution remains stable for up to 24 h, beyond which it should be discarded.

Liquid polymers hold significant importance in various industries, available in diverse concentrations and polymeric materials. Their intrinsic ability to form true solutions obviates the need for continuous stirring, enhancing convenience.

5.9.3.5 Inorganic Chemical Conditioning

Inorganic chemical conditioning is predominantly utilized in the vacuum or pressure filtration dewatering processes. The chemicals most commonly employed for this purpose are lime and ferric chloride. Although used less frequently, ferrous chloride, ferrous sulfate, and aluminum sulfate are also applied in specific scenarios.

Ferric Chloride

Ferric chloride is frequently utilized alongside lime for sludge conditioning, with lime being added subsequently (Zhao et al., 2020). When ferric chloride dissolves in water, it undergoes hydrolysis, producing positively charged iron complexes. These complexes effectively neutralize the negative surface charges on sludge solids, promoting their aggregation. Furthermore, ferric chloride reacts with the bicarbonate alkalinity in the sludge, forming hydroxides that act as flocculants.

$$2FeCl_3 + 3Ca(HCO_3)_2 \rightarrow 2Fe(OH)_3 + 3CaCl_2 + 6CO_2$$

Overall, the use of ferric chloride in combination with lime provides an efficient method for sludge conditioning. The hydrolysis of ferric chloride in water creates positively charged iron complexes, which effectively neutralize the negative surface charges of sludge solids, thereby facilitating their aggregation. Additionally, ferric chloride reacts with the bicarbonate alkalinity in the sludge, leading to the formation of hydroxides that function as flocculants. These flocculants are crucial in the coagulation and settling of sludge, ultimately enhancing the treatment and disposal.

Ferric chloride solutions are typically supplied in concentrations ranging from 30 to 40%. Diluting these solutions is not recommended, as it can lead to hydrolysis and the formation of ferric hydroxide precipitates. These solutions can be stored for long periods without degradation, although crystallization may occur at temperatures below − 1 °C (for example, a 45% $FeCl_3$ solution crystallizes at this temperature). Due to the highly corrosive nature of ferric chloride, special pumps, corrosion-resistant storage materials, and careful operational procedures are required for its handling.

Lime

Lime is often combined with ferric chloride primarily for pH control, odor management, and pathogen reduction (Farrell et al., 1974). The reaction between lime and bicarbonate ($CaCO_3$) produces a granular structure within the sludge, enhancing its porosity and reducing compressibility. Lime is available in two forms: quicklime (CaO) and slaked lime [$Ca(OH)_2$]. Quicklime must be slaked with water before use, resulting in the production of $Ca(OH)_2$. The slaking process generates significant heat, necessitating appropriate equipment and precautions to ensure the safety of plant workers. When selecting the slaking process, the calcium oxide (CaO) content in different types of quicklime should be considered. Slaked lime should be stored in dry areas to prevent hydration reactions with moisture in the air, which would render it ineffective.

In contrast, slaked lime does not require slaking, mixes easily with water, releases minimal heat, and does not necessitate special storage conditions. However, due to its higher cost and limited availability compared to quicklime, on-site slaking of quicklime may be more cost-effective for plants consuming more than 1–2 metric tons of lime per day.

Combining lime with ferric chloride facilitates sludge conditioning by enhancing the physical properties of the sludge. The reaction of lime with bicarbonate in the sludge forms calcium carbonate, creating a granular structure that improves sludge porosity and reduces compressibility. Quicklime, when slaked, produces $Ca(OH)_2$, which reacts exothermically, requiring proper equipment and safety measures to handle the heat generated. The calcium oxide content in quicklime varies, influencing the slaking process and effectiveness of the resulting $Ca(OH)_2$. Slaked lime, while easier to handle and store, is more expensive and less readily available than quicklime, making on-site slaking a more economical option for high-consumption plants.

5.9.4 Sludge Dewatering

Dewatering stands as a pivotal physical (mechanical) process utilized to diminish the moisture content of sludge for multifarious reasons:

(i) **Cost Reduction**: Dewatering curtails the expenses linked with transporting sludge to final disposal sites.
(ii) **Enhanced Handling**: Dewatered sludge presents superior manageability compared to thickened or liquid sludge, streamlining further processing.
(iii) **Incineration Preparation**: Pre-incineration, dewatering becomes imperative to expunge excess moisture, thereby augmenting the energy content of the sludge.
(iv) **Composting Facilitation**: Prior to composting, dewatering minimizes the necessity for additional bulking agents, simplifying the composting process.
(v) **Odor and Decomposition Control**: In certain scenarios, eliminating excess moisture is critical to eradicate odors and deter putrefaction, ensuring the sludge remains odorless and non-putrescible.
(vi) **Landfill Preparations**: Pre-landfilling, sludge dewatering becomes a prerequisite in monofills to mitigate leachate production at the landfill site, averting potential environmental concerns.

Diverse techniques are employed in the dewatering process. Some rely on natural evaporation and percolation, exemplified by drying beds and sludge lagoons, to extract moisture from solids. Others employ mechanical means, such as belt filter presses, vacuum filters, centrifuges, and recessed plate filter presses. Table 5.1 comprehensively discusses the advantages and disadvantages of these distinct sludge dewatering techniques.

Selecting the appropriate sludge dewatering processes can be intricate due to the myriad potential combinations of unit processes available for wastewater treatment. These processes encompass sludge thickening, stabilization, conditioning, dewatering, and ultimate use or disposal. It is imperative to meticulously evaluate the advantages and disadvantages of each technique to ensure optimal outcomes. By considering factors such as efficiency, cost-effectiveness, and environmental impact, informed decisions regarding the most suitable dewatering method for a given scenario can be made.

Table 5.1 Advantages and disadvantages of different sludge dewatering techniques

Dewatering method	Advantages	Disadvantages
Vacuum filter	• No need of skilled person • Low maintenance requirements for continuously operating equipment	• Highest consumption of energy per unit of sludge dewatered • Continuous operational attention required • Filtrate may have high suspended solids content depending on filter medium
Solid bowl centrifuge	• Clean appearance, minimum odour problems, fast startup and shutdown capabilities • Easy to install • Produces relatively dry sludge cake • Low capital cost to capacity ratio	• Scroll wear potential a high-maintenance problem • Requires grit removal and possibly a sludge grinder in the feed stream • Skilled maintenance personnel required • Moderately high suspended solids content in concentrate
Imperforate basket centrifuge	• Same machine can be used for both thickening and dewatering • Chemical conditioning may not be required • Clean appearance, minimal odour problems, fast startup and shutdown capabilities • Very flexible in meeting process requirements • Not affected by grit • Excellent results for difficult sludges	• Limited size capacity • Except for vacuum filters consumes more energy per unit of sludge dewatered • Skimming stream may produce significant recycled load • For easy dewatered sludges, has highest capital cost (cost-to-capacity-ratio) • For most sludges, produce lowest cake solids • Vibration
Belt filter press	• Low energy requirements • Relatively low capital and operating costs • Less complex mechanically and easier to maintain • High-pressure machines are capable of producing very dry cake • Minimal efforts requires for system shutdown	• Hydraulically limits throughout • Requires sludge grinder in feed stream • Very sensitive to incoming sludge feed characteristics • Short media life as compared to other device using cloth media • Automatic operation generally not advised

(continued)

Table 5.1 (continued)

Dewatering method	Advantages	Disadvantages
Recessed plate filter press	• Highest solid concentration • Low suspended solids in filtrate	• Batch operation • High equipment cost • High labour cost • Special support structure requirements • Large foot area requires for equipment • Skilled maintenance personnel required • Additional solids due to large chemical addition require disposal
Sludge drying beds	• Lowest capital cost where land is readily available • Small amount of operation attention and skill required • Low energy consumption • Low to no chemical requirements • Less sensitive to sludge variability • Higher solids content than mechanical methods	• Require large area of land • Requires stabilized sludge • Design requires consideration of climatic effects • Sludge removal is labour intensive
Sludge lagoons	• Low energy consumption • No chemical consumption • Organic matter is further stabilized • Low capital cost where land is available • Least amount of skill required for operation	• Potential for odour problems • Potential for groundwater pollution • More land intensive than mechanical methods appearance may be unsightly • Design requires consideration of climatic effects

5.9.5 Sludge Drying Beds

Sludge drying beds serve as a prevalent method for dewatering digested sludge. Upon completion of the drying process, the sludge is extracted and either disposed of in a landfill or repurposed as a soil conditioner. The utilization of drying beds offers several notable advantages, including cost-effectiveness, minimal maintenance requirements, and the ability to attain a high solids content in the resultant dried product (Haseltine, 1951).

Typically, five types of drying beds are employed in this process:

(i) Conventional sand
(ii) Paved
(iii) Artificial media

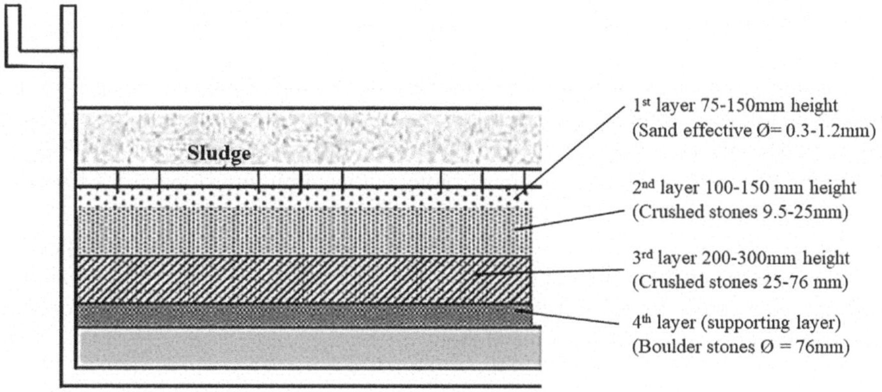

Fig. 5.3 Details of the draining medium (Goncalves et al., 2004)

(iv) Vacuum-assisted and
(v) Wetware beds.

5.9.5.1 Main Characteristics

Water is eliminated through the processes of evaporation and percolation, facilitated by a tank typically rectangular in shape, constructed with masonry or concrete walls and a concrete floor. Within the tank, several components aid in the drainage of sludge water.

5.9.5.2 Draining Medium

This medium permits the percolation of liquid from the sludge, passing through layers of sand at the top and gravel at the bottom (Fig. 5.3). These layers are arranged with increasing grain sizes, ranging from 0.3 mm in the upper part (sand) to 76 mm in the lower part (gravel). The total depth of these layers is approximately 0.50 m.

5.9.5.3 Supporting Layer

Constructed using hard burnt brick or other durable materials capable of withstanding the removal of dry sludge, the supporting layer is typically arranged with 20–30 mm joints filled with coarse-grained sand. It ensures proper distribution of the sludge, prevents clogging of the draining medium pores, and facilitates the removal of dewatered sludge without disturbing the layers of the draining medium.

5.9.5.4 Drainage System

This system comprises 100 mm pipes laid across the tank floor, featuring open or perforated joints to drain all the liquid percolating through the layers of the draining medium. The distance between the pipe drains should not exceed 3 m. The floor of the drying bed must be level and impermeable, with a minimum slope of 1% towards the main draining collector.

The removal of water from sludge involves a well-designed tank with specific components that enable efficient drainage. The careful arrangement of the draining medium, supporting layer, and drainage system ensures optimal performance and effective removal of liquid from the sludge.

5.9.5.5 General Aspects of Sludge Dewatering in Drying Beds

When the biodegradable solids content in sludge is well digested, natural drying can occur, resulting in characteristics conducive to quick dewatering. An example of such sludge is the bottom sludge from stabilization ponds, which typically exhibits these characteristics and has achieved sufficient biological stability for liquid–solid separation without prior treatment.

Sludge drying beds can be constructed as open-air structures or covered to shield against rainfall. Drying is conducted as a batch process, with the sludge sequentially routed to multiple drying beds (van Haandel & Lettinga, 1994). According to Hess (1973), digested sludges subjected to high hydraulic pressures, either in clarifiers or sludge digesters, may contain interstitial water saturated with gases such as CO_2 or methane. This sludge type may float in drying beds due to density differences between the digested sludge and water. During most of the dewatering period, water readily percolates through the draining bed until the sludge settles and transforms into a thick, pasty mass. At this juncture, percolation virtually ceases, and drying occurs through natural evaporation.

Sludge drying beds can be constructed as open-air structures or covered to protect against rainfall. Drying is carried out as a batch process, with:

Dry Sludge with Scarce and Thin Cracking: Indicates well-digested sludge with low water content.

A Large Number of Medium-Sized Cracks: Indicates digested sludge with high water content.

Small Quantities of Wide Cracks: Indicates poorly digested sticky sludge requiring extended drying periods.

Besides sludge physical properties, climatic conditions significantly influence the effectiveness of this process. Natural drying, for instance, can substantially reduce pathogenic organisms through sunlight exposure (van Haandel & Lettinga, 1994).

Once solids content reaches around 30%, prompt removal of sludge from the drying bed is advisable to forestall potential difficulties during later removal.

Allowing dry sludge to linger in the drying beds for extended periods can foster vegetation growth, indicative of poor plant management.

Hence, considering both sludge physical characteristics and prevailing climatic conditions is vital for optimal performance in this process. By doing so, we can effectively mitigate pathogenic organism presence and prevent complications during sludge removal. Additionally, timely sludge removal from the drying bed is crucial to uphold proper plant management and prevent unwanted vegetation growth.

5.9.6 *Centrifuges*

Centrifugation stands as a remarkably efficient process employed to segregate solid and liquid components through the utilization of centrifugal force. This potent force facilitates the swift settling of sludge solid particles during the initial stage, called clarification. Here, particles settle at a markedly accelerated rate compared to gravity alone.

Subsequent to clarification, the compaction stage ensues, wherein the sludge undergoes prolonged centrifugation, leading to the expulsion of a portion of its capillary water. This process effectively diminishes the moisture content of the sludge, rendering it more manageable and easier to handle.

Upon completion of the compaction stage, the resulting solid cake is meticulously extracted from the process. This cake, constituting the concentrated solid component of the sludge, can then undergo further processing or be disposed of as necessitated.

Similar to sludge thickening, centrifugation operates on the principle of sedimentation, relying on the density disparity between solid particles and the surrounding liquid. This sedimentation process is accurately described by Stokes' equation, furnishing a mathematical expression for the settling velocity of a solid particle in a fluid.

In essence, centrifugation presents a highly effective and efficient approach for separating solid and liquid components across various industries. Its capacity to harness the force of centrifugal force ensures rapid and precise separation, culminating in heightened productivity and enhanced quality of the final product.

$$V_s = \frac{(\rho_p - \rho_f) g d^2}{(18\mu)} \tag{5.1}$$

where

- Vs is particle settling velocity (in m/s units)
- ρ_p is density of the particle (in kg/m^3 units)
- ρ_f is density of the carrier fluid (in kg/m^3 units)
- g is acceleration due to gravity (in m/s^2 units)
- d is diameter of the particle (in m units) and
- μ is viscosity of the carrier fluid (in Pa s or kg/ms units).

The equation above demonstrates that the settling velocity of a particle is directly proportional to the disparity between the densities of the particle and the liquid, as well as to the square of the particle's diameter. Conversely, it is inversely related to the viscosity of the liquid.

The acceleration resulting from centrifugation (G) is typically expressed as a multiple of the gravitational constant, denoted as g (= 9.81 m/s^2). The centrifugal acceleration acting on a particle within a liquid confined in a cylinder can be calculated using the following formula:

$$G = \omega^2 R = \left(\frac{2\pi N}{60}\right)^2 R \tag{5.2}$$

where

G centrifugal acceleration of the particle (m/s^2)
ω angular velocity (rad/s)
R radius (m)
N rotation speed (revolutions/min).

The settling velocity (V), of a solid particle in a centrifuge with radius (R), can be determined by substituting the gravitational constant (g), in Eq. (5.1) with the value of (G) as defined in Eq. (5.2).

In the case of a sludge dewatering centrifuge, the centrifugal acceleration is 500–3000 times higher than the gravitational constant. This substantial increase in acceleration results in a settling velocity of 10 m/h, which is 50 times greater than the natural thickening velocities observed in sludges.

The magnitude of the forces involved in this process is so substantial that it causes the inner bond forces among particles to break apart. This phenomenon allows for a more efficient separation than what could be achieved through a simple static settling process.

By harnessing the power of centrifugal acceleration, the sludge dewatering centrifuge offers a superior method for separating solid particles. This enhanced separation capability is made possible by the increased settling velocity and the subsequent disruption of inner bond forces among particles.

5.9.6.1 Types of Centrifuges

Centrifuges prove to be versatile machines capable of serving both sludge thickening and dewatering. While the fundamental operating principle remains consistent, centrifuges can be installed in series to accomplish varying stages of thickening and dewatering. The equipment's capacity varies, typically ranging from 2.5 to 180 m^3/h of incoming sludge flow.

Two primary types of centrifuges including: vertical and horizontal shaft centrifuges are employed in sludge dewatering. These differ concerning sludge feeding, centrifugal force intensity, and the discharge mechanism for cake and liquid.

Horizontal shaft centrifuges are predominantly utilized for thickening and dewatering sludges, offering benefits such as higher cake solids content and semi-continuous feeding capability. In contrast, vertical shaft centrifuges see lesser use due to lower cake solids content and associated limitations.

Horizontal centrifuges typically adopt a solid-bowl design, featuring moving parts including the rotating bowl and the rotating conveyor/scroll. These parts are commonly crafted from stainless steel or carbon steel. Essential components of a centrifuge encompass the support basis, bowl, conveyor scroll, cover, differential speed gear, main drive, and feeding pipe.

The support basis, often fabricated from steel or cast iron, incorporates vibration insulators to minimize vibration transmission. The cover encapsulates all moving parts, serving to reduce odors and noise while collecting both the centrifuged liquid (centrate) and the dewatered sludge. The bowl, with a cone-cylinder shape, exhibits varying characteristics dependent on the manufacturer, with the length-to-diameter ratio ranging from 2.5:1 to 4:1 and diameters spanning from 230 to 1800 mm. The differential speed gear enables adjustment of rotational speed differences between the bowl and the screen conveyers.

Both the bowl and the conveyor scroll rotate in the same direction at high speed, with the scroll speed slightly differing from the bowl speed, thus creating a conveying effect. Positioned inside the bowl core, the scroll maintains a radial aperture of 1–2 mm, facilitating passage of the centrifuged liquid. Typically, velocities ranging from 800 to 3200 rpm result in cakes with solid levels exceeding 20% and a clarified centrate. Enhanced centrifugation speed can lead to reduced polyelectrolyte consumption, improved solids capture, and potentially higher cake solid content, although at increased maintenance costs due to bearing abrasion.

The centrifuge reduction gear-box introduces rotational differences of 1–30 rpm between the bowl and the screw conveyor. The sludge feeding rate, speed difference, and bowl rotational speed are key parameters controlling the solids retention time within a centrifuge. High solids cakes are achieved through a combination of extended retention time, low speed differential, and compatible sludge feeding rate.

The attainment of sludge thickening hinges upon the sludge type and initial solids concentration (Butler et al., 1997). Generally, dryer cakes are obtained when minimizing the speed difference between the bowl and the screw conveyor. Horizontal centrifuges can be categorized based on the feeding flow direction and cake withdrawal method, namely co-current or counter-current. These centrifuges vary regarding sludge feeding points, centrate removal (liquid phase), and the direction of solid and liquid phases within the unit. In co-current centrifuges, both solid and liquid phases traverse the entire longitudinal axis of the centrifuge bowl before discharge. Conversely, in counter-current models, sludge is fed on the opposite side of the centrate release point, with the solid phase directed towards the end of the conic section by the screw conveyor while the liquid phase follows the opposite path (Fig. 5.4).

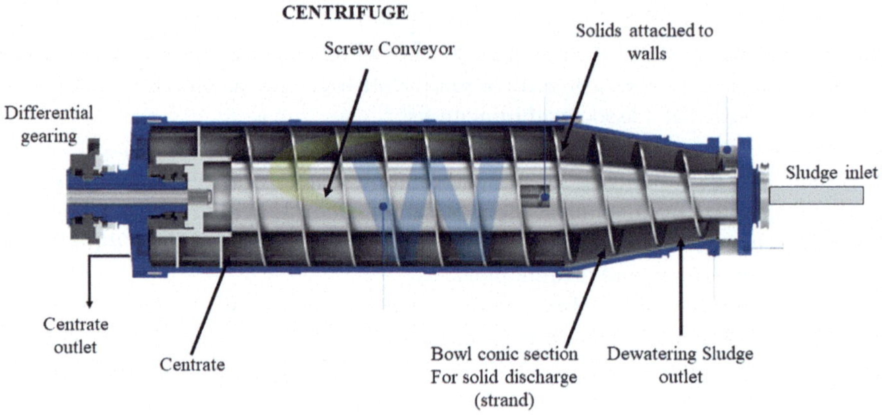

Fig. 5.4 Countercurrent horizontal centrifuge

5.9.7 Filter Press

Filter presses were initially developed for industrial use and have since been modified to be suitable for wastewater sludge dewatering operations (To et al., 2016). These presses operate through batch feeding, which requires skilled operators. However, their reliability is a major asset. Filter presses offer several advantages, including the ability to produce a cake with a higher solids concentration compared to other mechanical equipment. They also have a high solids capture rate and produce high-quality liquid effluent, known as filtrate. Additionally, filter presses require low chemical consumption for sludge conditioning.

- Cake with higher solids concentration than any other mechanical equipment
- High solids capture
- Quality of the liquid effluent (filtrate)
- Low chemical consumption for sludge conditioning.

5.9.7.1 Working Principle

The operating cycle of the filter press typically spans from 3 to 5 h and encompasses three primary stages.

Filling

Pumped sludge is introduced into the empty spaces between consecutive filter plates. The filling process typically lasts around 5–10 min, although it may extend up to 20 min in certain instances. The pressure exerted by the filling pump immediately initiates the separation of solids and liquids in the filter cloths.

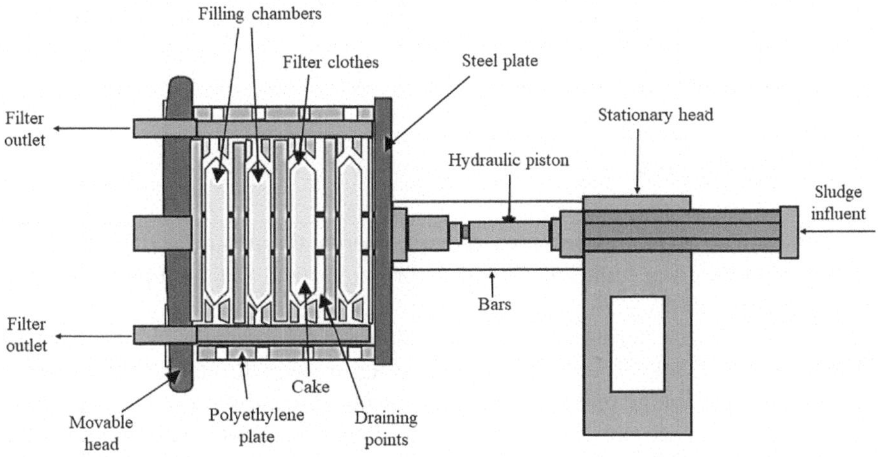

Fig. 5.5 Operating diagram of filter press

Filtration Under Maximum Pressure

During this phase, the applied pressure can reach up to 170 kPa that facilitates efficient separation of solids from liquids.

Cake Discharge

Cake discharge is a pivotal step in the filtration process, and its duration varies based on factors such as the flow rate of sludge feeding pump, the sludge type and solids content, the filterability of influent sludge, and the filter cloth's condition (Dick et al., 1980) (Fig. 5.5).

In this process, the liquid sludge is pumped into the recessed plates enveloped by filter cloths. As the sludge is pumped, the pressure between the plates increases, causing the solids (filter cake) to adhere to the filter media while the liquid sludge passes through the filter cloth. Subsequently, a hydraulic piston pushes the steel plate against the polyethylene plates, compressing the cake. Support bars on both the movable and stationary heads facilitate this compression.

The filtrate, or the liquid that has passed through the filter cloths, is collected through draining points and filtrate channels. Typically, the filtrate contains less than 15 mg SS/L. Conversely, the compressed cake is easily removed from the filter as the pneumatic piston moves back, separating the plates. Once the cake falls from the plate, it is ready to be directed to storage or its final destination.

Modern filter presses have become increasingly automated, reducing the need for manual labor. However, certain limitations must be considered when utilizing filter presses, including equipment weight, initial costs, and the regular replacement of filter cloths, which restrict their use to medium-sized and large wastewater treatment plants.

5.9.8 Belt Presses

Belt presses, also known as belt-filter presses, operate through three distinct zones: the gravity-dewatering zone, the low-pressure zone, and the high-pressure zone (Hwang & Min, 2003).

In the gravity-dewatering zone, positioned at the equipment's entrance, sludge is applied onto the upper belt, allowing free water to seep through the cloth pores. Subsequently, the sludge moves to the low-pressure zone, also known as the wedge zone, where gentle compression between the upper and lower belts releases the remaining free water.

The high-pressure zone, comprising several rollers of varying diameters arranged in series, further compresses the sludge between the belts, resulting in the release of interstitial water. Scrapers positioned on the upper and lower belts remove the dewatered sludge. Before receiving fresh diluted sludges, high-pressure water jets wash the upper and lower belts. It is crucial for the cloth washing water to have a minimum pressure of 6 kg/cm^2 and sufficient flow to effectively eliminate any attached sludge and polyelectrolyte residues from the cloth.

However, belt presses exhibit some drawbacks. Their open design renders them prone to aerosol emission, high noise levels, and potential foul odor emission, contingent upon the type of sludge being processed. Additionally, the considerable number of bearings required mandates regular maintenance and replacement. Nonetheless, belt presses offer the advantages of low initial costs and reduced electric power consumption.

In recent times, the emergence of decanter-type centrifuges has intensified competition between suppliers of both types of dewatering equipment. Despite their higher initial costs, centrifuges are currently preferred (Fig. 5.6).

5.10 Sludge Disposal

On a global scale, the expansion of metropolitan cities due to urbanization has presented numerous challenges in the final disposal of sewage sludge and its alternatives. This situation necessitates the consideration of processes such as incineration and wet air oxidation. Regardless of the chosen technology, all treatment and disposal processes come with their own set of advantages and disadvantages, particularly concerning potential contamination of soil, water bodies, and the atmosphere. The combustion of sludge, in particular, raises serious concerns regarding atmospheric pollution and the safe disposal of residual ashes.

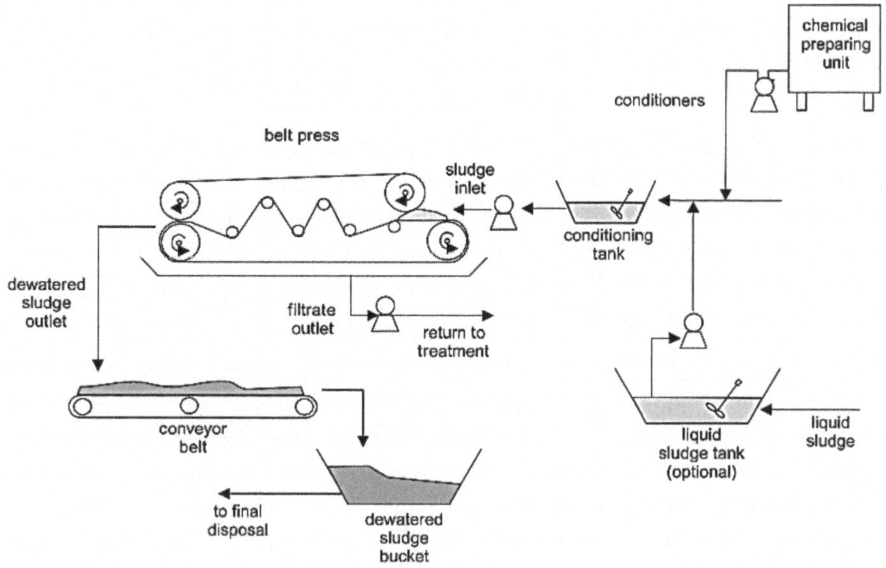

Fig. 5.6 Schematics diagram of belt press

In response to the increasing difficulties in maintaining landfills as a final sludge disposal route, many countries are turning to the incineration process. This shift is driven by factors such as the scarcity of landfill space, rising disposal costs, legislative constraints, and incentives for sludge recycling.

Various methods for sludge transformation and disposal include thermal drying, wet air drying, incineration, and land disposal (Zhang et al., 2014). Among these options, only landfills can be categorized as a definitive disposal method due to the fact that the other alternatives, despite their effectiveness in removing water, still generate residues that require final disposal. It is crucial to consider landfills solely as a last resort when other solutions, such as utilizing wastewater sludge, are not feasible.

The majority of sludges produced by municipal wastewater treatment plants possess physical and chemical properties that make them valuable for agricultural or industrial purposes. This opens up the possibility of adopting a more constructive approach towards sludge management. In certain circumstances, sludge can even be regarded as a valuable commodity rather than a useless byproduct.

5.11 Thermal Drying

Thermal drying is an incredibly versatile process that can be easily adapted to produce pellets for various purposes such as agricultural reuse, sanitary landfill disposal, or incineration (Nanda & Berruti, 2021). By applying heat, this process effectively evaporates moisture from sludge, resulting in the production of pellets that can be utilized as fuel for boilers, industrial heaters, cement kilns, and more. The concentration of solids in these pellets typically ranges from 65 to 95%, with a water content of 5–35%.

There are numerous advantages associated with thermal sludge drying. It significantly reduces the volume of sludge, which in turn leads to reduced storage and freight costs. The final product is stabilized and can be easily transported, stored, and handled. It is free from pathogenic organisms and maintains the beneficial characteristics of soil amendment derived from sewage sludge. It can be safely reused for agronomic purposes without any restrictions, or alternatively, it can be incinerated or disposed of in landfills. The flexibility of thermal drying allows for accommodation in small-sized packages, enhancing convenience and practicality.

However, it is important to acknowledge the limitations of thermal drying processes. The production of liquid effluents must be properly managed to prevent any adverse environmental impacts. The release of gases into the atmosphere is a concern that needs to be addressed through appropriate emission control measures. There is a potential risk of foul odors and disturbing noise associated with the process, which should be mitigated to ensure a pleasant and safe working environment.

Thermal drying processes can be classified into three categories: indirect, direct, and mixed. Indirect processes yield pellets with a solid concentration of up to 85%. For solid contents exceeding 90% and the potential production of organic mineral fertilizers, direct drying processes are recommended. The liquid effluent accounts for less than 1% of the total treatment plant flow and can be recycled to the plant headworks, provided there is sufficient capacity to handle the additional organic load. However, during thermal drying of anaerobic sludge, the treatment of the liquid effluent may encounter issues with excess ammonia nitrogen.

Both direct and indirect drying processes release gaseous emissions that have the potential to produce unpleasant odors. The odor associated with the dried product, although less intense, resembles the original sludge odor. To minimize the release of these gases, it is highly advisable to isolate the drying unit, preferably in a negative pressure environment.

5.12 Wet Air Oxidation

Wet air oxidation (WAO) originated in Norway for treating paper industry residues before being adapted for sewage sludge treatment in the United States during the 1960s. However, its initial application for sewage sludge did not meet expectations, leading to its modification for treating high-toxicity industrial liquid wastewater (Jansson, 2022). WAO is now recommended when the effluent is too dilute for incineration yet too toxic or refractory for biological treatment (Fig. 5.7).

The process relies on oxidizing dissolved or particulate organic matter in a liquid at temperatures from 100 to 374 °C, where water cannot exist in liquid form. High

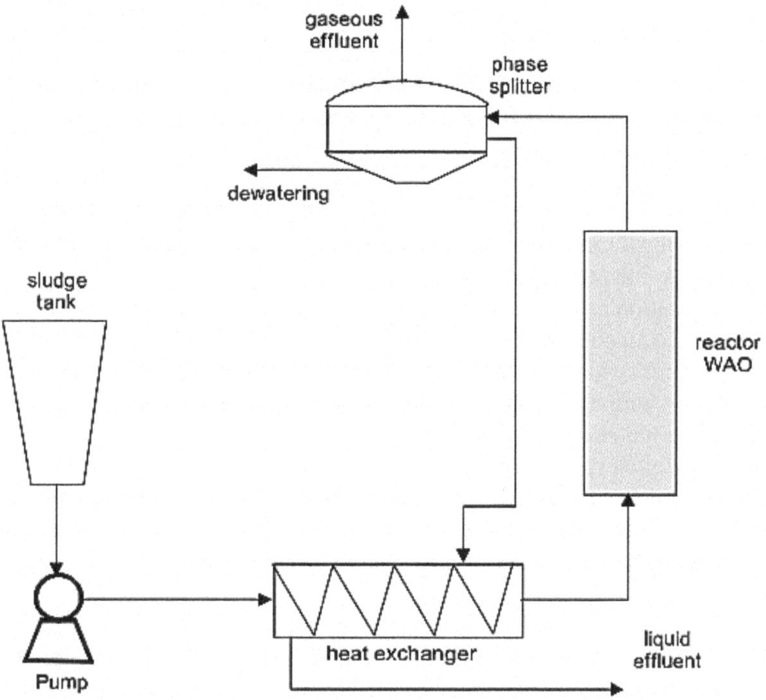

Fig. 5.7 Conventional wet air oxidation system with a vertical reactor

solubility of oxygen in aqueous solutions at elevated temperatures accelerates oxidation. WAO efficiently destroys organic matter in effluents with 1–20% solids concentrations, generating sufficient internal heat to sustain the reaction without external energy. An upper limit of 20% solids concentration prevents excessive temperature rise.

Advancements in technology and stricter environmental regulations have renewed interest in WAO for sewage sludge stabilization, especially for plants serving large metropolitan areas. Sewage sludge organic matter is categorized as easily oxidizable or not under WAO, with the former including common constituents like proteins, lipids, sugars, and fibers.

Key control variables in WAO are temperature, pressure, air/oxygen supply, and solids concentration, with variations such as low-pressure oxidation, intermediate pressure oxidation, and high-pressure oxidation. Low-pressure oxidation aims to reduce sludge volume and enhance dewaterability, while intermediate and high-pressure oxidation aim to oxidize volatile organic matter into CO_2 and water.

Challenges in industrial-scale WAO include foul odors, corrosion of equipment, power consumption for startup, high COD in liquid effluent, and high metal content in residual ashes. Despite its effectiveness, efficient operation and maintenance are necessary for successful implementation.

The vertical reactor wet air oxidation system (Fig. 5.7), efficiently directs influent sludge to the Wet Air Oxidation (WAO) reactor, where it undergoes temperature elevation via a heat exchanger. The WAO reactor effluent is then skillfully guided through a phase splitter, allowing effective dewatering of the sludge while redirecting the liquid component back through the heat exchanger to transfer heat to incoming sludge, optimizing system performance. Rigorous purification measures, including treatment by an electrostatic precipitator and filtration for solid particle and odor removal, ensuring the environmental compliance before the treated gaseous effluents are released into the atmosphere is necessary.

Wet air oxidation processes utilize either air or pure oxygen as the oxygen supply, with compressed air commonly used in wastewater treatment plants. While the capital costs for a wet air oxidation process may be higher than those for incineration when treating effluent with up to 20% solids, the operational cost is significantly lower due to reduced external energy requirements.

Advantages of wet air oxidation include its versatility in treating various organic sludge types and the production of a sterile, non-putrescible solid residue suitable for mechanical dewatering. However, limitations such as low nitrogen concentration and relatively high metals content may restrict its applicability for land application in agricultural reuse.

The gaseous output from wet air oxidation comprises nitrogen, oxygen, carbon dioxide, and hydrocarbons, with foul odor release influenced by the degree of oxidation achieved. Significant chemical oxygen demand (COD) removal occurs, with a large portion transformed into low-molecular-weight volatile acids. The liquid phase, containing lower organic content, can undergo further biological treatment, achieving high removal efficiencies for COD and BOD.

Integration of wet air oxidation technology with the activated sludge process, termed Deep Shaft Technology, has revolutionized sludge treatment by eliminating the need for pumps, heat exchangers, and high-pressure reactors, reducing capital costs substantially (Sandford & Gallo, 1978). This approach is particularly advantageous in space-limited situations.

5.13 Inciniration

Incineration is a highly efficient method for sludge stabilization, resulting in significant volume reduction, with residual ashes typically comprising less than 4% of the initial dewatered sludge volume. Incinerators are designed to handle sludge from multiple treatment plants, often accommodating populations exceeding 500,000 and with capacities exceeding 1 tonne per hour (Zhang et al., 2021a, 2021b, 2021c).

The incineration process involves the combustion of organic substances and pathogenic organisms in the presence of excess oxygen. To ensure environmental safety, sophisticated filter systems are employed to greatly reduce pollutant emissions. Regular monitoring of gases released into the atmosphere is essential for maintaining operational efficiency and safety.

The design of incinerators requires meticulous consideration of mass and energy balances. Although self-sustaining combustion occurs when solids concentration exceeds 35%, dewatered cakes containing 20–30% total solids can be burned with auxiliary fuels such as low-sulfur content boiler fuel. The calorific value of the sludge, determined by its combustible components (carbon, sulfur, and hydrogen), plays a crucial role in reducing fuel consumption (Table 5.2).

Complete combustion of sludge produces water vapor, carbon dioxide, sulfur dioxide, and inert ashes. Achieving optimal combustion requires an appropriate fuel-to-oxygen ratio, typically exceeding the stoichiometric value of 4.6 kg of air per kg of O_2. An excess of 35–100% more air is generally necessary to ensure complete combustion, with the exact amount dependent on sludge characteristics and incinerator type.

The quantity of oxygen required for the full combustion of organic matter can be ascertained by identifying the organic compounds and assuming that all carbon and hydrogen is converted into carbon dioxide and water.

$$\mathrm{C}a\mathrm{H}b\mathrm{O}_c\mathrm{N}d + (a + 0.25b - 0.5d)\mathrm{O}_2 \rightarrow a\mathrm{CO}_2 + 0.5c\mathrm{H}_2\mathrm{O} + 0.5d\mathrm{N}_2$$

Two types of incinerators are currently in use for sewage sludge:

- *Multiple chamber incinerator*
- *Fluidized bed incinerator.*

Multiple chamber incinerators and fluidized bed incinerators offer distinct advantages and considerations in sludge treatment. The choice between these options depends on factors such as operational costs, air quality, and the specific requirements of the wastewater treatment plant.

Multiple chamber incinerators are divided into three combustion zones: the higher zone for final moisture removal, the intermediate zone for combustion, and the lower cooling zone. Supplementary fuel may be required in the intermediate chamber. In contrast, fluidized bed incinerators consist of a single-chamber vessel where organic sludge particles remain in contact with a fluidized sand bed until complete combustion is achieved.

The trend now favors fluidized bed incinerators due to their lower operational costs and better air quality through chimney emissions. These incinerators operate under autogenous conditions at temperatures above 815 °C, ensuring complete destruction

Table 5.2 Calorific power of different types of sewage sludge

Type of sludge	Calorific power (kJ/kg dry solids)
Raw primary sludge	23,300–29,000
Anaerobically digested sludge	12,793
Activated sludge	19,770–23

Source Adapted from WEF (1996)

of volatile organic compounds at a cost-effective price. Modern dewatering equipment can feed cakes with over 35% total solids to incinerators, making autogenous combustion feasible.

However, the use of incinerators in sludge treatment is limited to wastewater treatment plants serving large urban areas due to high costs and sophisticated operation. Restrictions on agricultural sludge reuse, such as excessive metal concentrations, long hauling distances, and urban landfill volume constraints, may favor incineration as a viable alternative.

To control atmospheric emissions, incinerators optimize the combustion process and use air filters. Pollutants released include nitrogen oxides (NO_x), incomplete combustion products (e.g., carbon monoxide, dioxins, furans), acidic gases (e.g., sulfur dioxide, hydrochloric acid), and volatile organic compounds (e.g., toluene, chlorinated solvents). Solids in emissions consist of fine particulate matter composed of metals and suspended solids, which are removed using electrostatic precipitators.

Despite significantly reducing sludge volume, incineration cannot be considered a final disposal method due to residual ashes requiring appropriate disposal. Ashes primarily contain inert inorganic matter with a concentration of metals. Disposal alternatives include final landfill disposal or co-incineration in cement kilns or thermoelectric power plants using mineral coal as fuel. Co-incineration reduces capital, operational, and maintenance costs by integrating with existing industrial processes.

5.14 Landfill Disposal

Landfill disposal is a method used to safely dispose of solid urban waste onto soil, with minimal environmental impact and no harm to public health. This technique employs engineering methods to confine the waste within the smallest possible area and volume, covering the disposed waste with a layer of soil after each working day or at smaller intervals if necessary.

When it comes to sludge disposal in landfills, nutrient recovery or practical use of the sludge is not a concern. Within landfill cells, anaerobic biodegradation occurs, leading to the production of various by-products, including methane.

There are two types of landfill disposal for sludge.

5.14.1 Exclusive (Dedicated) Sanitary Landfills

These are specifically designed and constructed to receive sewage sludge, incorporating special features to accommodate the sludge's specific properties and comply with environmental regulations. Typically, these landfills require thermally dried sludges or cakes with high solids content (> 30%).

5.14.2 Co-disposal with Urban Solid Waste

In this approach, wastewater sludge is disposed of in a landfill alongside municipal solid waste. Mixing sludge with urban waste can accelerate the biodegradation process, depending on the nitrogen content and sludge inoculation potential. However, a drawback is that it may reduce the landfill's lifespan if a significant amount of sludge is present.

Various technologies are available for the construction, operation, and maintenance of both types of landfills, offering alternatives to traditional sludge disposal on land. Sludge from domestic wastewater treatment is generally classified as a non-hazardous residue, although it may be considered a non-inert residue. Extensive testing has shown that sludge from municipal treatment plants does not pose a hazardous waste threat. However, heavily contaminated sludge from industrial effluents may require disposal at a licensed hazardous waste landfill.

Landfills provide a flexible solution, accommodating varying volumes of sludge, absorbing excess demands from other disposal methods, and operating independently from external factors. When selecting a landfill as the final disposal route, the characteristics of the sludge, such as stabilization or pathogen levels, are not primary concerns. Instead, the availability of suitable land in close proximity to the wastewater treatment plant is crucial.

Site selection should involve a multidisciplinary study to identify the most environmentally and economically viable option. Obtaining approval from the environmental agency and addressing the concerns of the neighboring population during the design and construction phases are essential. Collaborative approaches can prioritize community well-being and minimize potential negative impacts.

5.15 Area Selection and Environmental Impact Consideration

The primary task in a sanitary landfill project is to conduct a comprehensive evaluation of the environmental impacts associated with various stages of implementation and operation. This assessment is crucial for determining the necessary protective measures to control and minimize any adverse effects. The selection of a suitable landfill site is of utmost importance as it can eliminate or reduce several impacts if the chosen location possesses favorable characteristics.

Improperly designed or operated landfills can lead to pollution in various ways, including air pollution from foul odors, toxic gases, or the release of particulate matter; contamination of surface water bodies from leachate drainage or sludge transportation through runoff; and soil and groundwater pollution from infiltrated liquids.

The process of selecting a landfill site involves several crucial steps:

(1) **Macro-region Selection**: Consider accessibility and waste generation points within different macro-regions to strategically identify suitable locations.
(2) **Identification of Legal Constraints**: Identify and address all legal constraints within the macro-regions to ensure compliance with regulations and guidelines.
(3) **Preliminary Evaluation of Macro-regions**: Thoroughly evaluate remaining macro-regions, considering topographic requirements like dry valleys and hillsides with slopes below 20%, as well as geological and hydrogeological aspects such as low-permeability lithology and minimal soil deformability. Ideally, select sites near the watershed limit with a water table depth greater than 1.50 m.
(4) **Preliminary Selection Using Aerial Photos**: Utilize aerial photos to identify potential sites that align with selective criteria, aiding in narrowing down options.
(5) **Field Survey and Preliminary Selection**: Conduct a comprehensive field survey to gather additional information and validate the preliminary selection, ensuring all relevant data is considered.
(6) **Development of Technical Studies**: Compare and evaluate potential areas using technical studies to analyze factors like environmental impact, infrastructure requirements, and long-term sustainability.
(7) **Environmental Licensing**: Once the selected landfill site is determined, obtain necessary environmental licensing to ensure compliance with regulations and guarantee the site meets all criteria for safe and responsible waste disposal.

References

Ahammad, S. Z., Graham, D. W., & Dolfing, J. (2020). Wastewater treatment: Biological. In *Managing water resources and hydrological systems* (pp. 561–576). CRC Press.

Beckett, R., & Le, N. P. (1990). The role of organic matter and ionic composition in determining the surface charge of suspended particles in natural waters. *Colloids and Surfaces, 44*, 35–49. https://doi.org/10.1016/0169-3402(90)80004-L

Butler, R. C., Finger, R. E., Pitts, J. F., & Strutynski, B. (1997). Advantages of cothickening primary and secondary sludges in dissolved air flotation thickeners. *Water Environment Research, 69*(3), 311–316. https://doi.org/10.2175/WER.69.3.8

Cieślik, B. M., Namieśnik, J., & Konieczka, P. (2015). Review of sewage sludge management: Standards, regulations and analytical methods. *Journal of Cleaner Production, 90*, 1–15. https://doi.org/10.1016/j.jclepro.2014.11.066

Di Costanzo, N., Cesaro, A., Di Capua, F., & Esposito, G. (2021). Exploiting the nutrient potential of anaerobically digested sewage sludge: A review. *Energies, 14*(23), 8149. https://doi.org/10.3390/en14238149

Dick, R. I., Ball, R. O., & Novak, J. T. (1980). Sludge dewatering. *Critical Reviews in Environmental Science and Technology, 10*(4), 269–337. https://doi.org/10.1080/10643388009388434

Environmental Protection Agency. (1978). *Sludge treatment and disposal* (EPA-625/4-78-012).

Farrell, J. B., Smith, J. E., Jr., Hathaway, S. W., & Dean, R. B. (1974). Lime stabilization of primary sludges. *Journal (Water Pollution Control Federation), 46*, 113–122.

Gawdzik, J., Długosz, J., & Urbaniak, M. (2015). General characteristics of the quantity and quality of sewage sludge from selected wastewater treatment plants in the Świętokrzyskie province. *Environment Protection Engineering, 41*(2).

Ghangrekar, M. M., Asolekar, S. R., & Joshi, S. G. (2005). Characteristics of sludge developed under different loading conditions during UASB reactor start-up and granulation. *Water Research, 39*(6), 1123–1133. https://doi.org/10.1016/j.watres.2005.01.010

Ginisty, P., Mailler, R., & Rocher, V. (2021). Sludge conditioning, thickening, and dewatering optimization in a screw centrifuge decanter: Which means for which result? *Journal of Environmental Management, 280*, 111745. https://doi.org/10.1016/j.jenvman.2020.111745

Goncalves, A., Esteves, A. M., Carvalho, M., & Machado, Â. (2004). Incorporation of sludges from a water treatment plant in cement mortars. In *International RILEM Conference on the Use of Recycled Materials in Buildings and Structures* (p. 843).

Gupta, V. K., Ali, I., Saleh, T. A., Nayak, A., & Agarwal, S. (2012). Chemical treatment technologies for waste-water recycling—An overview. *RSC Advances, 2*(16), 6380–6388. https://doi.org/10.1039/C2RA00581H

Haleem, N. (2024). *Development of innovative flocculation technologies for agricultural water treatment*.

Haseltine, T. R. (1951). Measurement of sludge drying bed performance. *Sewage and Industrial Wastes,* 1065–1083.

Hess, M. L. (1973). Lagoas de estabilização. *Capítulo 6: Lagoas anaeróbias* (2nd ed.). CETESB (in Portuguese).

Hwang, S., & Min, K. S. (2003). Improved sludge dewatering by addition of electro-osmosis to belt filter press. *Journal of Environmental Engineering and Science, 2*(2), 149–153. https://doi.org/10.1139/s02-014

Jansson, K. (2022). *Development of advanced oxidation processes for the Finnish pulp and paper industry water treatment*.

Knocke, W. R., Hamon, J. R., & Dulin, B. E. (1987). Effects of coagulation on sludge thickening and dewatering. *Journal—American Water Works Association, 79*(6), 89–98. https://doi.org/10.1002/j.1551-8833.1987.tb04966.x

Nanda, S., & Berruti, F. (2021). A technical review of bioenergy and resource recovery from municipal solid waste. *Journal of Hazardous Materials, 403*, 123970. https://doi.org/10.1016/j.jhazmat.2020.123970

Quintero-García, O. J., Pérez-Soler, H., & Amezcua-Allieri, M. A. (2023). Enzymatic treatments for biosolids: An outlook and recent trends. *International Journal of Environmental Research and Public Health, 20*(6), 4804. https://doi.org/10.3390/ijerph20064804

Raheem, A., Sikarwar, V. S., He, J., Dastyar, W., Dionysiou, D. D., Wang, W., & Zhao, M. (2018). Opportunities and challenges in sustainable treatment and resource reuse of sewage sludge: A review. *Chemical Engineering Journal, 337*, 616–641. https://doi.org/10.1016/j.cej.2017.12.173

Sandford, D. S., & Gallo, T. (1978). *Application of deep shaft technology to the treatment of food processing wastewater* (Vol. 188, pp. 341–358). Environmental Protection Technology Series EPA, 600/2-78.

Smith, S. J., Keane, C., Ahrens, L., & Wiberg, K. (2023). Integrated treatment of per- and polyfluoroalkyl substances in existing wastewater treatment plants: Scoping the potential of foam partitioning. *ACS ES&T Engineering, 3*(9), 1276–1285. https://doi.org/10.1021/acsestengg.3c00185

To, V. H. P., Nguyen, T. V., Vigneswaran, S., & Ngo, H. H. (2016). A review on sludge dewatering indices. *Water Science and Technology, 74*(1), 1–16. https://doi.org/10.2166/wst.2016.258

Van Haandel, A. C., & Lettinga, G. (1994). *Anaerobic sewage treatment*. Wiley.

Wang, X., Jin, P., Zhang, A., Gao, J., Zhang, B., & Hou, Y. (2020). Effect of mechanical elutriation on carbon source recovery from primary sludge in a novel activated primary tank. *Chemosphere, 240*, 124820. https://doi.org/10.1016/j.chemosphere.2019.124820

Zhang, F., Peng, Y., Wang, Z., Jiang, H., Ren, S., & Qiu, J. (2021a). Achieving synergetic treatment of sludge supernatant, waste activated sludge and secondary effluent for wastewater treatment plants (WWTPs) sustainable development. *Bioresource Technology, 337*, 125416. https://doi.org/10.1016/j.biortech.2021.125416

Zhang, L., Xu, C., Champagne, P., & Mabee, W. (2014). Overview of current biological and thermochemical treatment technologies for sustainable sludge management. *Waste Management & Research, 32*(7), 586–600. https://doi.org/10.1177/0734242X14540737

Zhang, Q., He, C., Ren, J., & Goodsite, M. E. (Eds.). (2021b). *Waste to renewable biohydrogen: Volume 1: Advances in theory and experiments*. Academic Press. https://doi.org/10.1016/B978-0-12-819173-9.00001-8

Zhang, S., Wang, F., Mei, Z., Lv, L., & Chi, Y. (2021c). Status and development of sludge incineration in China. *Waste and Biomass Valorization, 12*, 3541–3574. https://doi.org/10.1007/s12649-021-01424-5

Zhao, X., Jiang, J., Zhou, Z., Yang, J., Chen, G., Wu, W., ... Lou, Z. (2020). Applying organic polymer flocculants in conditioning and advanced dewatering of landfill sludge as a substitution of ferric trichloride and lime: Mechanism, optimization, and pilot-scale study. *Chemosphere, 260*, 127617. https://doi.org/10.1016/j.chemosphere.2020.127617

Zhou, X., Jiang, G., Wang, Q., & Yuan, Z. (2014). A review on sludge conditioning by sludge pre-treatment with a focus on advanced oxidation. *RSC Advances, 4*(92), 50644–50652. https://doi.org/10.1039/C4RA08004C

The manufacturer's authorised representative in the EU is Springer Nature Customer Service Centre GmbH, Europaplatz 3, 69115 Heidelberg, Germany. If you have any concerns regarding our products, please contact ProductSafety@springernature.com

Printed and bound by CPI Group (UK) Ltd, Croydon, CR0 4YY
01/10/2026
02225915-0010